This book provides an up-to-date account of the fast-moving field of lipase research. It covers both the lipases proper (triglyceride lipases) and the phospholipases. It gives a comprehensive picture of the state of our knowledge of these enzymes, with a strong bias towards the fields that today are attracting the greatest attention: their detailed molecular structure, their mechanism of action, their position in the evolution of enzymes, and their application both in the laboratory and in industry.

The book begins with a survey of known amino-acid sequences of lipases and an analysis of their kinships with the sequences of other, related protein families. A survey of lipases from plants is followed by several chapters, each of which looks at a topical aspect of the structures and mechanisms of action of representative lipases from animals, bacteria and fungi. On the applied side, current methods for isolating and purifying lipases are reviewed, and further chapters survey the value of lipases for chemical synthesis in particular and for industry in general. Finally, the medical aspects of some lipases are discussed, and the book ends with a glimpse into the future of lipases as potentially valuable 'engineered proteins'.

It is hoped that this book will appeal to those working in universities, in research institutes and in companies specialising in biotechnology, and will be a useful reference book in university departments of biochemistry and chemistry, for post-graduate students working on lipase-related projects, and for advanced research workers seeking to keep abreast of current developments within the lipase field.

The articles in this book were commissioned by the Protein Engineering programme of Nordisk Industrifond, Oslo.

Lipases – their structure, biochemistry and application

Lipases

their structure, biochemistry
and application

Edited by

PAUL WOOLLEY

Chemistry Department, Aarhus University

and

STEFFEN B. PETERSEN

Magnetic Resonance Centre, SINTEF, Trondheim

CAMBRIDGE
UNIVERSITY PRESS

CAMBRIDGE UNIVERSITY PRESS
Cambridge, New York, Melbourne, Madrid, Cape Town,
Singapore, São Paulo, Delhi, Tokyo, Mexico City

Cambridge University Press
The Edinburgh Building, Cambridge CB2 8RU, UK

Published in the United States of America by Cambridge University Press, New York

www.cambridge.org
Information on this title: www.cambridge.org/9780521207997

First published 1994
First paperback edition 2011

A catalogue record for this publication is available from the British Library

ISBN 978-0-521-44546-7 Hardback
ISBN 978-0-521-20799-7 Paperback

Additional resources for this publication at www.cambridge.org/9780521207997

Contents

Foreword

The idea for this book on lipases was prompted by the fact that standard works on these enzymes tend to treat them from a mainly physiological and classical biochemical standpoint, whereas in recent years there has been an especially rapid growth in our knowledge of the structures of lipases (including phospholipases). In addition, lipases are developing rapidly as objects of industrial interest. It was therefore felt that there was a need for a book in which, on the one hand, the biochemistry contains a strong structural element, and, on the other, account is taken of the increasing industrial importance, both realised and anticipated, of these enzymes.

The Nordisk Industrifond programme in Protein Engineering, of which the editors are co-ordinator and chairman respectively, has provided a base for the preparation of this book. Not only are lipases a target of the programme, but the participant groups also include many leading research groups in the lipase field. One of the activities of the Nordisk Industrifond programme is the organisation of symposia on topical aspects of protein research. Since the inception of the programme in 1989, many such meetings have been held in many forms. The present symposium, like that of Plato, took place on paper rather than in the flesh. It aims to be a collation of state-of-the-art reviews at research level. In commissioning articles, the editors have attempted to strike a balance between contributions from within and from outside the Nordisk Industrifond programme. However, the criterion for the selection of topics has first and foremost been a balanced treatment of present-day knowledge of lipases, even though the constraint of length has resulted in the omission of some potentially interesting topics. The book does not attempt to cover the literature comprehensively, but is intended rather to give a broad and up-to-date picture of the field, and at the same time to be a source-book for those looking for other publications that give more detailed treatment of individual topics.

It is hoped that this book will appeal to those working in universities, research institutes and in companies specialising in biotechnology, and will be a useful reference book in university departments of biochemistry and chemistry, for post-graduate students working on lipase-related projects, and for advanced research workers seeking to keep abreast of current developments within the lipase field.

P.W.
S.B.P.

November, 1992.

Nordisk Industrifond

Nordisk Industrifond (the Nordic Fund for Technology and Industrial Development) has organised a number of biotechnology programmes under the auspices of the Nordic Council of Ministers. The programmes aim to support research, dissemination of expertise and training of research workers in the Nordic community and to promote scientific exchanges with countries outside it. Collaboration between academic and industrial research centres at a 'pre-competitive' level is especially encouraged. The participating research groups come from all the five Nordic countries: Denmark, Finland, Iceland, Norway and Sweden.

One of these programmes is in Protein Engineering. Projects supported by this programme include work on lipases, cellulases and other enzymes. The articles in this book were commissioned by the programme as part of its efforts towards the dissemination of expertise through seminars, symposia and other activities.

Further information about the work of Nordisk Industrifond can be obtained from its secretariat at: Nedre Vollgate 8, 0158 Oslo 1, Norway.

List of Contributors

Chapter 1 – *Sequence comparisons within the lipase family*
Allan Svendsen
Novo Nordisk A/S, Novo Allé, 2880 Bagsværd, Denmark.

Chapter 2 – *A sequence analysis of lipases, esterases and related proteins*
Steffen B. Petersen, Finn Drabløs
MR Center, Natural Science Section, SINTEF UNIMED, 7034 Trondheim, Norway.

Chapter 3 – *Lipases from plants*
Kumar D. Mukherjee
Federal Centre for Cereal, Potato and Lipid Research, Institute for Biochemistry and Technology of Lipids, H.P. Kaufmann Institute, Piusallee 68, 48147 Münster, Germany.

Matthew J. Hills
John Innes Centre for Plant Science Research, Department of Brassica and Oilseeds Research, Colney Lane, Norwich, Norfolk NR4 7UJ, England.

Chapter 4 – *Three-dimensional structures of two lipases from filamentous fungi*
David M. Lawson, Andrzej M. Brzozowski, Guy G. Dodson, Rod E. Hubbard
Department of Chemistry, University of York, York YO1 5DD, England.

Birgitte Huge-Jensen, Esper Boel
Novo Nordisk A/S, 2880 Bagsværd, Copenhagen, Denmark.

Zygmunt S. Derewenda
Department of Biochemistry, University of Alberta, Edmonton, Alberta, Canada TG6 2H7.

Chapter 5 – *Structural aspects of phospholipase C*

Edward Hough, Sissel Hansen

Protein Crystallography Group, Institute of Mathematical and Physical Sciences, University of Tromsø, 9000 Tromsø, Norway.

Chapter 6 – *Phospholipase A₂: mechanism and structure*

Hubertus M. Verheij

Department of Enzymology and Protein Engineering, State University of Utrecht, CBLE, University Centre De Uithof, Padualaan 8, 3584 CH Utrecht, The Netherlands.

Bauke W. Dijkstra

Laboratory of Biophysical Chemistry, University of Groningen, Nijenborgh 4, 9747 AG Groningen, The Netherlands.

Chapter 7 – *Structure and mechanism of human pancreatic lipase*

Fritz K. Winkler, Klaus Gubernator

Pharma Research-New Technologies, F. Hoffmann-La Roche Ltd., 4002 Basel, Switzerland.

Chapter 8 – *Kinetics of triglyceride lipases*

Mats Martinelle, Karl Hult

Department of Biochemistry and Biotechnology, The Royal Institute of Technology, 100 44 Stockholm, Sweden.

Chapter 9 – *Gastric lipases: cellular, biochemical and kinetic aspects*

Frédéric Carrière, Youssef Gargouri, Hervé Moreau, Stephane Ransac, Ewa Rogalska, Robert Verger

Centre de Biochimie et de Biologie Moléculaire, CNRS Marseille, France.

Chapter 10 – *Lipase inhibitors*

Shamkant Patkar, Fredrik Björkling

Novo Nordisk A/S, Novo Allé, 2880 Bagsværd, Denmark.

Chapter 11 – *Substrates for phospholipase C and sphingomyelinase*

Ulrich Massing, Hansjörg Eibl

Max-Planck-Institut für biophysikalische Chemie, Am Faßberg, 37077 Göttingen, Germany.

Chapter 12 – *Isolation and purification of lipases*

M. Raquel Aires-Barros, M. Ângela Taipa, Joaquim M.S. Cabral

Laboratório de Engenharia Bioquímica, Instituto Superior Técnico,
1000 Lisboa, Portugal.

Chapter 13 – *Industrial applications of lipases*

Evgeny N. Vulfson

Department of Biotechnology and Enzymology, AFRC Institute of Food
Research, Earley Gate, Whiteknights Road, Reading RG6 2EF, England.

Chapter 14 – *Lipases and phospholipases in organic synthesis*

Jochem Kötting, Hansjörg Eibl

Max-Planck-Institut für biophysikalische Chemie, Am Faßberg,
37077 Göttingen, Germany.

Chapter 15 – *Medical aspects of triglyceride lipases*

Gunilla Bengtsson-Olivecrona, Thomas Olivecrona

Department of Medical Biochemistry and Biophysics, University of Umeå,
901 87 Umeå, Sweden.

Chapter 16 – *Protein engineering of lipases*

Richard Bott, A.J. Poulose

Genencor International Inc., 180 Kimball Way, South San Francisco,
CA 94080, U.S.A.

John W. Shield

Geobiotics Inc., 3505 Breakwater Avenue, Hayward, CA 94545, U.S.A.

Editors

Paul Woolley

Kemisk Institut, Aarhus Universitet, 8000 Århus C, Denmark.

Steffen B. Petersen

MR Center, Natural Science Section, SINTEF UNIMED, 7034 Trondheim,
Norway.

1.

Sequence comparisons within the lipase family

ALLAN SVENDSEN

At least 21 sequences of triacylglycerol lipases have been published. Of these, nine are from mammalian species and twelve are from microbial species. The mammalian lipase sequences are the group showing the clearest homology.* In contrast, within the microbial lipase group few relatively clearly homologous sequences are known. The mammalian sequences are briefly discussed here. Alignments of both mammalian and microbial lipases are shown. Some subfamilies within the microbial lipases are identified. Use of a 'forced' alignment to the known structure of *Mucor miehei* lipase reveals a previously undetected homologous stretch in the sequence of lipases.

1. BACKGROUND

Numerous attempts have been made to extract structural and functional information from sequences. When standard alignment methods (for a review, see reference 1) are applied to microbial lipase sequences, rather inconclusive results are obtained with regard to aligned residues and plausible gap positions. Alignment methods taking three-dimensional structural information into account have been published (2), as have methods that allow local sequence similarity within a sequence (3).

The aligning of lipase sequences can lead to the identification of essential lipase residues. If sequences can be aligned, expensive and demanding X-ray work can be avoided, and protein engineering work can be initiated at an earlier stage and on a reasonably rational basis.

The lipases considered here are all triacylglycerol lipases known to hydrolyse triglycerides (TG) to diglycerides (DG), monoglyceride (MG) or glycerol and fatty acids. Phospholipases will not be considered in the context of the review,

---

* *Editorial note*: In this book the term "homology" is used in its biological sense, implying relationship by evolution (72, 73). When similarity between sequences is observed empirically, on the basis of identical amino-acid residues at corresponding positions in aligned sequences, the fraction of identical residues is termed the *degree of identity* or *degree of similarity*.

though some of these could be of interest for comparison, on account of their dual activities. For example, guinea-pig pancreatic lipase (4) and *Staphylococcus hyicus* lipase have a phospholipase activity much higher than their triacylglycerol activity (5). Most of the phospholipases are totally different in sequence from triacylglycerol lipases.

The first published lipase sequence was that of porcine pancreatic lipase; its protein sequence was completed in 1981 (6). The primary structures of all the other lipases have been determined by recombinant-DNA techniques. Antonian reviewed twelve lipase sequences in 1988 (7). Most of the known lipase sequences have been published within the last four years. Their number will probably double in a few years. Ideas gained from comparison of lipase sequences, as in this chapter, will therefore need to be subject to constant revision. An example is the suggestion of an active-site residue at Asp104 in *Geotrichum candidum* lipase by homology with lipase from *Candida cylindracea* and acetylcholine esterase from torpedo (8); the crystal structure of *G. candidum* lipase later showed the active residue to be Glu354 (9).

In this chapter, homology will be considered in broad terms. One approach is the Dayhoff matrix, describing evolutionary and amino-acid physicochemical relationships (10).

2. HOMOLOGY WITHIN THE LIPASE FAMILY

Degrees of similarity between the different lipase groups are very low. One of their few known common features is the GxSxG consensus sequence (where x is any residue), common to most lipases. This consensus sequence is also observed in other esterases, and has frequently been termed the substrate-binding site.

2.1 Mammalian lipases

There are four groups, the lingual, the hepatic, the gastric and the pancreatic lipases (Figure 1). The gastric lipases have a region of rather weak homology around residues 240 and 260 (human pancreatic lipase numbering) (11). This region is called the flap region (11), a loop in the molecule that is placed on top of the active site. The highly conserved tryptophan in this region is absent from the gastric lipase sequence (Figure 2).

The mammalian lipase sequences considered in this chapter include triacylglycerol lipases (EC 3.1.1.3) from the alimentary canal. These lipases have approximately 450 amino-acid residues. The tracylglycerol lipase sequences have a catalytic domain of around 340 residues, and a probably non-catalytic domain of around 110 residues, as seen in human pancreatic lipase (11).

Triacylglycerol lipase sequences published include dog pancreatic (DPL; reference 12), rat lingual (RLL; 13), rat hepatic (RHL; 14), human hepatic (HHL;

		1	2	3	4	5	6	7	8	9
RHL	1	100	85	49	50	49	48	41	44	43
HHL	2		100	52	53	53	51	47	42	39
PHL	3			100	83	92	82	40	40	39
MPL	4				100	81	77	44	46	39
DPL	5					100	82	42	44	40
HPL	6						100	41	40	42
HGL	7							100	88	40
RLL	8								100	41
BSAL	9									100

Figure 1. Similarity of mammalian lipases. Numbers are per cent similarity. For abbreviations, see text. The gene sequences were converted to amino-acid sequences by the program TRANSLATE. The percentage similarity is optimised to the diagonal in order to see connections between 'families'. Families are defined as having >50% similarity and are shown boxed.

15–17), human pancreatic (HPL; 18), human gastric (HGL; 19), mouse pancreatic (MPL; 20) and porcine pancreatic (PPL; 6, 21).

The mammalian lipases have clear homology with lipoprotein lipases (LPL, EC 3.1.1.34; references 14, 22–24). Homology with hepatic and pancreatic lipases was found, as well as with *Drosophila* vitellogenins. The most pronounced similarity was detected in the N-terminal part of the proteins. Within these regions of the lipase, the putative lipid-binding region in found.

Published LPL sequences from cDNA include human (25), bovine (26), guinea-pig (27, 28), mouse (29), chicken (30) and a fragment from rat (31). The human LPL gene has ten exons (24, 32), the guinea-pig gene has nine (28). Bovine LPL was later characterised by protein methods, revealing the N- and C-terminal residues (Asp and Gly respectively), glycosylation sites, cystine bridges and active-site Ser as residue 134 (33). The carboxyl ester lipase from human pancreas has been shown to contain the consensus sequence GxSxG (34). A lipase found in human milk, milk bile-salt-activated lipase (BSAL; 35, 36), was shown to be identical with the carboxyl ester lipase found earlier (34).

2.2 Microbial lipases

Microbial sequences are known from both bacteria and fungi. They vary in length from around 258 to 544 amino-acid residues, and are compared in Figure 3.

The known bacterial lipase sequences are from *Staphylococcus aureus* (37) and *S. hyicus* (38), with 690 and 641 residues, respectively. The lengths quoted for

```
               X                        Y
        42    ▬▬                    ▬▬▬▬▬              ?   89
BSAL    TYGDEDCLYL NIWVPQGRK.  ..QVSRDLPV MIWIYGGAFL MGSGHGANFL
RLL     EVVTEDGYIL GVYRIPHGKN NSENIGKRPV VYLQHGLIAS AT..NWIANL
HGL     EVVTEDGYIL EVNRIPYGKK NSGNTGQRPV VFLQHGLLAS AT..NWISNL
DPL     TNKNPNNFQT LLPSDPSTIE ASNFQTDKKT RFTIHGFINK GE.ENWLLDM
HPL     TNENPNNFQE VA.ADSSSIS GSNFKTNRKT RFIIHGFIDK GE.ENWLANV
MPL     TNENPNNYQI ISATDPATIN ASNFQLDRKT RFIIHGFIDK GE.EGWLLDM
PHL     TNQNQNNYQE LV.ADPSTIT NSNFRMDRKT RFIIHGFIDK GE.EDWLSNI
HHL     GETNQ..GCQ IRINHPDTLQ ECGFNSSLPL VMIIHGWSVD GVLENWIWQM
RHL     KDESDRLGCQ LRPQHPETLQ ECGFNSSHPL VMIIHGWSVD GLLETWIWKI


        90                                            130
BSAL    NNYLYDGEEI ATRGNVIVVT FNYRVGPLGF LSTGDANLPG NYGLRDQHMA
RLL     PNNSLAFMLA DAGYDVWLGN SRGNTWSRKN VYYSPDSVEF WAFSFDEMAK
HGL     PNNSLAFILA DAGYDVWLGN SRGNTWARRN LYYSPDSVEF WAFSFDEMAK
DPL     CKNMFKVEE. ........VN CICVDWKKGS QTSYTQAANN VRVVGAQVAQ
HPL     CKNLFKVES. ........VN CICVDWKGGS RTGYTQASQN IRIVGAEVAY
MPL     CKKMFQVEK. ........VN CICVDWKRGS RTEYTQASYN TRVVGAEIAF
PHL     CKNLFKVES. ........VN CICVDWKGGS RTGYTQASQN IRIVGAEVAY
HHL     VAALKSQPAQ P.......VN VGLVDWITLA HDHYTIAVRN TRLVGKEVAA
RHL     VGALKSRQSQ P.......VN VGLVDWISLA YQHYAIAVRN TRVVGQEVAA


        131                                           175
BSAL    IAWVKRNI.A AFGGDPNNIT |LFGESAGGAS| VSLQTLSPYN K...GLIRRA
RLL     YDLPATINFI VQKTGQEKIH |YVGHSQGTTI| GFIAFSTNPT L..AKKIKTF
HGL     YDLPATIDFI VKKTGQKQLH |YVGHSQGTTI| GFIAFSTNPS L..AKRIKTF
DPL     MLSMLS...A NYSYSPSQVQ |LIGHSLGAHV| AGEAGSRTPG ...LGRITGL
HPL     FVEFLQ...S AFGYSPSNVH |VIGHSLGAHA| AGEAGRRTNG T..IGRITGL
MPL     LVQVLS...T EMGYSPENVH |LIGHSLGSHV| AGEAGRRLEG H..VGRITGL
PHL     FVEVLK...S SLGYSPSNVH |VIGHSLGSHA| AGEAGRRTNG T..IERITGL
HHL     LLRWLE...E SVQLSRSHVH |LIGYSLGAHV| SGFAGSSIGG THKIGRITGL
RHL     LLLWLE...E SMKFSRSKVH |LIGYSLGAHV| SGFAGSSMGG KRKIGRITGL


        176                                           220
BSAL    ISQSGVALSP WVIQKN.... ..PLFWAKKV AEKVGCPVGD AARMAQCLKV
RLL     YALAPVATVK YTQSPLKKIS FIPTFLFKLM FGKKMFLPHT YFDDFLGTEV
HGL     YALAPVATVK YTKSLINKLR FVPQSLFKFI FGDKIFYPHN FFDQFLATEV
DPL     DPVEASFQGT PEEVRLD... ..PTDADFVD VIHTDAAPLI PFLGFGTSQQ
HPL     DPAEPCFQGT PELVRLD... ..PSDAKFVD VIHTDGAPIV PNLGFGMSQV
MPL     DPAEPCFQGL PEEVRLD... ..PSDAMFVD VIHTDSAPII PYLGFGMSQK
PHL     DPAEPCFQGT PELVRLD... ..PSDAKFVD VIHTDAAPII PNLGFGMSQT
HHL     DAAGPLFEGS APSNRLS... ..PDDASFVD AIHTFTREHM GLSVGIK.QP
RHL     DPAGPMFEGT SPNERLS... ..PDDANFVD AIHTFTREHM GLSVGIK.QP


                              Z
        221                   ▬▬▬▬▬▬▬▬▬▬▬▬▬▬▬▬▬▬      270
BSAL    TDPRALTLAY KVPLAGLEYP MLHYVGFVPV IDGDFIPADP INLYANAADI
RLL     CSREVLDLLC SNTLFIFCGF DKKNLNVSRF DVYLGHNPAG TSVQDFLHWA
HGL     CSREMLNLLC SNALFIICGF DSKNFNTSRL DVYLSHNPAG TSVQNMFHWT
DPL     MGHLDFFPNG GEEMPGCKKN ALSQIVNLDG IWEGTRDFVA CNHLRSYKYY
HPL     VGHLDFFPNG GVEMPGCKKN ILSQIVDIDG IWEGTRDFAA CNHLRSYKYY
MPL     VGHLDFFPNG GKEIPGCQKN ILSTIVDING IWEGTRNFAA CNHLRSYKYY
PHL     VGHLDFFPNG GKQMPGCQKN ILSQIVDIDG IWEGTRDFVA CNHLRSYKYY·
HHL     IGHYDFYPNG GSFQPGCHFL ELYRHIAQHG FNAITQTIK. CSHERSVHLF
RHL     IAHYDFYPNG GSFQPGCHFL ELYKHIAEHG LNAITQTIK. CAHERSVHLF
```

Figure 2. Tentative alignment of mammalian lipases. Only the homology of residues 42–270 (HPL numbering) is shown. Segment B (Figure 4) is indicated with a bar marked X. The novel segment seen as box 1 in Figure 6 is indicated with a bar marked Y. The Trp in the lid region of *Mucor miehei* lipase are marked with a '?'. Sequences of ten residues including the GxSxG consensus sequences of the mammalian lipases are boxed. The so-called flap region of HPL is indicated by a bar marked Z.

		1	2	3	4	5	6	7	8	9	10	11	12
Morax lip. 3	1	100	40	43	42	44	41	45	44	43	38	44	41
Morax lip. 1	2		100	46	42	40	37	41	40	41	42	38	42
P. putida	3			100	46	44	35	39	41	45	42	42	36
C. cylind.	4				100	63	47	48	48	44	47	43	44
G. candidum	5					100	43	48	48	44	46	40	40
P. glumae	6						100	89	61	47	49	42	40
P. cepacia	7							100	60	43	49	41	40
P. fragi	8								100	52	49	41	40
S. hyicus	9									100	69	46	45
S. aureus	10										100	42	45
Humicola sp.	11											100	56
M. miehei	12												100

Figure 3. Similarity of microbial lipases. The sequences are from the NBRF protein data bank. Morax lip. 1 (Moraxella lip1) pre-form, *S. hyicus* lipase truncated to residues 245–622, *S. aureus* lipase truncated to residues 281–667 by homology with *S. hyicus*, *P. fragi* lipase pre-form, *G. candidum* lipase minus residues 1–19, *Mucor miehei* lipase truncated to residues 95–364, and from the EMBL gene bank Morax lip. 3 truncated to the pre-form (nucleotides 658-1605), *P. cepacia* lipase truncated to nucleotides 364–1458 and *C. cylindracea* lipase truncated to nucleotides 13–1614. In the special case of *C. cylindracea* all CTG codons were altered by hand to Ser instead of Leu. Lipase sequences from *P. putida*, *Humicola lanuginosa* and *P. glumae* were taken from patent applications. Methods used are the same as in Figure 1.

lipases found by screening of gene libraries are probably too great, as was seen for the *S. hyicus* preprotein, which was found to be 641 residues long when expressed in *S. carnosus*. In *S. hyicus*, it is processed between residues Thr245 and Val246 to give a protein of 46 kDa with an N-terminal sequence Val-Lys-Ala-Ala-Pro-Glu. Corresponding shortening will give *S. hyicus* lipase a length of 377 residues, and *S. aureus* lipase around 380 residues. The patent literature further contains sequences from *Bacillus* species.

The bacterial genus *Pseudomonas* is represented by many sequences. Most sequences from *Pseudomonas* are 258–320 residues in length. *Pseudomonas* lipase sequences have been described from *P. fragi*, with 277 amino-acid residues (39), *P. cepacia*, the mature protein containing 320 residues (40), *P. glumae*, 320 residues in length (41) and *P. putida*, 258 residues (42). A sequence from *P. fragi* with only 135 residues (43) and further sequences of *Pseudomonas* species from the patent literature are not included here.

Two bacterial sequences from the psychrotrophic organism *Moraxella* TA144 are known, lip1 (44) and lip3 (45). The length of the preproteins are 319 and 315 amino-acid residues, respectively.

Two fungal sequences have been found in *Geotrichum candidum*, termed I and II (8, 46). *G. candidum* lipase II is 544 residues long, the longest of the microbial lipases. These sequences contain the GxSxG consensus sequence in their N-terminal region. The X-ray structure of *G. candidum* lipase has recently been published (9), and shows a single structural domain.

Other known fungal lipase sequences are from the phycomycete fungus *Rhizomucor miehei*, previously classified as *Mucor miehei* (49), from *Humicola lanuginosa* (50) and from *Penicillium camembertii* U-150 (51). These sequences are 269 residues long. Two sequences from *Rhizopus* species can be found, one in the patent literature and one in the data bank SWISSPROT. The latter will be mentioned in the discussion.

Lipase lip1 from the asporogenic yeast *Candida cylindracea* (the strain deposited as *C. rugosa*) was found to contain a codon (CTG) normally coding for Leu, coding for Ser in the wild type (47). Other sequences from the fungal genus *Candida* are known. These are *C. antarctica*, components A and B. Twenty N-terminal residues have been reported from these (48), and the B component lipase sequence is in preparation (I.G. Clausen *et al.*, unpublished results).

The overall lengths of the lipases seem to be very similar (around 300 residues), except for the very long sequences of *G. candidum* and *C. cylindracea* lipases. They probably comprise one structural domain, as seen in *M. miehei* (52).

2.3 The active-site residues

The active-site residues have been found to be a triad of Ser, Asp and His, as in serine proteases (11, 52). The active-site Ser residue is contained in the consensus sequence GxSxG, known as the substrate-binding site from many esterases. In *Geotrichum candidum* it was found that the triad consisted of Ser, Glu and His (9). The order in sequence of the active-site residues in lipases is Ser...Asp...His (53). This is seen in the X-ray structures of lipases from *Mucor miehei* (Ser144, Asp203, His258; reference 52), *G. candidum* (Ser217, Glu354, His463; 9) and human pancreatic lipase (Ser152, Asp176 and His263; 11).

Bovine milk lipoprotein lipase was found to contain DFP-sensitive serine in the context $W-X_{1-3}-S$ (54). This sequence was then suspected to be involved in the enzymic mechanism. The sequence was not found in *G. candidum* (8, 46), and was only found in some of the mammalian lipases (55).

3. HOMOLOGY BETWEEN FAMILIES

Attention has frequently been drawn to similarities between the lipase GxSxG consensus sequence and other proteins. The human lecithin–cholesterol acyltransferase is one example (49). *Candida cylindracea* has been shown to be a member of the cholinesterase family of serine esterases (56). This was found also to be true for the homologous *Geotrichum candidum* (8). Bell *et al.* (56) found sequence homology between this and cholinesterases, esterases, lysophospholipases and thyroglobulins but not other lipases. Within this group, three highly conserved regions were found (Figure 4; reference 56). One segment, termed C, contained the consensus sequence GxSxG, and an extended homologous region between this and the N-terminus was found. Another segment, B, is highly conserved in cholinesterases, lysophospholipases and thyroglobulins. Segment A had a large number of hydrophobic residues, and was therefore suggested to be involved in catalysis or substrate-binding.

Sussman *et al.* 1991 (57) showed that the overall structure of torpedo acetylcholinesterase resembled closely that of *G. candidum* lipase (9). *G. candidum* lipase I and *C. cylindracea* lipase show 45% overall identity (46), suggesting that they have the same structural folding. The bile-salt-activated lipase and acetylcholinesterase share an identical stretch of 11 residues (EDCLYLNIWVP; reference 34), indicating a close relationship. This stretch is homologous to the B segment found by Bell *et al.* (56) and mentioned above.

Despite the weak overall homology between hepatic and pancreatic lipases, they may fold in a similar way. Twelve and ten Cys residues are conserved in the pancreatic and hepatic lipases, respectively. Eight of these are at the same positions, suggesting the same folding in these lipases. In *G. candidum* and *C. cylindracea* lipases, it was found that four cysteine residues were in identical positions, suggesting that structures are conserved among evolutionarily close species (46).

The homology between HHL, RHL and LPL (mouse, bovine and human) is emphasized by the finding of two N-glycosylation sites conserved among these sequences. However, they are not found in dog and pig pancreatic lipases (58).

Homology has been found between mouse LPL and fatty-acid-binding protein (28) in exon IV, whereas pancreatic and hepatic lipases resemble the fatty-acid-binding proteins to a much lesser extent. However, the comparison is weakened by the fact that LPL and fatty-acid-binding proteins are intracellular proteins while the other lipases are extracellular. It was supposed that a segment with the sequence NRCNNVG could be a connection between two functional domains. Homology between the central core of LPL and vitellogenins was observed (23).

Kirchgessner *et al.* (24) have proposed that exon-shuffling may have been involved in the evolution of human lipoprotein lipase, hepatic lipase and pancreatic lipase, in which they include the *Drosophila* yolk protein 1.

4. AUTOMATIC SEQUENCE ALIGNMENT

Sequences were taken from the Protein Identification Resource data bank (PIR, at NBRF; release 29.0), from the EMBL gene bank (release 28.0) and from the database SWISSPROT (release 19.0). Searches were done with the string

SEGMENT A

Lipase:	*Candida cylindracea*		NEAFLGIPFAEPPVGNLRFKDPVPYSG
Cholinesterase:	acetyl	Fruit fly	ISAFLGIPFAEPPVGNMRFRRPEPKKP
	acetyl	Electric ray	VHVYTGIPYAKPPVEDLRFRKPVPAEP
	butyryl	Human	VTAFLGIPYAQPPLGRLRFKKPQSLTK
Esterase:	Fruit fly		YYSYESIPYAEPPTGDLRFEAPEPYKQ
	Rabbit		VAVFLGVPFAKPPLGSLRFAPPQPAES
Lysophospholipase:	Rat		VDIFKGIPFATAKTLENPQRHPGWQGT
Thyroglobulin:	Bovine		VDQFLGVPYAAPPLGEKRFRAPEHLNW
	Rat		VYQFLGVPYAAPPLAENRFQAPEVLNW

SEGMENT B

Lipase:	*Candida cylindracea*		SPNSEDCLTKQVVEP---
Cholinesterase:	acetyl	Fruit fly	REMSEDCLYLNIWVPSPR
	acetyl	Electric ray	TNVSEDCLYINVWAPAKA
	butyryl	Human	TDLSEDCLYLNVWIPAPK
Esterase:	Fruit fly		LVGEEDCLTVSVYKPKNS
	Rabbit		LKFSEDCLYLNIYTPADL
Lysophospholipase:	Rat		TYGQEDCLYLNIWVPQGR
Thyroglobulin:	Bovine		PGVSEDCLYLNVFVPQNM
	Rat		PQISEDCLYLNVFVPENL

SEGMENT C

Lipase:	*Candida cylindracea*		WVADNIAAFGGDPTKVTIFGESAGSMM
Cholinesterase:	acetyl	Fruit fly	WVHDNIQFFGGDPKTVTIFGESAGGAS
	acetyl	Electric ray	WLKDNAHAFGGNPEWMTLFGESAGSSS
	butyryl	Human	WVQKNIAAFGGNPKSVTLFGESAGAAS
Esterase:	Fruit fly		WIKQNIASFGGEPQNVLLVGHSAGGAS
	Rabbit		WVQDNIANFGGDPGSVTIFGESAGGQS
Lysophospholipase:	Rat		WVKRNIAAFGGDPDNITIFGESAGGAI
Thyroglobulin:	Bovine		WVKTHIQAFGGDPRRVTLAADRGGADI
	Rat		WVKTHIGAFGGDPQRVTLAADRGGADV

Figure 4. Segments A, B and C found in the cholinesterase sequences. Reproduced with permission from Bell *et al.* 1989 (56).

'lipase'. For triacylglycerol lipases 33, 27 and 21 entries, respectively, were found. The database entry codes for the sequences used are given in Appendix 1.

The automatic alignment was done on VAX computer with the UWGCG package version 7.0 (59). The alignment program used was GAP, using the Needleman–Wunsch algorithm with a gap penalty of 3 and an extension penalty of 0.1. Because of the fact that the scores are obtained from non-optimised alignments, the quality of the alignments generated by GAP was checked by quantitative comparison with the results of ten similar alignments in which one of the sequences was randomised. Degrees of similarity less than 50% were defined to be non-significant.

4.1 Mammalian sequences

Three families were detected within the mammalian lipases, the hepatic, the pancreatic and the gastric/lingual family (Figure 1). The hepatic and pancreatic lipase sequences have a per cent similarity of around 50, whereas gastric/lingual lipase sequences have a low similarity to the other groups.

4.2 Microbial sequences

This group had an overall similarity of 40%. Four families are suggested, which have similarity higher than 50% (Figure 2). These are the *Candida cylindracea* family (fI; the cholinesterase family), the *Pseudomonas* family (fII), the *Staphylococcus* family (fIII) and the *Mucor* family (fIV). The two *Moraxella* lipase sequences had less than 50% per cent similarity to each other, as well as to any of the other sequences. Interestingly, the *Pseudomonas putida* sequence has higher similarity to *Moraxella* lip1 lipase and to *C. cylindracea* lipase than to the *Pseudomonas* lipase family.

4.3 Comparison of mammalian and microbial sequences

Similarity searches between mammalian and microbial lipases resulted in an average similarity much lower than 50% (data not shown). Two exceptions were observed: similarity of bile-salt-activated lipase (BSAL) to *Geotrichum candidum* lipase (54%), and of BSAL to *Candidum cylindracea* lipase (51%). The homology with *G. candidum* lipase was especially pronounced around the GxSxG consensus sequence and in the N-terminal part. This homologous region includes a homology with segment B (56) in the cholinesterase family, and a novel homologous part of microbial lipases comprising a segment of β strand in *Mucor miehei* and *G. candidum* lipases.

4.4 Dot-plots

Dot-plots of microbial lipases reveal the homology within the same groups as seen with the GAP program. A selection of dot-plots giving homologies within the 'subfamilies' of lipases is shown in Appendix 2. The dot-plots were carried out on a VAX computer using UWGCG package version 7.0 (59). The COMPARE and DOTPLOT programs were used with a window of 30 and a stringency of 12.

The dot-plots of sub-families fIII and fIV show clear diagonals with acceptably low levels of background noise. The stretch of diagonal that is lacking in the N-terminal region in the comparison with either of the other two sequences in sub-family fIV is noticeable. Family fII shows a higher degree of noise around the diagonals. Especially pronounced is the visual square of repeats, seen in the dot-plot between *Pseudomonas cepacia* and *P. glumae*. In the other two dot-plots, only one corner of the square of repeats is observed. The reason for this is the absence of homology in the C-terminal part, between residues 180 and 220 in the *P. fragi* lipase. The two microbial lipases in sub-family fI also have noticeably high noise. For convenience, the homology of BSAL with *Geotrichum candidum* lipase is shown. The N-terminal region is the most clearly homologous part.

5. FORCED ALIGNMENT – X FIXED POINTS ALIGNMENT

In the search for distant homologies, a more intuitive method was adopted. With the Dayhoff comparison matrix in mind, together with the normal gap probabilities in loop regions, one can align the lipase sequences with a guiding sequence(s) of known structure(s).

The ease or difficulty in aligning the sequences then depends upon picturing the relationships between them. Irrespective of whether the relationship is close or distant, or indeed of whether any parts of the sequence appear homologous at all, possible structural relationships are examined.

The method used here can be called the 'x fixed points' method, where x denotes any reasonable starting point such as homologous stretches, putative active-site residues or other supposed common structural features.

From the known structures of lipases we can extract some structural features that are alike. There are four major structural, functional or sequence homologies in the known lipases (Figures 5 and 6). (i) The consensus sequence GxSxG around the active-site serine (box 3). (ii) The strand–helix motif around the active serine residue. This fold is seen in *Mucor miehei* (52), *Geotrichum candidum* (9) and human pancreatic lipase (11). Parts of the lipases' β-sheet structures with parallel strands are seen in all three known lipase structures (15). (iii) A third feature is the buried active site covered by a lid or lids (box 2). These lids seem to be rather variable in structure and position in sequence, but most of them contain a tryptophan residue, supposedly in close contact with the active-site serine, as seen

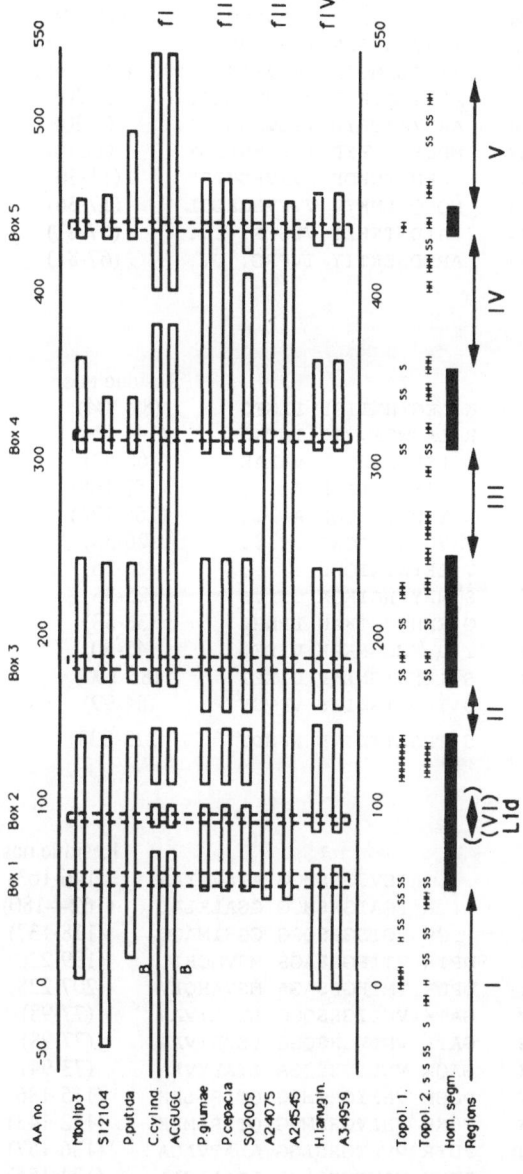

Figure 5. Tentative comparison of microbial lipase sequences. Comparison was done with the LINEUP program from the UWGCG package (59). Correspondingly placed stretches are shown by boxes. The topologies of *Mucor miehei* and *Geotrichum candidum* lipases are Topol. 1 and Topol. 2, respectively. Segments that fall in the same region (Hom. segm.) are shown as thick lines. Regions clearly showing homology or used as 'fixed points' are indicated by dashed boxes (shown in detail in Figure 6). Family relationships found by automatic alignment are shown to the right using family numbering (fI, fII, fIII and fIV). Highly variable regions are shown with arrows in the bottom line. Regions I–VI are discussed in the text. The segment B (see Figure 4) is indicated in the relevant sequences. Gaps of less than 10 residues are not shown. The *Mucor miehei* lid region is indicated as LID. Other codes are listed in Appendix 1.

Box 1

			Residue nos.
Morax lip. 1	AENDNVTGEP	LLLIHGFGGN	(60-79)
Morax lip. 3	GLLGAIAVVP	GYVSYENSIK	(110-129)
P. putida	DLGQGGVRHP	VILWGNGTGA	(41—60)
C. cylind.	ITKSIAMGKP	IIHVSVNY..	(139-158)
G. cand. lip. A	VKESLEMGQP	VVFVSINY..	(146-165)
P. glumae	ADTYAATRYP	VILVHGLAGT	(1-20)
P. cepacia	AAGYAATRYP	IILVHGLSGT	(1-20)
P. fragi	MDDSVNTRYP	ILLVHGLFGF	(1-20)
S. hyicus	NPENPKNKDP	FVFVHGFTGF	(11-30)
S. aureus	NQVQPLNKYP	VVFVHGFLGL	(17-36)
H. lanuginosa	LALDNTNKLI	VLSFR.....	(67-81)
M. miehei	ARGDSEKTIY	IVFRG.....	(67-81)

Box 2

		?	Residue nos.
Morax lip. 3	K	LEGYHLIIPD LLGFG	(89-104)
Morax lip. 1	R	LASWGFVVIT INTNS	(144-159)
P. putida	L	LSHWASHGFV VAAAE	(68-83)
C. cylind.	R	VSSWGFLAGD E....	(157-168)
G. cand. lip. A	R	TGPYGFLGGD A....	(165-176)
P. glumae	V	VDYWYGIQSD	(26-36)
P. cepacia	V	LEYWYGIQED	(26-36)
P. fragi	S	HHYFHGIKQA	(25-35)
S. hyicus	G	ENHWGGTKAN LRNHL	(38-53)
S. aureus	Y	PNYWGGNKFK VIEEL	(45-60)
H. lanuginosa	S	IENWIGNLNF DLKEI	(85-100)
M. miehei	S	IRNWIADLTF VPVSV	(84-99)
R. delemar	S	FRSAITDIVF NFSDY	(223-238)

Box 3

	*		Residue nos.
Morax lip. 3	.... THVGGNSMGG	AISVAYAA	(146-163)
Morax lip. 1	...R LGAIGWSMGG	GGALKLAT	(162—180)
P. putida	..GR VGTSGHSQGG	GGSIMAGQ	(118-137)
C. cylind.	DPTK VTIFGESAGS	MSVMCHIL	(199-220)
G. cand. lip. A	DPDK VMIFGESAGA	MSVAHQLV	(207-228)
P. glumae	GATK VNLIGHSQGG	LTSRYVAA	(77-98)
P. cepacia	GATK VNLVGHSQGG	LSSRYVAA	(77-98)
P. fragi	GAQR VNLIGHSQGA	LTARYVAA	(73-94)
S. hyicus	PGHP VHFIGHSMGG	QTIRLLEH	(115-186)
S. aureus	PGKK VHLVGHSMGG	QTIRLMEE	(122-143)
H. lanuginosa	PDYR VVFTGHSLGG	ALATVAGA	(136-157)
M. miehei	PSYK VAVTGHSLGG	ATALLCAL	(134-155)

Box 4

		*	Residue nos.
Morax lip. 3	YDFVMYKP	PYI	(201-211)
Morax lip. 1	DDRIAETK	KYA	(233-243)
P. putida	EDTIAFPY	LNA	(176-186)
C. cylind.	DEGTFFGT	SSL	(340—350)
G. cand. lip. A	DEGTILAP	VAI	(353-363)
P. glumae	QDALAALR	TLT	(158-168)
P. cepacia	QDALAALQ	TLT	(158-168)
P. fragi	QNALNALN	ALT	(154-164)
S. hyicus	YDLTREGA	EKI	(198-208)
S. aureus	YDLTLDGS	AKL	(256-266)
H. lanuginosa	NDIVPRLP	PRE	(199—209)
M. miehei	RDIVPHLP	PAA	(202-212)

Box 5

		*	Residue nos.
Morax lip. 3	NDV	GHVPMVEA	(278-288)
Morax lip. 1	NNG	SHFCPSYR	(260-270)
P. putida	RYV	SHFEPVGS	(202-212)
C. cylind.	LGT	FHSNDIVF	(445-455)
G. cand. lip. A	LGT	FHGSDLLF	(459-469)
P. glumae	YHW	NHLDEINQ	(272-282)
P. cepacia	YKW	NHLDEINQ	(292-302)
P. fragi	YPL	DHLDTINH	(256-266)
S. hyicus	KGW	DHSDFIGN	(352-362)
S. aureus	QGW	DHVDFIGV	(360-370)
H. lanuginosa	DIP	AHLWYFGL	(254-264)
M. miehei	SVL	DHLSYFGI	(253-263)

Figure 6. Homologous sequences or sequences used as 'fixed points' in the alignment procedure. The box numbers refer to the boxes in Figure 5. Under box 2 a segment from *Rhizopus delemar* lipase is shown. The numbers in the sequences are shown, using the sequences defined in Figure 3, except that the *P. cepacia* sequence is numbered as the mature protein, with the 44-residue propeptide deleted (40). Box 1: the hydrophobic stretch in front of the lid in *M. miehei* lipase. Box 2: the 'fixed point' lid with the Trp residue. Box 3: the 'fixed point' consensus sequence, GxSxG. Box 4: the 'fixed point' active site Asp/Glu or Xxx. Box 5: the 'fixed point' active-site residue histidine.

in HPL (11) and *M. miehei* lipase (52). (iv) Finally, the triad residues are in the order Ser...Asp/Glu...His, lying about 50 residues from one another (boxes 4 and 5; references 9, 11, 52, 57).

By using these common features as guides, 'fixed points', to align the sequences, a new feature was found. A hydrophobic stretch is seen immediately before the lid region in *M. miehei* lipase (Figures 5 and 6, box 1). This segment is a β strand in both *M. miehei* and *G. candidum* lipases.

Because of difficulties in aligning the Asp/Glu residue, the forced fixed-point alignment tends to allow other residues than Asp in the catalytic triad. In this alignment it can be Asn as well. Glu was recently proposed to be an active-site residue in *G. candidum* lipase (9) and in a homologous acetylcholine esterase (57).

The alignment indicates at least five regions (I–V; Figure 5) with great variability and a smaller region (VI) around the lid region in *M. miehei* lipase. When superimposed on the *M. miehei* structure, three of these regions show that the lipid contact zone varies most (variable regions III, IV, VI). The lipid contact zone can be defined as the side (face) of a lipase that changes conformation on interfacial activation. Two such variable regions are at the N and C termini (variable regions I and V). Variable region IV is the large insert in the C-terminal part of the sequence, as seen for lipases from *Pseudomonas* species and *G. candidum* and *Candida cylindracea*. Variable region III is another large insert between residues 250 and 325 in the comparison (Figure 5).

6. DISCUSSION

In this review, the lipases were assumed to have common structural features. Tentative alignment of all the lipase sequences seems to be possible, identifying some homologous parts, but also pointing out some parts where homology cannot be detected. These may well be the parts of the lipid contact zone responsible for specificity.

The novel sequence homology (box 1 in Figures 5 and 6) contains in most of the sequences a proline followed by a number of hydrophobic residues. The mammalian sequences, except for the pancreatic lipases, also contain this segment. The segment is shown to be part of a sheet structure in both *Geotrichum candidum* and *Mucor miehei* lipases. The rather different paths taken by these two enzymes after this β strand suggest a rather speculative relationship. But the fact that this motif is also seen in all the microbial lipases and in some mammalian lipases suggests at least some common structural feature.

The second 'fixed point' (box 2), around the lid Trp88 of *M. miehei* lipase, seems to be very variable in sequence, even within the sub-families (fI, fII, fIII and fIV). The importance of this sequence in comparison is therefore low.

A high degree of identity around the so-called consensus site is observed (box 3). A conserved basic residue adjacent to a conserved valine is positioned seven and six residues to the N-terminal side of the active-site serine, thus indicating structural or functional importance of these two residues.

The alignments of the sequences in boxes 4 and 5 are very tentative. Active-site Glu in *Candida cylindracea* lipase is clearly identified by its homology with *G. candidum* lipase. The active-site histidine of *C. cylindracea* is located in the same way. The two aspartic acid and one asparagine in the *Pseudomonas* sequences are chosen as they are the only aligning residues of interest, except perhaps for the glutamine just before the aspartic acid and asparagine residues. This suggests an asparagine in the active site of *P. fragi* lipase. The active-site histidine within the *Pseudomonas* sequences are suggested by homology. At least within sub-families fII and fIII, a consensus sequence of HxDxI is observed.

Recently Krejci *et al.* (61) proposed on the basis of sequence alignment that the active-site residue in acetylcholinesterase from *Torpedo californica* is Asp397. To test this, they mutated Asp397 to Asn, and the activity of the mutant enzyme was less than 5% of the wild-type value. The active-site Glu354 proposed by Schrag *et al.* (9) is nearly totally conserved in their alignment, except for five sequences that have Asp at this position.

A segment represented by SEDCLYLNVW from fruit-fly acetylcholinesterase is also found in BSAL, rat lingual lipase and human gastric lipase. The relationship between BSAL and the cholinesterase family is also indicated by a dot-plot, showing a clear diagonal comprising the N terminus of the BSAL (Appendix 2).

The sequence of *Rhizopus delemar* lipase (SWISSPROT database) has 71% and 53% similarity to *M. miehei* and *Humicola lanuginosa* lipases, respectively. Interestingly, the *Rhizopus delemar* lipase lacks the otherwise conserved tryptophan in the lid region (homology with *M. miehei* lipase); see box 2.

From the comparison of microbial lipases (Figure 5), it is seen that the central to N-terminal part of the lipase sequences have few and small insertions, and therefore probably have the greatest similarity. This could raise the question of whether the C-terminal parts of at least the microbial lipases are responsible for the lipases' specificity, as they have many loops in contact with the supposed buried active site.

Note added in proof

Since this manuscript was written, several new lipase sequences have been published. One is that of *Rhizopus niveus* lipase (62), and this sequence is identical with that of *Rhizopus delemar* (63). Another is *Bacillus subtilis* lipase (64). This lipase is very small (19 kDa), and it has the PVVMV motif at the beginning of the sequence, as is also seen for the *Pseudomonas* family. The consensus site in this lipase is not GxSxG but AxSxG. This consensus site has also been found in *Candida antarctica* component B lipase (I.G. Clausen, unpublished results). The C-terminal part (after the consensus site) is rather short, indicating a new family with a small, variable sub-domain. In the *Pseudomonas* family, the sequence of *P. auroginosa* lipase has been published (65). It also contains the PIVLA motif close to its N terminus. The paper describes the conserved Asp, His and Trp, to suggest the active-site residues. The suggestions for active asparagine do not fit with the proposals in this chapter. Another lipase from the genus *Pseudomonas* is the *P. fluorescens* B52 lipase (66). This lipase has a molecular weight of 50kDa. Two other lipase sequences of the same size have been published, *P. fluorescens* lipA (67) and *P. fluorescens* SIK W1 lipase (68). These three homologous sequences have no obvious PILVV-type motif. Their most pronounced similarity to other *Pseudomonas* species lipases is the consensus sequence. They indicate another new family. These lipase sequences will be analysed in more detail. The *Pseudomonas nov.* species 109 (69) lipase sequence contains the PIVLA motif, indicating that the mature lipase N terminus starts a few residues before that sequence. Among sequences that are homologous to lipases are the *P. fluorescens* carboxylesterase (70), which, like the *Pseudomonas* lipase family, has the PLIIN sequence motif close to the N-terminus. It still has a very short C-terminal part (as for *Bacillus subtilis* lipase) and has a total molecular weight of 23kDa.

APPENDIX 1. LIST OF ENTRY NUMBERS

PIR (NBRF) protein data bank

Mammalian sequences

A27442, rat hepatic lipase; A33533, human hepatic lipase; LIPG, pig (hepatic) lipase; A34671, mouse pancreatic lipase; A34494, human pancreatic lipase; LIDG, dog pancreatic lipase; S07145, human gastric lipase; LIRTT, rat lingual lipase; A37916, bile-salt-activated lipase; LIP$RHIDL, *Rhizopus delemar* lipase.

Microbial sequences

S12104, *Moraxella* lip1 lipase; A24075, *S. hyicus* lipase; A24545, *S. aureus* lipase; S02005, *P. fragi* lipase; ACGUGC, *G. candidum* lipase; A34959, *Mucor miehei* lipase.

EMBL gene data bank

Mbolip3, *Moraxella* lip3 lipase; Pselipaa, *P. cepacia* lipase; Ysalipase, *C. cylindracea* lipase.

APPENDIX 2. ALIGNMENT WITHIN LIPASE SUB-FAMILIES

A selection of dot-plots is shown. The sequences are identified in Appendix 1.

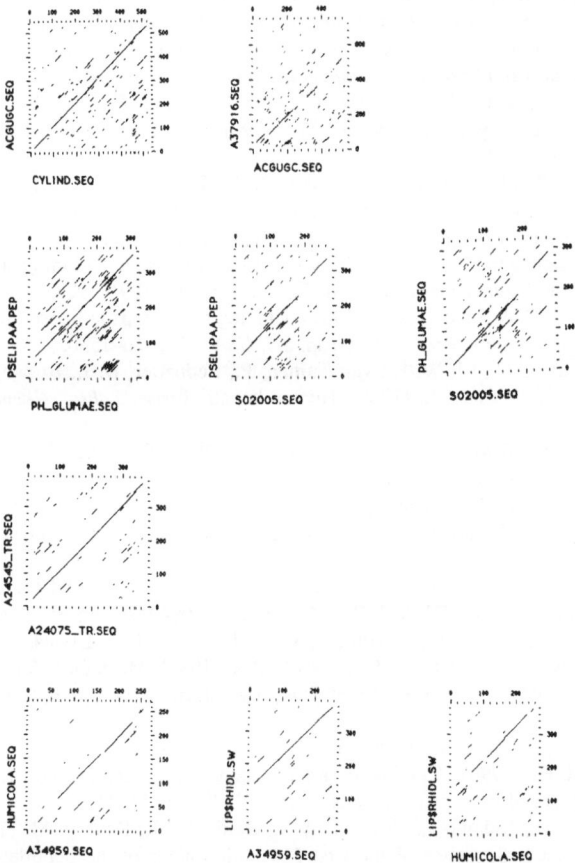

REFERENCES

1. Argos, P., Vingron, M. & Vogt, G. (1991). Protein sequence comparison: methods and significance. *Prot. Engin.*, 4, 375–383.
2. Taylor, W.R. (1986). Identification of protein sequence homology by consensus template alignment. *J. Mol. Biol.*, 188, 233–258.
3. Pearson, W.R. & Lipman, D.J. (1988). Improved tools for biological sequence comparison. *Proc. Natl. Acad. Sci. USA*, 85, 2444–2448.
4. Verger, R. (1984). Pancreatic lipases. In *Lipases*, ed. Borgström, B. & Brockman, H.L., pp. 83–150. Amsterdam: Elsevier.
5. van Oort, G.M., Deveer, A.M.Th.J., Dijkman, R., Tjeenk, M.L., Verheij, H.M., de Haas, G.H., Wenzig, E. & Götz, F. (1989). Purification and substrate specificity of *Staphylococcus hyicus* lipase. *Biochemistry*, 28, 9278–9285.
6. Caro, J. De, Boudouard, M., Bonicel, J., Guidoni, A., Desnuelle, P. & Rovery, M. (1981). Porcine pancreatic lipase. Completion of the primary structure. *Biochim. Biophys. Acta*,

671, 129–138.

7. Antonian, E. (1988). Recent advances in purification, characterisation and structure determination of lipases. *Lipids*, 23, 1101–1106.

8. Shimada, Y. Sugihara, A., Iizumi, T. & Tominaga, Y. (1990). cDNA cloning and characterisation of *Geotrichum candidum* lipase II. *J. Biochem.*, 107, 703–707.

9. Schrag, J.D., Li, Y., Wu, S. & Cygler, M. (1991). Ser-His-Glu triad forms the catalytic site of the lipase from *Geotrichum candidum*. *Nature*, 351, 761–764.

10. Dayhoff, M.O., Barker, W.C. & Hunt, L.T. (1983). *Methods Enzymol.* 91, 524–545.

11. Winkler, F.K., D'Arcy, A. & Hunziker, W. (1990). Structure of human pancreatic lipase. *Nature*, 343, 771–774.

12. Mickel, F.S., Weidenbach, F., Swarovsky, B., La Forge, K.S. & Sceele, G.A. (1989). Structure of the canine pancreatic lipase gene. *J. Biol. Chem.*, 264, 12895–12901.

13. Docherty, A.J.P., Bodmer, M.W., Anga l, S., Verger, R., Rivière, C., Lowe, P.A., Lyons, A., Emtage, J.S. & Harris, T.J.R. (1985). Molecular cloning and nucleotide sequence of rat lingual lipase cDNA. *Nucl. Acids Res.*, 13, 1891–1903.

14. Komaromy, M.C. & Schotz, M.C. (1987). Cloning of rat hepatic lipase cDNA: evidence for a lipase gene family. *Proc. Natl. Acad. Sci. USA*, 84, 1526–1530.

15. Datta, S., Luo, C-C., Li, W.-H., van Tuinen, P., Ledbetter, D.H., Brown, M.A., Chen, S-H., Liu, S.-w. & Chan, L. (1988). Human hepatic lipase. *J. Biol. Chem.*, 263, 1107–1110.

16. Cai, S.-J., Wong, D.M., Chen, S.-H. & Chan, L. (1989). Structure of the human hepatic triglyceride lipase gene. *Biochemistry*, 28, 8966–8971.

17. Ameis, D., Stahnke, G., Kobayashi, J., McLean, J., Lee, G., Buscher, M., Schotz, M.C. & Will, H. (1990). Isolation and characterisation of the human hepatic lipase gene. *J. Biol. Chem.*, 265, 6552–6555.

18. Lowe, M.E., Rosenblum, J.L. & Strauss, A.W. (1989). Cloning and characterisation of human pancreatic lipase cDNA. *J. Biol. Chem.*, 264, 20042–20048.

19. Bodmer, M.W., Angal, S., Yarranton, G.T., Harris, T.J.R., Lyons, A., King, D.J., Pieroni, G., Rivière, C., Verger, R. & Lowe, P.A. (1987). Molecular cloning of a human gastric lipase and expression of the enzyme in yeast. *Biochim. Biophys. Acta*, 909, 237–244.

20. Grusby, M.J., Nabavi, N., Wong, H., Dick, R.F., Bluestone, J.A., Schotz, M.C. & Glimcher, L.H. (1990). Cloning of an interleukin-4 inducible gene from cytotoxic T lymphocytes and its identification as a lipase. *Cell*, 60, 451–459.

21. Bianchetta, J.D., Bidaud, J., Guidoni, A.A., Bonicel, J.J., Rovery, M. (1979). Porcine pancreatic lipase. Sequence of the first 234 amino acids of the peptide chain. *Eur. J. Biochem.*, 97, 395–405.

22. Ben-Avram, C.M., Ben-Zeev, O., Lee, T.D., Haaga, K., Shively, J.E., Goers, J., Pedersen, M.E., Reeve, J.R. & Schotz, M.C. (1986). Homology of lipoprotein lipase to pancreatic lipase. *Proc. Natl. Acad. Sci. USA*, 83, 4185–4189.

23. Persson, B., Bengtsson-Olivecrona, G., Enerbäck, S., Olivecrona, T. & Jörnvall, H. (1989). Structural features of lipoprotein lipase. *Eur. J. Biochem.*, 179, 39–45.

24. Kirchgessner, T.G., Chuat, J-C., Heinzmann, C., Etienne, J., Guilhot, S., Svensson, K., Ameis, D., Pilon, C., d'Auriol, L., Andalibi, A., Schotz, M., Galibert, F. & Lusis, A.J. (1989). Organisation of the human lipoprotein lipase gene and evolution of the lipase gene family. *Proc. Natl. Acad. Sci. USA*, 86, 9647–9651.

25. Wion, K.L., Kirchgessner, T.G., Lusis, A.J., Schotz, M.C., Lawn, R.M. (1987). Human lipoprotein lipase complementary DNA sequence. *Science*, 235, 1638–1641.

26. Senda, M., Oka, K., Brown, W.V., Qasba, P.K. & Furuichi, Y. (1987). Molecular cloning and sequence of a cDNA coding for bovine lipoprotein lipase. *Proc. Natl. Acad. Sci. USA*, 84, 4369–4373.

27. Enerbäck, S., Semb, H., Bengtsson-Olivecrona, G., Carlsson, P., Hermansson, M.-L.,

Olivecrona, T. & Bjursell, G. (1987). Molecular cloning and sequence analysis of cDNA encoding lipoprotein lipase of guinea-pig. *Gene*, 58, 1–12.

28. Enerbäck, S. & Bjursell, G. (1989). Genomic organisation of the encoding guinea-pig lipoprotein lipase; evidence for exon fusion and unconventional splicing. *Gene*, 84, 391–397.

29. Kirchgessner, T.G., Svensson, K.L., Lusis, A.J. & Schotz, M.C. (1987). The sequence of cDNA encoding lipoprotein lipase. A member of a lipase gene family. *J. Biol. Chem.*, 262, 8463–8466.

30. Cooper, D.A., Stein, J.C., Strieleman, P.J. & Bensadoun, A. (1989). Avian adipose lipoprotein lipase: cDNA sequence and reciprocal regulation of mRNA levels in adipose and heart. *Biochim. Biophys. Acta*, 1008, 92–101.

31. Reynolds, M.V., Awald, P.D., Gordon, D.F., Gutierrez Hartmann, A., Rule, D.C., Wood, W.M. & Eckel, R.H. (1990). Lipoprotein lipase gene expression in rat adipocytes is regulated by isoproterenol and insulin through different mechanisms. *Mol. Endocrinol.*, 4, 1416–1422.

32. Deeb, S.S. & Peng, R. (1989). Structure of the human lipoprotein lipase gene. *Biochemistry*, 28, 4131–4135.

33. Yang, C-Y., Gu, Z-W., Yang, H-X., Rohde, M.F., Gotto, A.M. & Pownall, H.J. (1989). Structure of bovine milk lipoprotein lipase. *J. Biol. Chem.*, 264, 16822–16827.

34. Reue, K., Zambaux, J., Wong, H., Lee, G., Leete, T.H., Ronk, M., Shively, J.E., Sternby, B., Borgström, B., Ameis, D. & Schotz, M.G. (1991). cDNA cloning of carboxyl ester lipase from human pancreas reveals a unique proline-rich repeat unit. *J. Lipid Res.*, 32, 267–276.

35. Baba, T., Downs, D., Jackson, K.W., Tang, J. & Wang, C.-S. (1991). Structure of human milk bile salt activated lipase. *Biochemistry*, 30, 500–510.

36. Nilsson, J., Bläckberg, L., Carlsson, P., Enerbäck, S., Hernell, O. & Bjursell, G. (1990). cDNA cloning of human milk bile-salt-stimulated lipase and evidence for its identity to pancreatic carboxylic ester hydrolase. *Eur. J. Biochem.*, 192, 543–550.

37. Lee, C.Y. & Iandolo, J.J. (1986). Lysogenic conversion of staphylococcal lipase is caused by insertion of the bacteriophage L54a genome into the lipase structural gene. *J. Bacteriol.*, 166, 385–391.

38. Götz, F. Popp, F., Korn, E. Schleifer, K.H. (1985). Complete nucleotide sequence of the lipase gene from *Staphylococcus hyicus* cloned in *Staphylococcus carnosus. Nucl. Acids Res.*, 13, 5895–5906.

39. Aoyama, S., Yoshida, N. & Inouye, S. (1988). Cloning, sequencing and expression of the lipase gene from *Pseudomonas fragi* IFO-12049 in *E. coli. FEBS Letters*, 242, 36–40.

40. Jørgensen, S., Skov, K.W. & Diderichsen, B. (1991). Cloning, sequence, and expression of a lipase gene from *Pseudomonas cepacia*: lipase production in heterologous hosts requires two *Pseudomonas* genes. *J. Bacteriol.*, 173, 559–567.

41. Batenburg, A.M., Egmont, M.R., Frenken, L.G.J. & Verrips, C.T. (1990). Lipase enzymes including mutant lipase enzymes, e.g., from *Pseudomonas* species, are produced and modified by recombinant DNA technique. The enzymes are applicable in detergent and cleaning compositions, with advantage for example of improved stability to proteolytic digestion. European patent application, EP O 407 225 Al.

42. Gray, G.L., Poulose, A.J. & Power, S.D. (1986). A substantially enzymatically pure hydrolase is provided which is secreted by and isolatable from *Pseudomonas putida* ATCC 52552. Cloning the gene expressing the hydrolase into a suitable expression vector and culturing, such as fermenting the *E. coli* strain JM101 harboring a plasmid designated pSNtacII, has been found to provide surprisingly high yields of the hydrolase. European patent application EP O 268 452 A2.

43. Kugimiya, W., Otani, Y., Hashimoto, Y. & Takagi, Y. (1986). Molecular cloning and nucleotide sequence of the lipase gene from *Pseudomonas fragi. Biochim. Biophys. Res.*

Commun., 141, 185–190.
44. Feller, G., Thiry, M. & Gerday, C. (1990). Sequence of the lipase gene from the Antarctic psychrotroph Moraxella TA144. Nucl. Acids Res., 18, 6431.
45. Feller, G., Thiry, M. & Gerday, C. (1991). Nucleotide sequence of the lipase gene lip3 from the Antarctic psychrotroph Moraxella TA144. Biochim. Biophys. Acta, 1088, 323–324.
46. Shimada, Y., Sugihare, A., Tominaga, Y., Iizumi, T. & Tsunasawa, S. (1989). cDNA molecular cloning of Geotrichum candidum lipase. J. Biochem., 106, 383–388.
47. Kawaguchi, Y., Honda, H., Taniguchi-Morimura, J. & Iwasaki, S. (1989). The codon CUG is read as serine in an asporogenic yeast Candida cylindracea. Nature, 341, 164–166.
48. Patkar, S.A., Björkling, F., Zundel, M., Schülein, M., Svendsen, A., Heldt-Hansen, H.P. & Gormsen, E. (1993). Purification of two lipases from Candida antarctica and their inhibition by various inhibitors. Indian J. Chem., 32B, 76–80.
49. Boel, E., Huge-Jensen, B., Christensen, M., Thim, L. & Fiil, N.P. (1988). Rhizomucor miehei triglyceride lipase is synthesized as a precursor. Lipids, 23, 701–706.
50. Boel, E. & Huge-Jensen, B. (1988). Recombinant Humicola lipase and process for production of recombinant Humicola lipases. European patent application 0 305 216A1.
51. Yamaguchi, S., Mase, T. & Takeuchi, K. (1991). Cloning and structure of the mono- and diacylglycerol lipase-encoding gene from Penicillium camembertii U-150. Gene, 103, 61–67.
52. Brady, L., Bzrozowski, A.M., Derewenda, Z.S., Dodson, E., Dodson, G., Tolley, S., Turkenberg, J.P., Christiansen, L., Huge-Jensen, B., Nørskov, L., Thim, L. & Menge, U. (1990). A serine protease triad forms the catalytic centre of a triacylglycerol lipase. Nature, 343, 767–770.
53. Gubernator, K., Müller, K. & Winkler, F.K. (1991). The structure of human pancreatic lipase suggests a locally inverted, trypsin-like mechanism. In Lipases: Structure, Mechanism and Genetic Engineering, GBF monographs, vol. 16, ed. Alberghina, L., Schmid, R.D. & Verger, R., pp. 9–16. Weinheim: VCH.
54. Reddy, N.M., Maraganore, J.M., Meredith, S.C., Heinrikson, R.L. & Kézdy, F.J. (1986). Isolation of an active-site peptide of lipoprotein lipase from bovine milk and determination of its amino acid sequence. J. Biol. Chem., 261, 9678–9683.
55. Kordel, M., Menge, U., Morelle, G., Erdmann, H. & Schmid, R.D. (1991). Comparative analysis of lipases in view of protein design. In Lipases: Structure, Mechanism and Genetic Engineering, GBF monographs, vol. 16, ed. Alberghina, L., Schmid, R.D. & Verger, R., pp. 421–424. Weinheim: VCH.
56. Bell, A.W., Dumas, F., Shaw, K., Thomas, D.Y. & Ceska, T. (1989). Active site residues of Candida cylindracea lipase. Abstract no. T170 from 3rd Symposium of Protein Society meeting Seattle WA, USA.
57. Sussman, J.L., Harel, M., Frolow, F., Oefner, C., Goldman, A., Toker, L. & Silman, I. (1991). Atomic structure of acetylcholinesterase from Torpedo californica: a prototypic acetylcholine-binding protein. Science, 253, 872–879.
58. Datta, S., Luo, C-C., Li, W-H., van Tuinen, P., Ledbetter, D.H., Brown, M.A., Chen, S.-H., Liu, S.-W. & Chan, L. (1988). Human hepatic lipase. J. Biol. Chem., 263, 1107–1110.
59. Devereux, J., Haeberli, P. & Smithies, O. (1984). A comprehensive set of sequence analysis programs for the VAX. Nucl. Acids Res., 12, 387–395.
60. Brzozowski, A.M., Derewenda, U., Derewenda, Z.S., Dodson, G., Lawson, D.M., Turkenberg, J.P., Björkling, F., Huge-Jensen, B., Patkar, S.A. & Thim, L. (1991). A model for interfacial activation in lipases from the structure of a fungal lipase inhibitor complex. Nature, 351, 491–494.
61. Krejci, E., Duval, N., Chatonnet, A., Vincens, P. & Massoulié, J. (1991). Cholinesterase-like domains in enzymes and structural proteins: functional and evolutionary relationships

and identification of a catalytically essential aspartic acid. *Proc. Natl. Acad. Sci. USA*, 88, 6647–6651.

62. Kugimiya, W. Otani, Y., Kohno, M. & Hashimoto, Y. (1992). Cloning and sequence analysis of cDNA encoding *Rhizopus niveus* lipase. *Biosci. Biotech. Biochem.*, 56, 716–719.

63. Haas, M.J., Allen, J. & Berka, T.R. (1991). Cloning, expression and characterization of a cDNA encoding a lipase from *Rhizopus delemar*. *Gene*, 109, 107–113.

64. Dartois, V., Baulard, A., Schanck, K. & Colson, C. (1992). Cloning, nucleotide sequence and expression in *Escherichia coli* of a lipase gene from *Bacillus subtilis* 168. *Biochim. Biophys. Acta*, 1131, 253–260.

65. Chibara-Siomi, M., Yoshikawa, K., Oshima-Hirayama, N., Yamamoto, K., Sogabe, Y., Nakatani, T., Nishioka, T. & Oda, J. (1992). Purification, molecular cloning and expression of lipase from *Pseudomonas aeruginosa*. *Arch. Biochim. Biophys.*, 296, 505–513.

66. Johnson, L.A., Beacham, I.R., MacRae, I.C. & Free, M.L. (1992). Degradation of triglyerides by a Pseudomonad isolated from milk: molecular analysis of a lipase-encoding gene and its expression in *Escherichia coli*. *Appl. Environ. Microbiol.*, 58, 1776–1779.

67. Tan, Y. & Miller, K.J. (1992). Cloning, expression and nucleotide sequence of a lipase gene from *Pseudomonas fluorescens* B52. *Appl. Environ. Microbiol.*, 58, 1402–1407.

68. Ihara, F., Okamoto, I., Nihira, T. & Yamada, Y. (1992). Requirement in trans of the downstream limL gene for activation of lactonizing lipase from *Pseudomonas* sp. 109. *J. Fermentation Bioengin.*, 73, 337–342.

69. Ihara, F., Kageyama, Y., Hirata, M., Nihira, T. & Yamada, Y. (1991). Purification, characterisation and molecular cloning of lactonising lipase from *Pseudomonas* species. *J. Biol. Chem.*, 266, 18135–18140.

70. Hong, K.W., Jang, W.H. & Choi, K.D. (1991). Characterisation of *Pseudomonas fluorescens* carboxylesterase: cloning and expression of the esterase gene in *Escherichia coli*. *Agric. Biol. Chem.*, 55, 2839–2845.

71. Hide, W.A., Chan, L. & Li, W.H. (1992). Structure and evolution of the lipase superfamily. *J. Lipid Res.*, 33, 167–178.

72. Reeck, G.R., de Haën, C., Teller, D.C., Doolittle, R.F., Fitch, W.M., Dickerson, R.E., Chambon, P., McLachlan, A.D., Margoliash, E., Jukes, T.H. & Zuckerkandl, E. (1987). Homology in proteins and nucleic acids. *Cell*, 50, 667.

73. Darwin, C. (1872). *The Origin of Species by Means of Natural Selection* (glossary by W.S. Dallas, pp. 434–435), 6th. edn. London: John Murray.

2.

A sequence analysis of lipases, esterases and related proteins

STEFFEN B. PETERSEN and FINN DRABLØS

This chapter describes the search for common sequence motifs in a large number of lipases, esterases and related proteins. The analysis used MULTIM, a novel program suite for semi-automatic multiple sequence alignment in protein engineering and protein sequence studies. With few exceptions, all the sequences contained the GxSxG motif (where x is any amino-acid residue). A classification of the contexts of the putative active serine showed that the proteins can be grouped into two major classes, one displaying the AGY and the other the TCN codon for the active-site serine. The AGY codon class includes the lipoprotein lipases, the hepatic and the pancreatic lipases. The comparison also reveals that the conserved regions identified by MULTIM in the lipoprotein lipase class are involved in the genetic disease familial hyperlipoproteinæmia. The TCN codon class embraces many of the esterases including human milk bile-stimulated lipase, lysophospholipase, hormone-sensitive lipase, acetyl- and butyrylcholinesterases and thioesterases. The mammalian gastric lipases appear to be distinct from the other mammalian lipases. The *Geotrichum* and *Candida rugosa* lipases were found to resemble the acetylcholinesterases, including the hormone-sensitive lipase from rat and lysophospholipase. Vitellogenins displayed significant similarity to the lipoprotein lipases. Likewise, neurotactins clearly share similarities with the esterases. The GxSxG motif is also found in proteases. However, evidence from both sequences and three-dimensional structure appear to rule out a distant evolutionary relationship between lipases and proteases.

1. SEQUENCES AND STRUCTURES ANALYSED

During the last decade, an ever-growing number of protein sequences has been determined, either directly or by deduction from the genetic sequence. Sequence comparison between proteins belonging to the same protein family can reveal single residues or clusters of residues that have been conserved and may help us in identifying key residues that are essential for retention of protein structure or function.

In many cases, the amino-acid sequences from a family of enzymes involved in degradation or production of key metabolites in the living cell have been determined for a range of organisms. One such example is the group of enzymes that embraces both lipases and esterases (for a general source, see reference 1). Here we compare more than 75 such protein sequences. Where possible, we have combined the sequence information with relevant three-dimensional structural information. We have also employed limited phylogenetic considerations and a novel method for making multiple alignments. The result is the identification of sub-classes of these enzymes that share functional and presumably also structural features.

The protein sequences considered here are all related to triglyceride lipases or esterases. The central goal of this comparison is to identify sequence information that can be used to identify new members of the triglyceride lipase class, the esterase class and the related classes identified here.

Lipases were first identified in 1856 by Claude Bernard. They hydrolyse insoluble oil droplets into soluble products. They resemble esterases in some respects, but differ significantly in their ability to act on insoluble neutral esters (2). The presence of lipoprotein lipases in blood plasma was first noted in 1943 by Hahn (3). Initially the enzyme was described as a 'clearing factor', and nine years later Anfinsen and co-workers ascribed the clearing factor to the presence of a lipolytic enzyme (4). Since then, there has been a growing interest in both the lipolytic enzymes and in their substrates, the acylglycerols. Both lipoprotein lipase and pancreatic lipase exhibit increased activity in the presence of co-factors. The co-factors bind to the lipid surface and enhance the affinity of the lipase for the lipid surface. In addition to this, the co-factors overcome the inhibitory effect that bile salts have on the lipase activity. For the human lipoprotein lipase, the co-factor is apolipoprotein C-II, and for the pancreatic lipase the co-factor is simply called colipase. No homology has been found between apolipoprotein C-II and colipase. The lipoprotein lipases, the pancreatic lipases and the hepatic lipases share a common gene organisation (5, 6). In the case of the hepatic lipase, a putative liver-specific element for apolipoprotein was identified. The pancreatic lipase contains four potential glycosylation sites, and all cysteines are totally conserved. Until recently it was believed that dietary lipids were degraded only in the intestine. However, it is now clear that human gastric lipase is the first in a series of lipases involved in dietary fat digestion in humans. No lipolytic activity was found at locations above the fundic mucosa in the stomach, where the gastric lipase is produced (7). Human and rabbit gastric lipases have been isolated, sequenced and characterised (8–10). No significant sequence homology has been found with other lipases outside the immediate neighbourhood of the putative active serine. In addition, it has been reported that the N-terminal tetrapeptide is essential for lipid-binding and lipase activity in the human gastric lipase (9).

Esterases hydrolyse carboxylic esters, and are found in animals, plants and bacteria (11). Although no clear differentiation seems possible between lipases and esterases, the latter prefer substrates that are short-chain fatty-acid esters or water-soluble molecules. Choline esterases can be divided into acetylcholinesterases and butyrylcholinesterases. Both groups of cholinesterases are found in human brain, muscle and blood. The active site of the human serum cholinesterase is known to possess both an esteratic site and an anionic site. The anionic site is believed to be necessary for the processing of positively charged substrates. Increased levels of human serum cholinesterases are typically found in obesity, diabetes mellitus, coronary artery disease, fatty liver *etc*. It is believed that human serum cholinesterase originates in the liver.

Cholinesterases are highly polymorphic proteins, capable of rapidly hydrolysing the neurotransmitter acetylcholine in the brain and involved in terminating neurotransmission in neuromuscular junctions and cholinergic synapses. Human acetylcholinesterase displays homology with bovine thyroglobulin (12). Human serum cholinesterase has three disulphide bonds and forms a tetrameric unit through the formation of disulphide bonds between subunits (13). Human serum butyrylcholinesterase has been reported to exhibit peptidase activity, and a protein fragment has been isolated that displays peptidase activity, but not butyrylcholinesterase activity (14).

Cutinases form a group of fungal enzymes that are capable of hydrolysing the insoluble lipid polyester matrix (*e.g.* cutin) that covers the stems and leaves of plants (15, 16). A key component of this material is a mixture of long (>18-C) hydroxy fatty acids. The cutinases constitute a class of enzymes used by fungi to invade the plant tissue. Recently, it was shown that fungi that do not produce cutinase are still capable of destroying plant cell walls (17).

Five three-dimensional structures determined by X-ray diffraction analysis are considered here: lipase from *Rhizomucor miehei* (18), human pancreatic lipase (19), *Geotrichum candidum* lipase (20), *Fusarium* cutinase (16) and acetylcholinesterase (21). These three-dimensional structures provide vital insight into the similarities and dissimilarities, in respect both of structure and of substrate-binding, that exist within this group of proteins. The active site in triglyceride lipases possesses a triad of residues similar to the catalytic triad in the trypsin class of proteases (18). In the *R. miehei* lipase, the active site contains a serine, a histidine and an aspartic acid, and is well shielded from the solvent by a protective loop that contains a tryptophan. During enzymic action, this loop is believed to move and thus facilitate access to the active site. A similar active site has been identified in human pancreatic lipase (19). The active site in esterases has been identified in the case of the *Geotrichum* lipase and contains a serine, a glutamic acid and a histidine, thus resembling the *R. miehei* active site, but with aspartic acid replaced by glutamic acid.

2. METHODS OF COMPARISON

2.1 Sequence databases

In a comparison of this kind it is of central importance to use the biological sequence databases. We use in particular SWISSPROT (release 20) and Protein Identification Resource (PIR, release 29). In addition, the EMBL nucleic-acid database (release 30) has been used to study the serine codons, and the local context around the serines.

2.2 MULTIM

Whenever a class of proteins such as the lipases are studied, it is of importance to be capable of identifying which parts of the amino-acid chain are partially or fully conserved. Any such conservation may indicate structural or functional importance. Several computer-based methods exist for aligning multiple biological sequences (22–25). In general, these methods apply a scoring scheme that is based on a pairwise comparison of amino acids between the sequences that are being aligned. The optimal alignment is the one that leads to the highest score. Insertions and deletions are introduced at the cost of a reduced total score (26). A central feature in all such multiple alignments is the choice of scoring scheme. Three principal schemes are in use. (1) Identity: non-zero scores are only given if identical residues are found in the two sequences. (2) Similarity: scores are given depending upon the degree of similarity that exist between the residues compared. (3) Genetic: scores are based upon the genetic distance between the two residues that are compared. The nucleic-acid triplets coding for two residues may differ at one, two or three positions, and the genetic score scheme reflects this as a decreasing score. The similarity scheme may be based on physico-chemical properties such as hydrophobicity; however, many different scales expressing differences in hydrophobicity exist (27).

MULTIM is a program suite that identifies, filters for significance and subsequently displays the position and extent of sequence motifs that occur in all members of a set of proteins. Initially, the user defines a sequence window size along with the minimum number of residues to be conserved, either by identity or similarity, within the sequence window. If such a specific pattern of residues exists in all the sequences in the set, we have a conserved motif. MULTIM will create all possible motifs for any window and test whether such motifs exist in all sequences. The extent of the conservation is defined by the user; currently, windows of up to 20 residues may be searched for, with a number of conserved residues that is less than or equal to the motif length. The motifs that MULTIM identifies may subsequently be evaluated for their statistical significance. Finally, the motifs are displayed graphically and their significance is displayed by a colour

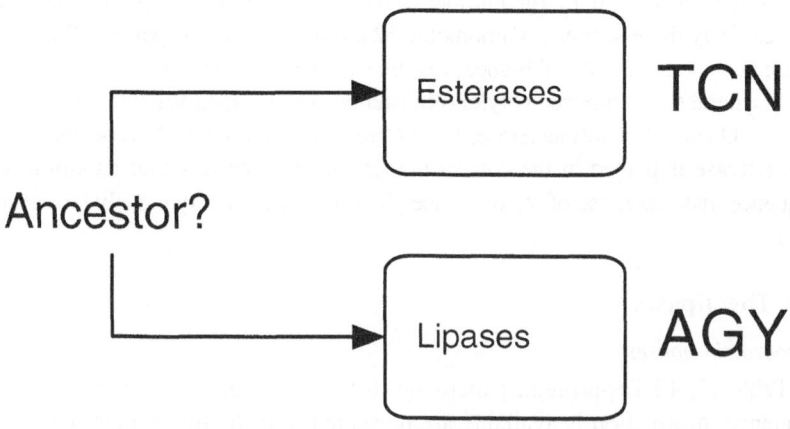

Figure 1. A classification of lipases and esterases determined from the nucleic acid sequence around the active serine. The tree has two branches, one predominantly based upon TCN codon usage for serine, the other upon the AGY codon.

representation. It is important to note that MULTIM does not rely on any particular type of gap-penalties; thus, insertions and deletions do not limit MULTIM's ability to detect conserved motifs. The software is available for Silicon Graphics workstations and IBM-compatible PCs equipped with a VGA graphics adapter.

3. RESULTS FROM THE COMPARISON

3.1 Codon usage for the active-site serine

An interesting property in the genetics of serine residues is the fact that two very different types of codon may be used, either TCN (where N is any nucleotide) or AGY (where Y is a pyrimidine). A double mutation is needed to change a TCN-type serine to an AGY-type serine. Since the probability for such a mutation is low, it seems likely that Nature became stuck with a specific codon usage for the active-site serine once it had evolved. In order to investigate this further, we used the TREEALIGN program (28, 29) for phylogenetic analysis of 32 polynucleotide sequences. A fragment of 39 nucleic-acid residues symmetrically positioned around the codon for the active serine was used, and no gaps were allowed in the alignment. In the tree drawn by the program, the data set is divided into two well-defined branches (Figure 1). One branch consists of 13 sequences that use a TCN-type codon for the active serine. Many of these sequences have previously been classified as esterases. The other branch consists of 16 sequences that use an

AGY-type codon, and a large number of these have previously been classified as lipases. Only three sequences among the 32 did not fit into this pattern. Rat gastric lipase is placed in the AGY branch, together with most of the other lipases. Rabbit cholinesterase and mouse acetylcholinesterase are grouped together in the TCN branch. One acetylcholinesterase, from *Drosophila*, uses a TCN-type codon, and this esterase is placed in the correct branch, despite the fact that its amino-acid sequence matches those of the other acetylcholinesterases very closely (see section 3.3).

3.2 The lipases

Lipoprotein lipases

In Table 1*, 13 lipoprotein, pancreatic, hepatic and gastric lipases for which sequence information is available are presented. For the lipoprotein, pancreatic and hepatic lipases, the size (including signal and prosequence) ranges from 449 to 499 residues, and the local sequence around the putative active-site serine is almost totally conserved. This observation is further strengthened by the fact that the serine-codon usage in all 11 lipases is AGC. For the two gastric lipases, the size is 395 and 398 residues, and the context of the active-site serine maintains some similarity to the lipases; at a position five residues before the active serine, a histidine is found in both of the gastric lipases and in ten of the other lipases. The residue immediately preceding the active serine is a His in the two gastric lipases, and this pattern is found in the four pancreatic lipases. However, the serine codon in the gastric lipases is TCT. Finally, position of the putative active serine in the overall sequence of the gastric lipases is similar to the one found in the hepatic and pancreatic lipases.

A MULTIM analysis of the lipases revealed several interesting facts. Although hepatic lipase has evolved at a comparatively rapid pace (5), local regions of homology are very significant when one compares the hepatic with the lipoprotein lipases (data not shown). It is also noted that several motifs are found in the region of the sequence corresponding to the location of the small domain in human pancreatic lipase (data not shown; see section 3.5). In Figure 2 a MULTIM analysis of the lipoprotein, pancreatic and hepatic lipases is shown. Note that no significant homology is detected in the small domain. This can reflect the utilisation of different co-factors among the three types of enzyme, although the co-factors presumably bind to the same region. If this is true, it is likely that a hepatic co-factor may resemble the apolipoprotein C-II, since a MULTIM analysis shows homology in this region.

The active asparagine is found less than 30 residues after from the active serine

---

* The Tables and Figures 2–4 are collected at the end of this chapter (pp. 39–47).

(Table 2). Within the lipoprotein lipases the residues around this Asp are well conserved, but not totally conserved as was the case for the active serine. When the hepatic and pancreatic lipases are included in the comparison, the variation increases. A motif common to the lipoprotein, pancreatic and hepatic lipases is RITGLDPAxPxF. With two exceptions (dog pancreatic and human hepatic lipases), all the lipases contain proline at positions 1 and 4 after the asparagine.

The active histidine in the lipoprotein, hepatic and pancreatic lipases is located about 100 residues downstream from the active serine (Table 2). The 11 lipoprotein, hepatic and pancreatic lipases all share the motif CxHxR around the active His. The hepatic lipases are quite similar to the lipoprotein lipases, whereas the pancreatic lipases display a number of aromatic residues in this stretch that are absent in the lipoprotein and pancreatic lipases. Interestingly in this connection, Datta *et al.* (5) noted that all Cys residues are stringently conserved in the human pancreatic lipases.

Phylogenetic analysis of the lipoprotein lipases places the lipoprotein lipases in one group, the hepatic lipases in another and the pancreatic lipases in a third, independently of which organism they stem from. This indicates that an organism, ancestral to all the present-day organisms considered here, existed, in which the lipoprotein, hepatic and pancreatic lipases had already developed. However, since the three types of lipases display clear homology of sequence, as well as similarity of gene organisation (6, 30, 31), it is reasonable to assume that they developed by duplication of an ancestral gene. The present-day lipoprotein lipases show relatively few variations within the group, whereas both hepatic and pancreatic lipases display considerable differences (5). In addition, each of the latter two display clearer homology with the lipoprotein lipases than with each other. We therefore propose that a lipoprotein-lipase-like enzyme developed first, and that the other two lipases developed from duplicated genes. This may also indicate that only small variations in the lipid composition has occurred in the evolutionary period relevant for the comparison. If the substrate for these enzymes had changed significantly in the course of time, then the substrate-binding sites on the enzymes would also change.

On the basis of the MULTIM sequence analysis one can extract a conserved, but disperse sequence motif characteristic of the lipoprotein, hepatic and pancreatic lipases (numbers refer to human pancreatic lipase, including the signal sequence, while active-site residues are indicated with an asterisk):

```
 91 IHGFIDxxxx xxxxxxxxxx xxxxxxNxxx VDWxxxxxxx xxxxxxxxxx 140

141 xxxxxxxxxx xxxxxxxxxx xxxxxxGxSL GAHAAGxAGx xxxxxxxRIT 190
                                        *
191 GLDPAxxxxx xxxxxxRLxP xDAKFVDVIH Txxxxxxxxx xxxxxxxxxH 240
       *
241 xDxxPNGGxx xPGCxxxxxx xxxxxxxxxx xxxxFAxxxH xRS       283
                                         *
```

If one searches sequence databases such as PIR with this motif, one finds only lipoprotein-lipase-related sequences.

Disease-related mutations of human lipoprotein lipase

Deficiency in lipoprotein lipases is a relatively rare genetic disease (familial hyperlipoproteinæmia), in which plasma from a fasting patient appears milky because of large amounts of chylomicrons (32). As cDNA probes are available for human lipoprotein lipases, the mutations causing the abnormality are known in several cases: Gln133→STOP, Gly169→Glu, Ala203→Thr, Gly215→Glu, Ile221→Thr, Pro234→Leu and Arg270→His (32; numbers refer to human lipoprotein lipase, including the signal sequence). In the annotation of the MULTIM alignment of the lipoprotein family (Figure 2), the disease mutations have been highlighted as well. It is very interesting to note that in most cases the mutations causing disease are associated with conserved regions in this family of proteins (data not shown). The exceptions are positions 215 and 221, which possibly could have a functional or structural rôle specific for the lipoprotein lipases. It is known that the mutation Ala203→Thr causes abnormal heparin-binding in addition to loss of enzymic activity. Similarly, the Gly215→Glu also impairs heparin-binding.

For human hepatic lipase, it is reported that the mutation Thr405→Met leads to a total deficiency in hepatic lipase. This location is clearly outside the conserved sequence regions in the lipoprotein lipase family as identified by MULTIM, and may be considered as a position of distinct importance for the hepatic lipases. We suggest that Thr405 is involved in the binding of co-factor (6). If three-dimensional structural homology can be assumed to exist within the lipoprotein lipase family, position 405 is located in the small domain, while the large domain encompasses the active site in the human pancreatic lipase, which in all likelihood exhibits the same overall three-dimensional structure as the hepatic lipase.

3.3 The cholinesterase class

In most of the proteins in this class the sequence around the putative active serine is GESAG (Tables 4–6; reference 33). This motif is shared with the lysophospholipase, esterase P from rat, sterol esterase and acetyl- and butyrylcholinesterases. In addition to this, the *Geotrichum* lipase and the *Candida rugosa* lipase display the same motif, indicating that they may be better classified as esterases than as lipases, as was also pointed out by Schrag *et al.* (20). However, there are exceptions to this motif: in the hormone-sensitive lipase from rat the motif changes into GDSAG, maintaining a negative charge immediately before the serine. The deviation from the other esterases becomes even larger if one widens the context, where the other esterases conform to a shared motif: [VI]xxFGESAGxxV (where [VI] means that either V or I is accepted at this position). It is interesting to note that this region around active serine resembles in many ways the corresponding region in trypsin, CxGDSGGPxxC, where the two cysteines form a disulphide bridge. Finally, we note that the codon usage in the case of the *Drosophila* acetylcholinesterase is TCN, whereas the rest are based upon an AGY codon. It is known that *Drosophila* does not utilise methylation of its bases, which may have an influence on evolutionary events (34). The cholinesterases are stabilised by three intramolecular disulphide bridges: Cys65–Cys92, Cys252–Cys263, Cys400–Cys519 and one intermolecular disulphide bond that involves Cys571 in both molecules (numbers refer to human serum cholinesterase; reference 13).

Disease-related mutations in human serum cholinesterase

There is only one gene for human serum cholinesterase. Two mutations have been described, Asp70→Gly and Ala359→Thr. In the Asp70→Gly variant, an abnormal response to the muscle relaxant succinylcholine is reported, including a substantial prolongation of muscle paralysis with apnea (35). The Ala539→Thr mutation causes the enzyme to lose 33% of its activity, indicating that this location either is involved in binding or is structurally important for catalysis.

Insect esterases

Neither the mosquito esterase nor *Drosophila* esterase-P exhibits the esterase motif discussed above. Instead, these share the motif LxGHSAGxAS. A similar motif, LxGHSMGGQTI, is found in the *Staphylococcus* lipases; the N-terminal residues of these motifs are identical. In the case of the mammalian pancreatic lipases we also find a related motif: [LV]IGHSLG[AS]H[AV]A. On the basis of these data, we propose that the two insect esterases should be classified as lipases instead.

Human milk bile-stimulated lipase

Human milk bile-stimulated lipase contains an undecamer that is repeated 16 times close to the C terminus (36). The repeat sequence is GAPPVPPTGDS with few variations. The rôle of this repetitive domain is not known at present. It is interesting to note that the human cholesterol esterase is also bile-stimulated. The disperse conserved motif is seen not to include the repetitive domain; it thus seems reasonable to assume that the repetitive domain plays a structural rather that a substrate-binding rôle. This possibility is supported further by the fact that the repeat contains a large fraction of proline residues, which are known to exhibit low torsional flexibility owing to the fact that one of the torsional angles is fixed. It is, furthermore, unlikely that the 16 repeats form secondary-structural elements involving α helices, since proline is unable to participate in hydrogen bonds in the α helix. Collagen-like helix formation is of course still possible. Finally, the amino-acid sequence of bile-stimulated human lipase is approximately 150 amino-acid residues longer than those of the other members of the cholinesterase class with which it is compared in Figure 4. In rat lysophospholipase we find a repeat that closely resembles the repeat in human bile-stimulated lipase, indicating that these two proteins may be related.

Hormone-sensitive lipase

In rat hormone-sensitive lipase, the sequence around the active-site serine is GDSAG, which resembles the that of the cholinesterases. The glutamic acid immediately preceding the active-site serine is replaced by an aspartic acid, and in addition we find two cysteines close by, indicating the existence of disulphide bridges close to the active-site serine. This feature is also exhibited by the active-site serine context of trypsin. The active-site aspartic acid and histidine have been identified by homology involving considerations both of sequence and of distance. However, the degree of similarity is marginal.

Lysophospholipase

The lysophospholipase resembles in its function the phospholipases. It hydrolyses the ester group in phospholipids at the same point as phospholipases A_1 and B. The sequence around the putative active serine is GESAG, identical to the bile-stimulated, *Geotrichum* and *Candida rugosa* lipases and the acetyl- and butyrylcholinesterases. In addition, it is very similar to the bile-stimulated human lipase in the neighbourhoods of the putative active-site aspartic acid and histidine residues.

Features common to the esterase family

Figure 4 shows the MULTIM output on seven esterases: the 'lipases' from *Geotrichum*, *Candida rugosa*, one representative of the acetylcholine esterases (*Torpedo californica*), one of the cholinesterases (human) and human bile-stimulated lipase, as well as the lysophospholipase from rat. As is seen from the figure, the homology is observed in the N-terminal part of the sequences. Surprisingly, neither the Asp (or Glu) nor the His environment seems to be well conserved. The consensus sequence is given below (numbers follow the *Geotrichum* lipase II sequence and the active-site serine is indicated by an asterisk):

```
 37      FxGI Pxxxxxxxxx xxxxxxxxxx xxxxxxxxxx xxxxxxxxxx 80

 81 xxxxxxxxxx xxxxxxxxxx xxxxxxxxxx xxxxxEDCLx xxxxxxxxxx 130

131 xxxxxxxxxW IxGGxFxxGx xxxxxxxxxx xxxxxxxxxx xxxxxxxxxx 180

181 xxxxxxxxxx xxxxxxxGL xDQxxxxxWV xxNIxxFGGx PxxxxxFGES 230
                                                     *
231 AG                                                  232
```

If this consensus motif is used to search in the PIR sequence data bank, 29 highly significant 'hits' are found, the majority of which are listed as esterases, carb-oxylesterases, sterolesterases, esterase B1, and lysophospholipase. In addition, some related proteins are found: crystal protein precursor from slime mould (A34576), D2 protein precursor, also from slime mould (B34576), as well as three thyroglobulins from cow, human and rat (UIBO, S00014, UIRT). It seems very reasonable to assume that these proteins, in addition to their recognised function, also possesses esterase functionality, or have evolved from such enzymes. Finally, it should be noted that lipase 2 from *Moraxella* (37, 38) has significant sequence similarity with the cholinesterases.

3.4 The fungal and procaryote class

Staphylococcus *and* Pseudomonas *lipases*

The *Staphylococcus* lipase can be aligned with the *Pseudomonas* lipases and three regions of similarity are identified by MULTIM. The regions are separated by stretches of similar length in the two sets of fungal proteins, indicating that the sequences indeed have the same origin. The central match contains the putative active serine, and the match upstream from this a conserved LxGxD motif. Downstream from the active serine the motif NDGLVS is found. Since both motifs flanking the serine motif contain an aspartic acid, it is only by homology with other known serine hydrolases that the aspartate furthest downstream can be pointed out as the one most likely to be an active-site residue. Again by comparison with other serine hydrolases, a possible active-site histidine can be identified, this time by inspection of the sequences to find a histidine conserved in all of them (data not shown).

A comparison of the active-serine contexts reveals three out of four conform to the motif V[HN][AVIL][AVIL]GHSxGGxTxR. For the active aspartic acid, the *Pseudomonas* contexts look similar, and the same can be concluded from the *Staphylococcus* sequences; however no significant homology can be found between the two pairs. Around the active hisitidines three out of four conform to the motif WDHxD.

Geotrichum *and* Candida rugosa *lipases*

These three proteins are highly similar, as is shown in Figure 3. This has also been shown previously by Shimada *et al.* (39). Homologous regions are seen throughout the whole length of the sequences, indicating a very close relationship. The homology is particularly clear in the active-serine region. This is further supported by inspection of the amino-acid residues close to the active-site residues. Around the serine, the motif VxIFGESAGxMSV is conserved. Around the aspartic acid we find that the motif QxDEGT is shared among all three, and, as was reported by Schrag *et al.* (20), it is the glutamic acid and not the aspartic acid that is the active-site carboxylic acid. Finally, around the histidine PxLGTFHxxxxF is conserved. The codon usage is not known for *Geotrichum* lipase I, whereas it is TCN for *Geotrichum* lipase II and CTG in the *Candida rugosa* lipase (40).

The cutinases

Three cutinase sequences were available for the present study: those from *Fusarium solani sisi, Colletotrichum capsici* and *Colletotrichum gloeosporiodes*. The consensus sequence was extracted from the MULTIM analysis and is given below. Note that the GGYSQGxA motif is present in all three cutinases. Upstream from and adjacent to the putative active serine they all contain tyrosine, as do the lipoprotein lipases, while the downstream neighbour is glutamine, which is rather large and hydrophilic compared with the hydrophobic leucine that is found in the lipoprotein lipases. Some of the thioesterases also contain Tyr at the position immediately upstream from serine. The putative active-site aspartic acid is also observed. No conserved histidine is found with a window size of 6 residues, but one is identified at a window size of 9. The conserved sequence in cutinases as detected by MULTIM is (numbers refer to the *Fusarium* sequence):

```
 21  xxxxxLExRQ xxxxxxxxxx xxxxxxxxxV IFIYARxSTE xGNxGxxxxx   70

 71  xxxxxxxxxx xxxVWIQGVG GxYxAxxxxx xxPxGTSxxA IxxxxxLFxx  120

121  ANTKCPxxxx xxGGYSQGxA xxxxxxxxxx xxxxxxIxGT VLxxxTKNLQ  170

171  NxGRIPNYxx xxxxxxxxxx DxVCxGxLxx xxxxxxxxxx xxxxAPxFL   219
```

3.5 Three-dimensional structure

Rhizomucor miehei *and human pancreatic lipases*

Human pancreatic lipase (19) is structurally closely related to *Rhizomucor miehei* lipase (18) although protein sequence alignment fails to indicate any close relationship. In addition to the large main domain that is shared, the pancreatic lipase contains a small domain, the rôle of which is still unclear. The consensus motif extracted from the MULTIM analysis of the lipoprotein lipase class is located solely in what corresponds to the catalytic domain in the human pancreatic lipase. In addition, almost all of the conserved residues are positioned either at the edge of a defined secondary structural motif, or in loop regions; both observations indicate that the consensus motif is located at or close to the surface and may at least in part reflect common structural features in the substrate for the lipoprotein lipases. The conserved cysteine at position 254 is involved in disulphide bridge formation to cysteine 261.

Plate 1 highlights the conserved residues identified from the MULTIM analysis of the lipoprotein, hepatic and pancreatic residues (see Figure 2), with the X-ray structure of human pancreatic lipase used as reference.

Geotrichum *lipase (20)*

When information about secondary structural obtained from the X-ray structure determination is used to annotate the MULTIM output, a very interesting picture emerges: most of the residues identified as conserved in a subclass (such as the lipoprotein class) are located at the surface of the protein, either at the end of β sheets, at the interface between two concatenated secondary structures, or in loops. These residues are thus not involved in structural conservation of the protein, although the structure must be shared as well; rather, they define the residues involved in substrate-binding, and as such they provide information about how similar a structure their substrates must have. This is the only interpretation that can explain the fact that enzymes stemming from very diverse origins, ranging from fungi to humans, such as the acetylcholinesterase class, display conserved, but disperse sequence motifs. The MULTIM method can thus be an excellent tool for identifying residues that will determine substrate-binding.

Esterases

In Plate 2, the conserved residues identified by the MULTIM analysis of the esterase group of enzymes shown in Figure 3 are highlighted using the X-ray structure of the acetylcholine esterase (21) as reference.

3.6 Distant relationships

Since the three-dimensional structures of the lipases differ from those of the trypsins, it is not very likely that they are related by evolution. Since, however, they do display similar active-site configurations and local motifs around the active serine and aspartic acid, it seems more likely that they represent a case of convergent evolution, similar to what is observed between the trypsins and the subtilisins. Although analysis of the sequences close to the serine will not account for the presence or absence of similarities outside the region used for the study, it does have the potential of revealing relationships that can be substantiated using other methods such as the MULTIM analysis. It is relevant in this connection to point out that the consensus motif identified by MULTIM not only selected successfully both the obvious and expected members of the family, but also found some proteins that, unexpectedly, shared enough of the sequence motif: *e.g.*, neurotactins, thyroglobulins, slime-mould proteins D2 and crystal protein. Similarly, the lipase consensus motif identified vitellogenins as homologues, as pointed out earlier by Persson *et al.* (10).

Proteases

The lipases share the active-site composition with several other serine hydrolases such as the protease enzymes. The GxSxG motif is found in the lipases and esterases, as was pointed out by Brenner (41). However, no evidence for a clearer relationship is known to the authors. Benner proposes that the GxSxG should be considered to be an original active-site module, which evolution has retained in many different designs.

Phospholipases

The substrates for phospholipases are highly hydrophobic, as is the case for the lipases. However, the phosphoester bond differs in nature from the carboxylic ester bond, and in general the enzymic mechanisms can be expected to differ significantly. However, we have identified a phospholipase A_1 that contains the GxSxG motif, indicating that the phospholipases A_1 may constitute an esterase-like class. It is also known that the guinea-pig lipoprotein lipase can function both as a triglycerol lipase and as a phospholipase. Phospholipases A_2 and C are very different from the lipases considered in this chapter.

Other lipases

It will be interesting to compare the amino-acid sequences and three-dimensional structures for fish and plant lipases and esterases. Fish lipases may have been optimised for processing longer fatty-acid chains than mammalian enzymes are required to act on, and it will also be of value to obtain information about expected evolutionary relationships between the plant lipases and esterases and their animal counterparts. Finally, we note that no lipase or esterase capable of hydrolysing the membrane lipids of archæbacteria has yet been sequenced. Since these lipids are in many ways unique (42), we may obtain new and valuable information from enzymes that can degrade these lipids.

4. CONCLUSION

A large set of proteins in the categories lipases and esterases, as well as proteins related to these, have been analysed. Amino acids around the putative active-site serine are well conserved in all of these, but with distinct variations that appear to correlate well with the functional classification into lipases and esterases. A classification analysis clusters the two different types of putative active serine sequence contexts into distinct groups.

REFERENCES

1. Alberghina, L., Schmid, R.D. & Verger, R. (1991). *Lipases: Structure, Mechanism and Genetic Engineering* (GBF MONOGRAPHS vol. 16). Weinheim, Germany: VCH.
2. Okuda, H. (1991). Lipases. In *A Study of Enzymes*, ed. Kuby, S.A., pp. 579–593. Boca Raton FL, USA: CRC Press.
3. Hahn, P.F. (1943). Abolishment of alimentary lipemia following injection of heparin. *Science*, 98, 19–20.
4. Anfinsen, C.B., Boyle, E. & Brown, R.K. (1952). The role of heparin in lipoprotein metabolism. *Science*, 115, 583–586.

5. Datta, S., Luo, C.C., Li, W.H., Van Tuinen, P., Ledbetter, D.H., Brown, M.A., Chen, S.H., Liu, S.W. & Chan, L. (1988) Human pancreatic lipase: Cloned cDNA sequence, restriction fragment length polymorphisms, chromosomal localization and evolutionary relationship with lipoprotein lipase and hepatic lipase. *J. Biol. Chem*, 263, 1107–1110.

6. Sensel, M.G., Legrand-Lorans, A., Wang, M.E. & Bensadoun, A. (1990). Isolation and characterization of clones for the rat hepatic lipase gene upsteam regulatory region. *Biochim. Biophys. Acta*, 1048, 297–302.

7. Carrière, H., Moreau, H., Raphel, V., Laugier, R., Benicourt, C., Junien, J.L. & Verger, R. (1991). Purification and biochemical characterisation of dog gastric lipase. *Eur. J. Biochem.*, 202, 75–83.

8. Moreau, H., Gargouri, Y., Lecat, D., Junien, J.L. & Verger, R. (1988). Purification, characterization and kinetic properties of the rabbit gastric lipase. *Biochim. Biophys. Acta*, 960, 286–293.

9. Bernback, S. & Bläckberg, L. (1989). Human gastric lipase: The N-terminal tetrapeptide is essential for lipid binding and lipase activity. *Eur J. Biochem.* 182, 495–499.

10. Persson, B., Bengtsson-Olivecrona G., Enerback, S., Olivecrona, T. & Jornvall, H. (1989). Structural features of lipoprotein lipase: Lipase family relationships, binding interactions, non-equivalence of lipase cofactors, vitellogenin similarities and functional subdivision of lipoprotein lipase. *Eur. J. Biochem*, 179, 39–45.

11. Okuda, H. (1991). Esterases (including choline esterases). In *A Study of Enzymes*, ed. Kuby, S.A., pp. 563–577. Boca Raton FL, USA: CRC Press.

12. Soreq, H. & Gnatt, A. (1987). Molecular biological search for human genes encoding for cholinesterases. *Mol. Neurobiol.*, 1, 47–80.

13. Lockridge, O., Adkins, S. & La Du, B.N. (1987). Location of disulphide bonds within the sequence of human serum cholinesterase. *J. Biol. Chem.*, 262, 12945–12952.

14. Rao, R.V. & Balasubramanian, A.S. (1990). Localization of the peptidase activity of human serum butyrylcholinesterase in an approximately 50-kDa fragment obtained by limited α-chymotrypsin digestion. *Eur. J. Biochem.* 188, 637–643.

15. Martinez, C., Abergel, C., Cambillau, C., de Geus, P. & Lauwereys, M. (1991). Crystallographic study of a recombinant cutinase from *Fusarium solani pisi*. In *Lipases: Structure, Mechanism and Genetic Engineering*, GBF monographs, vol. 16, ed. Alberghina, L., Schmid, R.D. & Verger, R., pp. 67–70. Weinheim: VCH.

16. Martinez, C., de Geus, P., Lauwereys, M., Matthyssens, G. & Cambillau, C. (1992). *Fusarium solani* cutinase is a lipolytic enzyme with a catalytic serine accessible to solvent. *Nature*, 356, 615–618.

17. Stahl, D.J. & Schäfer, W. (1992) Cutinase is not required for fungal pathogenicity on pea. *Plant Cell* 4, 621–629.

18. Brady, L., Brzozowski, M., Derewenda, Z.S., Dodson, E., Dodson, G., Tolley, S., Turkenburg, J.P., Christiansen, L., Huge-Jensen, B., Nørskov, L., Thim, L. & Menge, U. (1990). A serine protease triad forms the catalytic centre of a triacylglycerol lipase. *Nature*, 343, 767–770.

19. Winkler, F.K., D'Arcy, A. & Hunziker, W. (1990). Structure of human pancreatic lipase. *Nature*, 343, 771–774.

20. Schrag, J.D., Li, Y., Wu, S. & Cygler, M. (1991). Ser-His-Glu triad forms the catalytic site of the lipase from *Geotrichum candidum*, *Nature*, 351, 761–764.

21. Sussman, J.L., Harel, M. Frolow, F., Oefner, C., Goldman, A., Toker, L. & Silman, I. (1991). Atomic structure of acetylcholinersterase from *Torpedo californica*: a prototypic acetylcholine-binding protein. *Science* 253, 872–879.

22. Taylor, W.R. (1987). Multiple sequence alignment by a pairwise algorithm. *Comput. Appl. Biosci.*, 3, 81–87.

23. Carillo, H. & Lipman, D. (1988). The multiple sequence alignment problem in biology. *SIAM J. Appl. Math.*, 48, 1073–1082.

24. Lipman, D.J., Altschul, S.F. & Kececioglu, J.D. (1989). A tool for multiple sequence alignment. *Proc. Natl. Acad. Sci. USA*, 86, 4412–4415.
25. Altschul, S.F. & Lipman D.J. (1990). Protein database searches for multiple alignments. *Proc. Natl. Acad. Sci. USA*, 87, 5509–5513.
26. Altschul, S.F. (1989). Gap costs for multiple sequence alignment. *J. Theor. Biol.*, 138, 297–309.
27. George, D.G., Barker, W.C. & Hunt, L.T. (1990). Mutation data matrix and its uses. *Methods Enzymol.*, 183, 333–351.
28. Hein, J. (1989). A new method that simultaneously aligns and reconstructs ancestral sequences for any number of homologous sequences, when the phylogeny is given. *Mol. Biol. Evol.*, 6, 649–668.
29. Hein, J. (1990). Unified approach to alignment and phylogenies. *Methods Enzymol.*, 183, 626–645.
30. Cai, S.J., Wong, D.M., Chen, S.H. & Chan, L. (1989). Structure of the human hepatic triglyceride lipase gene. *Biochemistry*, 28, 8966–8971.
31. Ameis, D., Stahncke, G., Kobayashi, J., McLean, J., Lee, G., Buscher, M., Schotz, M.C. & Will, H. (1990). Isolation and characterization of the human hepatic lipase gene. *J. Biol. Chem.*, 265, 6552–6555.
32. Wang, C.-S., Hartsuck, J. & McConathy, W.J. (1992). Structure and functional properties of lipoprotein lipase. *Biochim. Biophys. Acta*, 1123, 1–17.
33. Gibney, G., Camp, S., Dionne, M., MacPhee-Quigley, K. & Taylor, P. (1990). Mutagenesis of essential functional residues in acetylcholinesterase. *Proc. Natl. Acad. Sci. USA*, 87, 7546–7550.
34. Prytz, H., personal communication, 1992.
35. Lockridge, O. (1990). Genetic variants of human serum cholinesterase influence metabolism of the muscle relaxant succinylcholine. *Pharmacol. Ther.* 47, 35–60.
36. Bläckberg, L., Bjursell, G., Carlson, P. Enerbäck, S., Hernell, O. & Nilsson, J. (1991). cDNA cloning and sequencing of human milk bile-salt stimulated lipase. In *Lipases: Structure, Mechanism and Genetic Engineering*, GBF monographs, vol. 16, ed. Alberghina, L., Schmid, R.D. & Verger, R., pp. 203–206. Weinheim: VCH.
37. Feller, G., Thiry, M. Arpigny, J.L., Mergeay, M. & Gerday, C. (1990). Lipases from psychrotrophic antarctic bacteria. *FEMS Microbiol. Letters*, 66, 239–244.
38. Feller, G., Thiry, M. & Gerday, C. (1991). Nucleotide sequence of the lipase gene lip2 from the antarctic psychrotroph *Moraxella TA144* and site-specific mutagenesis of the conserved serine and histidine residues. *DNA Cell Biol.* 10, 381–388.
39. Shimada, Y., Sugihara, A. Iizumi, T. & Tomiuaga, Y. (1990). cDNA cloning and characterization of *Geotrichum candidum* lipase II. *J. Biochem.*, 107, 703–707.
40. Kawaguchi, Y., Honda, H., Taniguchi-Morimura J. & Iwasaki, S. (1989). The codon CUG is read as serine in an asporogenic yeast *Candida cylindracea*. *Nature*, 341, 164–166.
41. Brenner, S. (1988). The molecular evolution of genes and proteins: a tale of two serines. *Nature*, 334, 528–530.
42. Kates, M. (1992). Archaebacterial lipids: structure, biosynthesis and function. In *The Archaebacteria, Biochemistry and Biotechnology*, ed. Danson, M.J., Hough, D.W. & Lunt, G.G. (Biochemical Society Symposium no. 58), pp. 51–73. London: Portland Press.

Human hepatic lipase

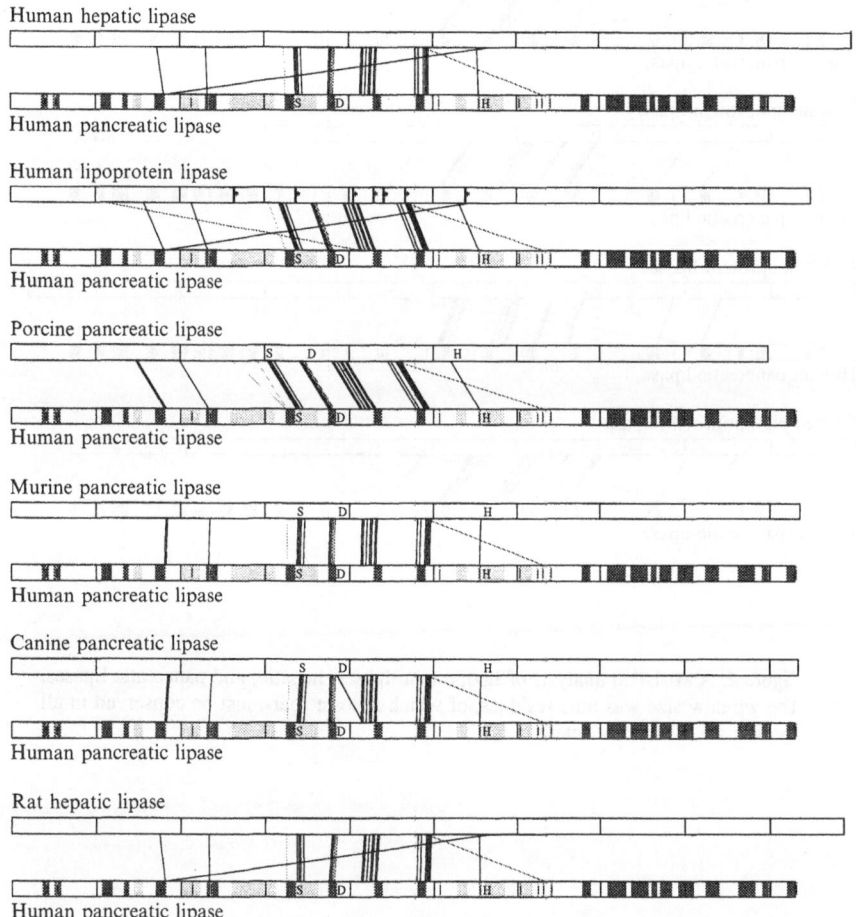

Human pancreatic lipase

Human lipoprotein lipase

Human pancreatic lipase

Porcine pancreatic lipase

Human pancreatic lipase

Murine pancreatic lipase

Human pancreatic lipase

Canine pancreatic lipase

Human pancreatic lipase

Rat hepatic lipase

Human pancreatic lipase

Bovine lipoprotein lipase

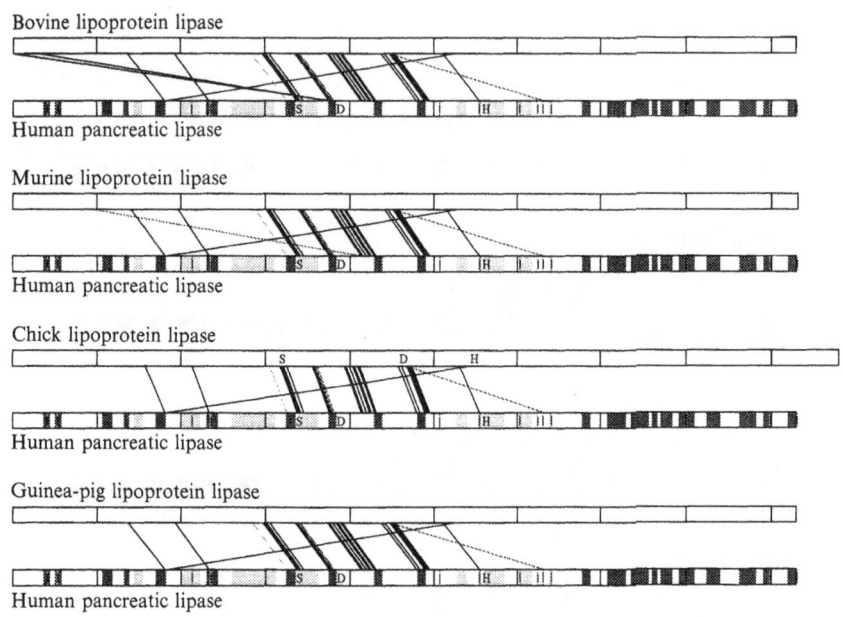

Human pancreatic lipase

Murine lipoprotein lipase

Human pancreatic lipase

Chick lipoprotein lipase

Human pancreatic lipase

Guinea-pig lipoprotein lipase

Human pancreatic lipase

Figure 2. A MULTIM analysis of lipoprotein lipases, hepatic, and pancreatic lipases. The window size was nine residues, of which at least four must be conserved in all species; see the text for details.

Geotrichum candidum lipase 1

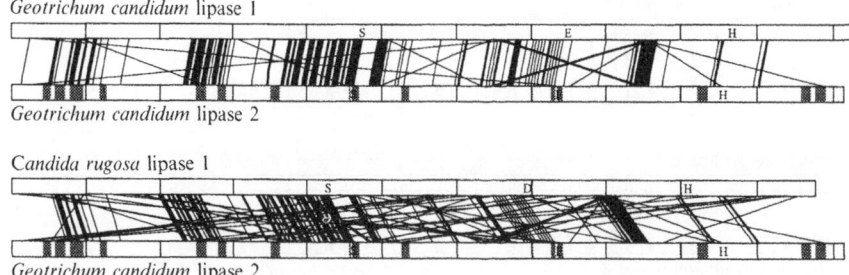

Geotrichum candidum lipase 2

Candida rugosa lipase 1

Geotrichum candidum lipase 2

Figure 3. A MULTIM analysis of the *Geotrichum* and *Candida rugosa* lipases using a window size of seven residues of which at least four were conserved; see the text for details.

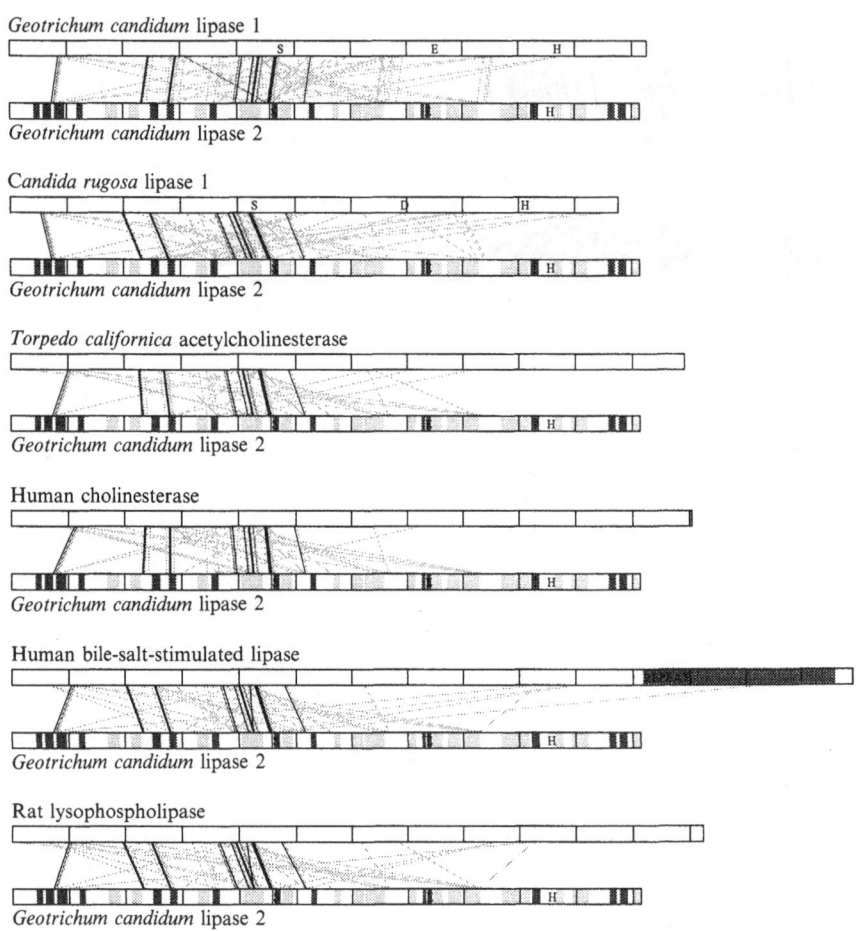

Figure 4. A MULTIM analysis of selected members of the esterase family. The window size is six residues, of which at least three are conserved.

Subclass	Code	Origin	Length	Signal	SER	Subsequence	Codon
Lipoprotein	LIPL$CAVPO	*Cavia porcellus*	465	17	149	VHLLGYSLGAHAA	AGC
	LIPL$BOVIN	*Bos taurus*	>465	>15	149	VHLLGYSLGAHAA	AGC
	LIPL$HUMAN	*Homo sapiens*	475	27	159	VHLLGYSLGAHAA	AGC
	LIPL$MOUSE	*Mus musculus*	>466	>19	151	VHLLGYSLGAHAA	AGC
	LIPL$CHICK	*Gallus gallus*	490	25	159	VHLLGYSLGAHAA	AGC
Pancreatic	LIPP$PIG	*Sus scrofa*	449	?	152	VHVIGHSLGSHAA	?
	LIPP$CANFA	*Canis familiaris*	467	17	171	VQLIGHSLGAHVA	AGC
	LIPP$HUMAN	*Homo sapiens*	465	16	169	VHVIGHSLGAHAA	AGC
	LIPP$MOUSE	*Mus musculus*	468	16	170	VHLIPHSLGSHVA	AGC
Hepatic	LIPH$RAT	*Rattus norvegicus*	494	22	169	VHLIGYSLGAHVS	AGC
	LIPH$HUMAN	*Homo sapiens*	499	22	168	VHLIGYSLGAHVS	AGC
Gastric	LIPG$HUMAN	*Homo sapiens*	398	19	172	LHYVGHSQGTTIG	TCT
	LIPG$RAT	*Rattus norvegicus*	395	18	171	IHYVGHSQGTTIG	TCT

Table 1. The lipoprotein, pancreatic, hepatic and gastric lipases. The subclass, the SWISSPROT entry code, the origin of the protein, the lengths of the total protein and the signal sequence, the position of the putative active serine, the relevant subsequence and the codon for the serine are shown. Note that the codon usage for the relevant serine in the gastric lipases differs from that in the rest of the lipases included in this table.

Subclass	Code	ASP	Subsequence	HIS	Subsequence
Lipoprotein	LIPL$CAVPO	173	SRITGLDPAGPNF	258	QLVKCSHERSIHL
	LIPL$BOVIN	173	NRITGLDPAGPNF	258	QLVKCSHERSVHL
	LIPL$HUMAN	183	NRITGLDPAGPNF	278	QLVKCSHERSIHL
	LIPL$MOUSE	175	NRITGLDPAGPNF	260	QLVKCSHERSIHL
	LIPL$CHICK	183	NRITGLDPAGPTF	278	QLVKCSHERSIHL
Pancreatic	LIPP$PIG	176	ERITGLDPAEPCF	263	DFVACNHLRSYKY
	LIPP$CANFA	194	GRITGLDPVEASF	281	DFVACNHLRSYKY
	LIPP$HUMAN	193	GRITGLDPAEPCF	280	DFAACNHLRSYKY
	LIPP$MOUSE	194	GRITGLDPAEPCF	281	NFAACNHLRSYKY
Hepatic	LIPH$RAT	195	GRITGLDPAGPMF	280	QTINCAHERSVHL
	LIPH$HUMAN	194	GRITGLDAAGPLF	279	QTIKCSHERSVHL
Gastric	LIPG$HUMAN	347	GKDLLADPQDVGL	363	LSNLLFHKEILAY
	LIPG$RAT	346	GNDILADPQDVAM	364	LPNLIYHKEIPFY

Table 2. Subsequences for the putative active-site Asp and His positions for the lipoprotein, pancreatic, hepatic and gastric lipases.

Subclass	Code	Origin	Length	Signal	SER	Subsequence	Codon
Rhizomucor	LIP$RHIMI	*Rhizomucor miehei*	363	24+70	238	VAVTGHSLGGATA	?
	LIP$RHIDL	*Rhizopus delemar*	392	26	268	VIVTGHSLGGAQA	TCA
Staphylococcus	LIP$STAHY	*Staphylococcus hyicus*	641	38	36	VHFIGHSMGGQTI	AGT
	LIP$STAAU	*Staphylococcus aureus*	690	37	412	VHLVGHSMGGQTI	AGT
Pseudomonas	LIP$PSECE	*Pseudomonas cepacia*	364	44	131	VNLVGHSQGGLSS	AGC
	LIP$PSEFR	*Pseudomonas fragi*	277	18	83	VNLIGHSQGALTA	AGC
Moraxella	LIP1$MORSP	*Moraxella* sp. (TA144)	319	?	189	LGAIGWSMGGGGA	TCA
	LIP2$MORSP	*Moraxella* sp. (TA144)	433	?	239	IVLSGDSAGGCLA	?
	LIP3$MORSP	*Moraxella* sp. (TA144)	315	?	142	THVGGNSMGGAIS	TCG
Cutinases	CUTI$FUSSO	*Fusarium solani pisi*	230	31	136	LIAGGYSQGAALA	AGC
					70	LGTLGPSIASNLE	AGC
	CUTI$COLCA	*Colletotrichum capsici*	228	?	140	VVAGGYSQGTAVM	AGC
					63	PGNMGISAGPIVA	AGC
	CUTI$COLGL	*Colletotrichum gloeosporioides*	224	?	136	IVSGGYSQGTAVM	AGC
					60	PGNMGISAGPIVA	AGC

Table 3. The fungal and procaryotic classes of lipases. See Table 1 for details. For the *Rhizomucor miehei* lipase, the length of the propeptide is indicated in addition to the signal sequence. For the cutinases, alternative sites with motifs similar to the putative active serine are also shown.

Subclass	Code	Origin	Length	Signal	SER	Subsequence	Codon
Hormone-sensitive	LIPS$RAT	*Rattus norvegicus*	757	?	423	ICLAGDSAGGNLC	AGC
Bile-salt-stimulated	BSSL$HUMAN	*Homo sapiens*	745	23	217	ITLFGESAGGASV	TCT
Lysophospho	LIPY$RAT	*Rattus norvegicus*	612	30	214	ITIFGESAGGAIV	TCT
Geotrichum	LIP1$GEOCN	*Geotrichum candidum*	563	19	236	VMIFGESAGAMSV	?
	LIP2$GEOCN	*Geotrichum candidum*	>557	>13	230	VMIFGESAGAMSV	TCC
Candida	LIP1$CANRU	*Candida rugosa*	>538	>4	213	VTIFGESAGSMSV	CTG

Table 4. Cholinesterase-like lipases. See Table 1 for details. The codon CTG for Candida rugosa is read as a Ser (reference 40).

Subclass	Code	Origin	Length	Signal	SER	Subsequence	Codon
Acetylcholine	ACES$DROME	*Drosophila melanogaster*	649	36	276	MTLFGESAGSSSV	TCG
	ACES$MOUSE	*Mus musculus*	614	31	234	VTLFGESAGAASV	AGT
	ACES$TORCA	*Torpedo californica*	596	21	221	VTIFGESAGGASV	AGT
	ACES$TORMA	*Torpedo marmorata*	599	24	224	VTLFGESAGRASV	AGT
	ACES$HUMAN	*Homo sapiens*	614	31	234	VTLFGESAGAASV	AGC
	ACES$BOVIN	*Bos taurus*	583	?	203	VTLFGESAGAASV	?
Choline	CHLE$HUMAN	*Homo sapiens*	602	28	226	VTLFGESAGAASV	AGT
	CHLE$RABIT	*Oryctolagus cuniculus*	581	8	205	VTLFGESAGAASV	AGT
Carboxyl	EST$ACICA	*Acinetobacter calcoaceticus*	290	?	131	IIISGDSCGANLH	TCA
	EST1$CULPI	*Culex pipiens*	540	?	191	VTLAGHSAGAASV	AGC
	EST1$RABIT	*Oryctolagus cuniculus*	539	?	195	VTIFGESAGGQSV	?
	EST1$RAT	*Rattus norvegicus*	549	18	221	VTIFGESAGGVSV	TCA
	EST1$HUMAN	*Homo sapiens*	>507	?	161	VTIFGESAGGESV	TCA
	EST2$RABIT	*Oryctolagus cuniculus*	532	?	201	VTIFGESAGGTSV	?
	EST6$DROME	*Drosophila melanogaster*	544	21	209	VLLVGHSAGGASV	TCC

Table 5. Acetylcholine, choline and carboxyl esterases. See Table 1 for details.

Subclass	Code	Origin	Length	Signal	SER	Subsequence	Codon
Esterase D	ESTD$HUMAN	Homo sapiens	>297	?	164	MSIFGHSMGGHGA	TCC
Juvenile Hormone esterase	ESTJ$HELVI	Heliothis virescens	564	19	220	ITIAGQSAGASAA	AGC
Esterase P	ESTP$DROME	Drosophila melanogaster	544	19	206	IVLIGHSAGGASA	TCT
	ESTP$RAT	Rattus norvegicus	565	18	221	VTIFGESAGGFSV	TCT
Tropin esterase	TPES$PSEPU	Pseudomonas putida	272	?	110	TTVIGHSMGSMTA	?
Sterol esterase	A33668(PIR)	Bos taurus	>597	?	212	ITLFGESAGGQSV	?
Gramicidin S biosynth. prot.	GRST$BACBR	Bacillus brevis	256	?	95	FAFLGHSMGALIS	AGC
Thioesterases	SAST$ANAPL	Anas platyrhynchos	251	?	90	FALFGHSFGSFVS	AGT
	SAST$RAT	Rattus norvegicus	263	?	101	FAFFGHSFGSYIA	AGT
	SAST$VIBAN	Vibrio anguillarum	252	?	92	TIIVGHSMGAQVA	AGT
Fatty acid synthetase	FAS1$YEAST	Saccharomyces cerevisiae	1845	?	274	KGATGHSQGLVTA	TCT
					1808	ATFAGHSLGEYAA	TCT
	FAS$RAT	Rattus norvegicus	2505	?	2228	YRVAGYSFGACVA	TCT
	FAS$MOUSE	Mus musculus	>838	?	635	HRIAGYSFGACVA	TCA
	FAS$CHICK	Gallus gallus	2446	?	2235	YRIAGYSFGACVA	TCT
Arginine esterase	ESTA$CANFA	Canis familiaris	260	?	212	DTCYGDSGGPLIC	TCA
Proteinase	RVVA$VIPRU	Vipera russelli siamensis	236	?	182	DTCHGDSGGPLIC	?
	RVVG$VIPRU	Vipera russelli siamensis	236	?	182	DTCHGDSGGPLIC	?
Enterochelin esterases	FES$ECOLI	Escherichia coli	374	?	255	TVVAQSFGGLSA	AGT
Cholesterol esterases	CHES$RAT	Rattus norvegicus	612	20	214	ITIFGESAGAASV	TCT
	CHES$HUMAN	Homo sapiens	745	23	217	ITLFGESAGASV	TCT
Methylesterase	CHEB$SALTY	Salmonella typhimuriun	349	?	164	LIAIGASTGGTEA	?

Table 6. Various esterases. See Table 1 for details.

3.

Lipases from plants

KUMAR D. MUKHERJEE and MATTHEW J. HILLS

What is special about lipases in plants? Lipolytic enzymes are widely distributed in the plant kingdom, yet the knowledge of such lipases from plants is still rather limited compared with those from mammalian systems and micro-organisms. Lipases from plants can be broadly classified into:

- triacylglycerol lipases (EC 3.1.1.3) or 'true' lipases that hydrolyse triacyl-glycerols, which are the major lipid constituents of storage tissues, such as seeds
- non-specific lipid acylhydrolases exhibiting combined action of various lipases, such as phospholipases A_1 (EC 3.1.1.32), A_2 (EC 3.1.1.4), B (EC 3.1.1.5), glycolipase, sulpholipase and monoacylglycerol lipase, which occur in diverse plant tissues
- phospholipases C (EC 3.1.4.3) and D (EC 3.1.4.4), the latter being more widely distributed in plants.

The occurrence, properties and physiological rôle of various lipases in plants have been covered extensively in several reviews (1–6). The major objective of this chapter is – as the title suggests – to focus on lipases from plants, with emphasis on:

- isolation of lipases from plants
- characteristics of lipases isolated from plants
- applications of lipases from plants.

1. TECHNOLOGICAL IMPORTANCE OF LIPASES

Commercial values of agricultural commodities, such as oilseeds and cereal grains, can be affected significantly by the action of lipolytic enzymes contained in such products. Improper storage and handling of oilseeds and cereal grains may initiate the activity of lipases that are generally not present in the dormant seed. Consequently, undesirable products, such as 'free' fatty acids, are formed that are prone to autoxidative processes which lead to severe impairment of the quality of the final products.

In the oil-milling industry, it is customary to carry out a moist heat treatment of the ground seeds in a so-called 'cooking' process prior to oil recovery. The objective of cooking is not only to rupture the membranes of the oil-bearing organelles for easier recovery of the oil, but also to prevent the formation of undesirable substances by deactivation of endogenous enzymes, including lipases, if present. A few examples will be given here of the technological implications of lipases in oilseeds and cereal grains, and measures to counteract their deleterious effects on lipids.

A major problem in the processing of oil palm is posed by post-harvest enzymic hydrolysis of the oil, which can lead to extensive formation of free fatty acids in the oil (7). It is believed that the lipids in the oil palm mesocarp are protected from the lipases by the membranes of the vacuoles (8), and these membranes are disrupted by bruising during harvesting and handling which leads to high lipolytic activity in the oil palm during storage (9). Although earlier studies report the absence of lipase in the oil palm mesocarp (10, 11) an active triacylglycerol lipase has been isolated later from the flesh of the ripe mesocarp (12). The lipase of the ripe oil palm fruit is sensitive to chilling inactivation at 8 °C, and the enzyme is completely inactivated by heating at 45 °C for 10 h or at 55 °C for 30 min (13). Similar findings have been reported by another group (14). Others have reported, however, that overnight exposure of oil palm fruits to a temperature of 5 °C may induce lipase-catalysed release of up to 70% of the total fatty acids (15). In industrial processing, the fresh fruit bunches of oil palm are sterilised at temperatures of 100 °C or more, in order to inactivate the lipase completely (16).

Like the lipase of oil palm, the endogenous lipase of the olive fruit can cause extensive lipolysis of the oil, especially if the fruit is bruised during harvesting (17).

Lipases also play a major rôle in the quality of rice bran oil, an important industrial commodity in rice-producing countries. The prime drawback of rice bran as a source of oil is the presence of highly active lipolytic enzymes in the bran (18–21). The lipolytic activity is observed as soon as the bran is removed from the rice, and lipolysis continues with prolonged storage (18), yielding high levels of free fatty acids in the oil, which becomes unsuitable for processing to an edible oil. Several methods have been applied to arrest the lipolysis in the bran, which include heating, treatment with hydrochloric acid (22) or exposure to microwave irradiation (23), the latter technique being found to yield rice bran oil containing only 2% free fatty acids.

Lipases and acylhydrolases have been observed in wheat, oats, barley and rice (19, 24) and studied in some detail because they cause problems of rancidity and loss of baking performance in flour. Although grains do not contain much stored oil compared with oilseeds, fatty acids released by the action of endogenous lipases on those triacylglycerols in the flour are oxidised by lipoxygenases and the products of the oxidation reactions cause the 'off' flavours (24).

Another lipase that has to be taken into consideration in the processing of oilseeds is phospholipase D. In industrial practice, crude hexane-extracted oils, such as soybean oils, are subjected to a 'degumming' process that involves hydration of phospholipids and their subsequent removal from the oil together with the mucilegenous substances. The non-hydratable phospholipids, such as phosphatidic acids and lysophosphatidic acids, if present in the degummed oil at high levels, contribute to undesirable taste, colour, odour and stability of the refined oil (25). It has been shown recently for soybeans that phospholipase D can remain fairly active in catalysing the hydrolysis of hydratable phospholipids, (*e.g.* phosphatidylcholines) to non-hydratable phospholipids, (*e.g.* phosphatidic acids), even at a temperature of 65 °C in the presence of water-saturated hexane, although in aqueous media this enzyme is readily inactivated by heat (26). These results show that non-hydratable phospholipids can be formed by the action of phospholipase D during oil extraction in the commercial processing of soybean.

2. ISOLATION AND CHARACTERISATION OF LIPASES, ACYLHYDROLASES AND PHOSPHOLIPASES FROM PLANTS

2.1 Triacylglycerol lipases

Many of the true lipases so far studied in plants have been found to be membrane-bound and have been examined whilst attached to the membranes. Exceptions, which have been dealt with in some detail, are lipases from maize, rape and castor bean. These have all been solubilised and been either purified to homogeneity or at least substantially purified.

Maize

Lipase from the scutella of maize seedlings has been purified to apparent homogeneity (27). About 50% of total lipase activity in homogenates of maize scutella is associated with the oil bodies (28) and is tightly bound since it resists washing with salt solutions. As with the oil-body lipases from castor and rape, the maize lipase catalyses autolysis of the triacylglycerols stored in the oil body, releasing free fatty acids (28). The lipase has a neutral pH optimum and, in contrast to the castor lipase, the maize enzyme is not inhibited by sulphydryl reagents. The maize enzyme acts on fatty acids esterified at both the primary and secondary positions of glycerol (6). It was found that upon removal of the triacylglycerols from the oil bodies with diethyl ether the lipase became partially solubilised. After removal of the ether, the de-lipidated oil bodies were centrifuged at $100\,000\,g$ for 90 min. Between 50 and 80% of the lipase activity remained in the supernatant, but the proteins were probably in an aggregated state, as they eluted in the void volume from a Sephadex G-200 column (27). The lipase in the supernatant fraction was further purified by ion-exchange chromatography and sucrose density-gradient centrifugation in the presence of deoxycholate as a detergent. The peak fraction of lipase activity was found to contain one polypeptide with a molecular weight of 65 kDa, although the native protein in detergent solution was estimated to consist of a tetramer with a molecular weight of 270 kDa (27). Antibodies were raised against the 65-kDa polypeptide and used to immunoprecipitate lipase synthesized by translation *in vitro* from mRNAs isolated from maize scutella. It was shown that the enzyme is synthesized on free rather than membrane-bound polyribosomes and that there is no co- or post-translational modification of the lipase (29). It was also shown that the appearance of lipase activity following germination of the maize seed was due to synthesis of the protein *de novo* rather than to activation of pre-existing protein. The turnover of lipase protein in the scutellum also appears to be slow, since once the maximal activity has been reached at 6 days after germination, it remains high, although the levels of lipase mRNA decrease rapidly after 6 days (29).

Oilseed rape

Density-gradient centrifugation studies have shown that the majority of the lipase activity in homogenates from cotyledons of oilseed rape seedlings is found in very light membrane fragments or protein/lipid vesicles (30, 31). A small proportion of the lipase was found to be associated with oil bodies (32), and there were some indications that the properties of the oil-body enzyme were different from those found in the particulate fraction (31). However, in one study as much as 50% of the lipase activity was found to be associated with oil bodies (33). The lipase in the particulate fraction is thought to be associated with fragments of oil bodies created during the hydrolysis of the oil, since they can be separated from the endoplasmic reticulum on the sucrose density gradients (30, 31). In some rape cultivars, almost all of the lipase-containing particles will pellet upon centrifugation at 100 000 g for 1 h, whereas in others as much as 70% of the lipase activity remains in the supernatant. The lipase in the 100 000-g supernatant is eluted in the void volume of a Sephacryl S-300 column, indicating that it was present in large aggregates. Analysis of the lipid content of the lipase-containing particles has shown them to contain 50% free fatty acid, 30% acylglycerols (mainly as di- and monoacylglycerols) and 20% phospholipids (34). The lipid-to-protein ratio is about 3:1. The presence of large proportions of neutral lipid supports the idea that the lipase is associated with particles derived from the storage oil bodies. The rape lipase has been partially purified in several laboratories (34–36), but not yet to homogeneity. It appears to be occur in very low abundance, and this has hampered efforts to isolate it. Detergents are required for the solubilisation of the lipase before column chromatography. It was shown that CHAPS and Triton X-100 inhibited rape lipase activity (31, 35) but the effect of Triton X-100 was reversible (35). Octylglucoside and deoxycholate were found to stimulate lipase activity (31) and deoxycholate was used routinely in lipase assays (30, 34). It is not clear why some detergents cause inhibition of lipase activity while others stimulate activity by up to tenfold. The Canadian group used a cultivar (Westar) from which the lipase pelleted at $150\,000\,g$. Triton X-100 was used to solubilise the lipase, and gel filtration and ion-exchange chromatography were used as purification steps (35). The peak activity from the gel-filtration column was associated with an M_r of about 250kDa. The pH optimum was found to be about 7.5, similar to that previously reported (31). A procedure for the purification of lipase from the $100\,000\text{-}g$ supernatant of rape cotyledon homogenates from another cultivar has been described which gave almost a 300-fold purification. The lipase in the $100\,000\text{-}g$ supernatant was precipitated with $MgCl_2$ and polyethyleneglycol and purified by gel filtration and ion-exchange chromatography (37). The rape lipase was estimated to have an M_r of about 300kDa (34). This is in reasonable agreement with that determined previously for rape (35) and that found for maize (27). The lipase has a strong affinity for hydroxyapatite and a 200- to 300-fold purification was achieved in this step (36). This gave a 9000-fold purification over the original $100\,000\text{-}g$ supernatant, but SDS-PAGE analysis revealed several silver-stained bands on the gel of which none was an obvious candidate for being the lipase polypeptide (36). The antibody raised against the alkaline lipase from castor (38) was used to probe a Western blot of the peak fractions from the hydroxyapatite column, and a band at 67 kDa was revealed (36). The fractions containing the 62-kDa protein which cross-reacted most strongly with the castor alkaline lipase antibody did not show lipase activity.

The pH optimum of the rape lipase was about 7.5, and it was found to hydrolyse fatty acids from all positions on the glycerol backbone; the first removed was either from the *sn*-1 or the *sn*-3 position (34). Activity against oleoyl (O) and palmitoyl (P) moieties was similar to that found with POP and OPO as triacylglycerol substrates. The rape lipase showed unusual selectivities against fatty acids containing *cis*-6 double bonds, and the biotechnological possibilities of this are described in a later section.

Castor bean

Three triacylglycerol lipases have been identified in the endosperm of germinating castor-bean seedlings. Two of these are located in the oil-body membrane and have acidic and neutral pH optima, respectively, whilst the third is found in the glyoxysomal membrane and has an alkaline pH optimum. The acid lipase was the subject of some study by Ory and co-workers in the 1960s (39, 40). The enzyme is firmly embedded in the oil-body membrane and fails to be solubilised by a wide range of non-denaturing detergents (authors' observations). It has an acidic pH optimum of about 4.2, and at this pH value the lipase can hydrolyse all of the endogenous substrate in less than 30 min (41). Some experiments were undertaken to isolate the acid lipase protein, but it was found to precipitate (40). It is thought that the acid lipase requires a glycoprotein and a lipid co-factor for catalytic activity (40). The lipid co-factor was found to be a cyclic tetramer of ricinoleic acid, where the 12-hydroxyl group is esterified to the carboxyl group of the ricinoleoyl moiety next to it. The lipid could only be extracted by dry butanol and could not be detected if water was present. Recent studies have shown that many lipases catalyse esterification reactions in dry organic solvents (42). This raises the possibility that the cyclic compound was formed by such esterification reactions during the butanol extraction procedure to form the cyclic estolide. The acid lipase will cleave fatty acids of C4 to C18 chain length from triacylglycerols, whilst diacylglycerols are hydrolysed faster than triacylglycerols (39). Methyl and butyl esters of fatty acids are hydrolysed but phospholipids are not. In a study using *iso*-propyl esters, it was shown that the acid lipase cleaves fatty acids from the *sn*-2 position (43).

The alkaline lipase in the glyoxysomal membrane has been purified to apparent homogeneity following its solubilisation with deoxycholate (38). The procedure involved column chromatography using ion-exchange and hydroxyapatite. A final purification factor of 272 was achieved and SDS-PAGE revealed one band with an apparent M_r value of 62 kDa. The alkaline lipase is active against triacylglycerols containing ricinoleic acid, but is most active against monoacylglycerols (38). Antibodies were raised against this polypeptide, and these were shown to immuno-inhibit the lipase activity. These antibodies were later shown to immuno-inhibit lipase from rape, maize and peanut (44).

It was not clear how triacylglycerol hydrolysis was catalysed in castor bean endosperm, as the acid lipase of the oil body is inactive at neutral pH values and the alkaline lipase of the glyoxysome is spatially separated from the substrate and is much more active on mono- than on triacylglycerols (5). However, studies on the pH dependence of autolysis of the oil bodies during early seedling growth showed that a neutral lipase associated with the lipid body membrane is synthesized during seedling development and it has sufficient activity to account for triacylglycerol hydrolysis rates found *in vivo* (41). The neutral lipase is stimulated 40-fold by 30 µM free Ca^{2+}, the effect of Ca^{2+} being stronger at more alkaline pH values. The neutral lipase is easily separated from the acid lipase by solubilisation with deoxycholate (authors' observations), indicating that the neutral lipase activity is due to the presence of a new protein rather than a shift in the pH curve of the acid lipase.

Peanut

Studies were made on peanut lipase, purified about 17-fold by ammonium sulphate precipitation and gel filtration from an acetone powder made from developing seeds (45). The enzyme was rather unstable, which precluded any further purification. The lipase had a pH optimum of 8.5 and a K_m of 0.26 mM with tributyrin as substrate. The molecular weight estimated from gel filtration on Sephadex and SDS-PAGE was 55 kDa. The lipase in these studies was soluble and found in developing rather than germinated seeds, so it is likely to be quite different from the alkaline glyoxysomal membrane lipase reported later, although this also has an alkaline pH optimum (46).

Vernonia anthelmintica

Ungerminated seeds of this plant contain a lipase with a pH optimum at about 7.5-8.0. Its intracellular location is not known. It was examined in detail after initial studies had suggested that it might act specifically at the *sn*-2 position (47). It was purified 30-fold from an acetone powder of the dry seed (48). The enzyme eluted in the void volume of a Sephadex G-200 column indicating a molecular weight of greater than 200 000 for the native enzyme, or its presence in large aggregates as was found for the maize and rape lipases in the absence of detergents (27, 34). The enzyme acted on both the primary and secondary ester positions of the triacylglycerols and showed no fatty acid preference, although activity against the natural substrate (trivernolin), which contains an epoxy fatty acid, was not examined (48).

Oil palm

Lipase activity has been observed in the mesocarp of developing oil palm fruits (12) and in the shoots of germinated seedlings (49). The lipase is synthesized during the period of fruit ripening (13) and was located in the oil body by histological staining methods (14). It is not clear whether lipase-catalysed degradation of triacylglycerols had occurred during incubation at low temperature in the latter study. The mesocarp oil-body lipase was partially purified from the oil-body fraction of mesocarp homogenates (12) following extraction of the triacylglycerols by diethyl ether. The lipase was precipitated by ammonium sulphate and subjected to ion-exchange and gel-filtration chromatography to give a threefold purification. The pH optimum of the lipase was 4.5, and it was most active against palm oil as substrate. The enzyme was very strongly inhibited by mercuric ions, glycylglycine, cholesterol and lecithin, and was stimulated to a small extent by phenol and EDTA (12). A study of germinated palm seeds showed that lipase activity could not be detected in extracts of the kernel, the site of triacylglycerol storage, and very little activity was found in the haustorium, but lipase was found in extracts of the shoot (49). The enzymes involved in fatty-acid oxidation were, however, found mainly in the haustorium. Translocation of radiolabelled triacylglycerols or fatty acids was measured at very low levels, so the mechanism of oil mobilisation in oil palm seedlings remains unclear (50). It was suggested that physiological regulators of lipase activity might cause inhibition during extraction of the kernel and haustorium, thus masking their presence (49).

Wheat

Wheat grains contain lipases, acylhydrolases and esterases; problems in distinguishing between these activities in wheat and other grains have been discussed previously (5). The wheat lipase activity increases during germination for the mobilisation of lipid reserves (51). Lipases present in the bran and isolated germ of dry wheat have very recently been purified and studied in some detail (52). The lipases were isolated in microsomal fractions, solubilised with Triton X-100 and purified by chromatography on ion-exchange and hydroxyapatite columns. The bran lipase was purified 240-fold from microsomes and was free from contaminating esterase activity. The lipase from the germ was separated from bran lipase when whole flour was used as the source material. The lipases had similar molecular weights of 67 kDa (germ) and 70 kDa (bran), but had different characteristics with regard to heat stability and pH optima (52). In common with the maize scutellum lipase, the wheat lipases were unaffected by inhibitors of sulphydryl groups. The wheat-bran lipase was extremely heat-stable, and retained 50% of activity after 2 h at 95 °C.

Rice

Two lipases have been purified to apparent homogeneity from rice bran and studied quite extensively. The enzymes are soluble and, unlike the oilseed lipases, do not require the use of detergents during purification. Rice lipase I is activated by Ca^{2+} and has a pH optimum of about 7.5. Its molecular weight is about 40 kDa (53). It appears to cleave preferentially fatty acids from the *sn*-1 and *sn*-3 positions. A rice bran lipase of 40 kDa was also purified by another group (20). The bran lipase II was purified by gel filtration and ion-exchange chromatography (19); the native enzyme has a molecular weight of 32 kDa by SDS-PAGE under non-reducing conditions (54) and an isoelectric point of 9.1. Rice bran lipase II also has a pH optimum of about 7.5 and is very much more active towards triacylglycerols containing short-chain fatty acids. Following cleavage of disulphide bridges, SDS-PAGE revealed the presence of two subunits with molecular weights of 14 kDa and 3.9 kDa, but the arrangement of the subunits is not known (54). Amino acid compositions of the rice-bran lipases were obtained, but no sequence data are available.

Barley

Lipase activity in barley aleurone cells is important in the mobilisation of triacylglycerols stored in lipid droplets to supply acyl groups for phospholipid synthesis and for oxidation (55). Although lipase has not been purified from barley, hydrolysis of triacylglycerols is induced by gibberellic acid (GA) by an unusual mechanism. Ultrastructural and biochemical studies showed that lipase in barley aleurone is associated with the protein bodies, but upon application of GA, the lipase is transferred to the oil bodies, probably by fusion of the protein-body and oil-body membranes (56). It is interesting to contrast this with other seed types such as rape or castor, where lipase is not found to be associated with protein bodies.

Oat

Lipase from the pericarp of oat was partially purified by precipitation and shown to have a pH optimum of 7.5 (57). Although the enzyme shows no positional specificity on triacylglycerols as substrate (58), it will only cleave one butyric acid group from tributyrin to yield dibutyrin. It will not hydrolyse either mono- or dibutyrin. The use of oat lipase as a biocatalyst is described later in this chapter.

Potato tuber

True lipase activity has been purified about 18-fold from potato tubers by ion-exchange and gel permeation chromatography (59). It is a soluble enzyme with an estimated M_r of 77 kDa and a pH optimum of about 7.5. It is most active against long-chain triacylglycerols and shows no activity against phospho- and galacto-lipids; it is thus distinct from the main tuber acylhydrolase described later in this chapter.

Lipase activities from several other plant species have been reported in extracts from germinated seeds, and in some cases parameters such as substrate selectivity were studied (60). However, there are no reported attempts to solubilise or purify these lipases. This material is covered in previous reviews (5). These lipases include that from soybean glyoxysomes (61), apple seedlings (62), *Cucumeriopsis edulis* (63), elm seedlings (60) and seeds of the Douglas fir (64). Non-seed-associated lipase has also been reported in the zone of maximum elongation of maize root tips (65) and recently characterised further (66).

Regulation of lipase activity

The activity of lipases in seedling extracts is sometimes much higher than the rate at which fatty acids are oxidised *in vivo* (5, 67) but free fatty acids do not accumulate (68). It is apparent that plants have a mechanism for the metabolic regulation of lipase activity *in vivo* (5). It was found that lipase from oilseed rape was inhibited by free fatty acids, in particular by erucic acid, which is present in high amounts in some cultivars. This suggested that lipase activity might be regulated *in vivo* by fatty-acid concentration (69). Studies with the neutral oil-body lipase from castor bean showed that fatty acids were unlikely to regulate that lipase activity, since they accumulated to quite high concentrations in the reaction buffer, and lipolysis continued for 4 h (70). However, when ATP and coenzyme A were included in the reaction mixture, lipolysis was almost totally inhibited within 15 min. An acyl-CoA synthetase, which is also present in the oil-body membrane, synthesized acyl-CoAs from the fatty acids released by the action of the lipase. Acyl-CoAs on their own inhibited lipase activity by about 30%. A mixture of acyl-CoA and free CoA together caused up to 95% inhibition of lipase activity (70). During periods of reduced acyl-CoA oxidation by the glyoxysome, a build up of acyl-CoAs would cause feedback inhibition of lipase activity, thus preventing the further hydrolysis of triacylglycerols.

Regulation of lipase activity by inhibitory proteins has also been suggested. There are reports of proteins partially purified from soybean (71) and sunflower (72) that are inhibitory to lipases from plant, animal and fungal sources. It was shown that these proteins penetrate the surface of the oil droplets and prevent efficient binding of the lipase (73). However, other proteins such as bovine serum albumin, mellitin and lactoglobulin also strongly inhibit lipase activity, by interacting with the surface of the oil emulsion (73). It should also be mentioned that oil bodies are completely coated with an oleosin protein annulus (74, 75), so the significance of the rôle of these surface active inhibitors *in vivo* is not clear (76). There have been suggestions that the oleosins might play a part in the process of lipolysis by acting as receptors for lipases (74, 77). It is interesting to note that one of the rape oleosins contains a region that is thought to span the interface between the aqueous and oil phases (78). The amino-acid sequence of this region shows 54% similarity, allowing for conservative substitutions, with the lipoprotein lipase-binding site of human apolipoprotein C-II (74).

2.2 Lipid acylhydrolases

Potato tuber

Non-specific acylhydrolase activities from potato tubers have been studied extensively and earlier work reviewed (5). The protein is known as patatin, a storage glycoprotein of the vacuole which represents 30–40% of total soluble protein in the tuber. It removes both acyl groups from a wide range of glycerolipids but not triacylglycerols (79). The pH optimum towards monogalactosyldiacylglycerols was about 5.0 and towards phosphatidylcholines (PC) was 8.5, but competition experiments showed that the active site for the substrates was the same. These results were confirmed later by using recombinant patatin (80). The acylhydrolase has been purified in several laboratories using different procedures (81–83). The M_r of patatin, estimated from translated nucleotide sequence of a cDNA cloned from a tuber cDNA library, is 40 kDa (84). Correlation of the amount of acylhydrolase inactivated by adding SDS with the increase in the amount of SDS-dissociated patatin located by staining on SDS-PAGE strongly suggested that patatin was the protein responsible for the acylhydrolase activities (83). However, acylhydrolase activity of patatin was unequivocally demonstrated only when pure patatin was synthesized as a recombinant protein using a baculovirus expression system (80). This ruled out the possibility that acylhydrolase activity in patatin purified from tubers was actually due to the presence of a copurifying contaminant. The recombinant patatin removes acyl groups from phospholipids, monoacylglycerols, *p*-nitrophenyl esters and galactolipids but not from di- or triacylglycerols as was previously reported (82). The enzyme also catalyses the transesterification of acyl groups in the presence of methanol, giving fatty-acid methyl esters, which confirmed previous observations with isolated tuber protein (81). Since it is present in such high amounts, patatin is clearly a storage protein, but the physiological significance of its acylhydrolase activity is not clear. It has been suggested that this might protect tubers from invasion by pathogens by releasing fatty acids after mechanical damage (80).

Leaf acylhydrolase

Galactolipase from *Phaseolus multiflorus*, which removes fatty acids from both the *sn*-1 and *sn*-2 positions of monogalactosyldiacylglycerols and digalactosyl-diacylglycerols was purified about 60-fold (85). It had an apparent M_r of 110 kDa and was inhibited by dithiothreitol (DTT). Another galactolipase was purified from *P. vulgaris* and was shown to be stimulated by DTT (86). Palmitoylated gauze chromatography was important for maintaining stability of the enzyme during purification (86). Two acylhydrolase activities were purified about 70-fold from *P. vulgaris* leaves (87). Acylhydrolase I showed a strong preference for phospholipids and had a pH optimum of 5.3, whereas acylhydrolase II was more active against galactolipids and showed a pH optimum of 7.3 (88). Both activities were inhibited by free fatty acids and Triton X-100, but sulphydryl inhibitors showed no effect. If methanol was included at 30% v/v in the reaction mixture, acyltransferase activities were very significant. An acylhydrolase from potato leaves has also been purified and shows a similar pattern of substrate specificity to the tuber acylhydrolase (89). It was shown that the major acylhydrolase activities in potato leaves are apparently regulated by calmodulin (90) and by protein phosphorylation/dephosphorylation (91). It was later shown that the acylhydro-lase was activated by the action of a protease (*ca.* 100 kDa). It was suggested that, as in potato tubers, the acylhydrolase might be activated upon mechanical damage to give protection against fungal infection (92).

2.3 Phospholipases

Phospholipase C

Phospholipase C (PLC) in plants has received attention recently because of its important rôle in the phosphoinositol cycle signal transduction pathway in the plasma membrane (93). PLC, which acts on phosphatidyl inositol 4,5-bisphosphate (PIP2), was found in a sedimentable fraction of celery extracts (94). As much as 80% of the activity was solubilised by deoxycholate, but the enzyme was rather unstable. A soluble PLC in the celery extracts acted only on phosphatidyl inositol (PI) and not on PIP2, and is probably not involved in the cycle (94). Plasma membrane PLC has recently been partially purified from wheat roots by use of octylglucoside as detergent and ion-exchange and hydroxyapatite chromatography (95). Optimal activity of the PLC was found between pH 6 and pH 7. It required micromolar concentrations of Ca^{2+} for activity, and Mg^{2+} in the millimolar range stimulated activity still further. PLC acting on PIP2 has also been partially purified from oat root plasma membranes, and preliminary evidence from gel-filtration studies suggest that it has an M_r value of 68 kDa (96).

Phospholipase D

Plants have proved to be a good source of phospholipase D (PLD), and that from cabbage is commercially available. It catalyses the hydrolysis of the base group such as choline or ethanolamine from phospholipids. Procedures for isolation of PLD with increasing degrees of purity from cabbage and peanut have been published over the past 30 years and have been reviewed (97). In the early studies it was found that the phospholipase D also catalysed transphosphatidylation reactions, and the ratio of this activity to the hydrolytic activity remained constant during purification of the cabbage (98) and the peanut enzymes (99), indicating that the same protein catalyses both reactions.

Phospholipase D from peanut was purified nearly 1200-fold, but required preparative gel electrophoresis to yield a single gel band showing PLD activity (100). Gel filtration of the protein gave an estimate of its M_r value of 200 kDa, but this tended to dissociate to subunits of $M_r \approx 25$ kDa (101). Activity was stimulated by SDS and other detergents, and by diethyl ether (102). The enzyme was unstable in dilute solution and lost activity at its pH optimum of 5.6. Phospholipase D from *Brassica oleracea* was purified to apparent homogeneity (103). The enzyme had a neutral pH optimum, but this varied depending on Ca^{2+} concentration from 6.0 in the presence of Ca^{2+} to 7.3 at low concentrations of Ca^{2+}. Anionic amphipathic compounds also caused changes in the pH optimum of the enzyme. The molecular weight was estimated to be about 113 kDa by SDS-PAGE. The stereochemistry of the cabbage PLD has been studied in some detail using sophisticated methods. Using PC as substrate, which had been labelled with different isotopes of oxygen, the structures of the products of PLD reactions were determined using nuclear magnetic resonance (NMR) spectroscopy (104). The cabbage PLD retains its configuration at the phosphorus in both transphosphatidylation and hydrolysis reactions.

Lysophospholipase in barley

Lysophosphatidyl choline (LPC) is the major phospholipid in starch grains in seeds of barley and other species. During germination, a specific lysophospholipase is synthesized and this catalyses the hydrolysis of the LPC which is released as the starch is mobilised. The enzyme was purified 4000-fold by solubilisation with NaCl, and chromatography on CM-cellulose, octyl-agarose, Con-A Sepharose and phenyl-agarose. This glycoprotein has an M_r of 36 kDa after removal of the carbohydrate group. The pH optimum is 8.5 and the enzyme is specific in cleaving the acyl group from LPC (105).

3. APPLICATIONS OF LIPASES FROM PLANTS

Although lipase preparations obtained from plants have been used as catalysts for hydrolysis and esterification of lipids in a few earlier studies (39, 106, 107), it was not until recently that several publications have appeared that describe the isolation of lipases from plants for their possible applications in various biotransformation reactions (32, 34, 108, 109). It has been suggested that lipases from oilseeds, owing to the technical simplicity of their isolation (32) and their unique substrate specificities (6) compared with most of the microbial lipases could be exploited suitably in lipid biotechnology.

3.1 Triacylglycerol lipases from rape

Simple procedures for the isolation of triacylglycerol lipases from germinating cruciferous oilseeds, such as rape (*Brassica napus*) and mustard (*Sinapis alba*), with the object of using such enzyme preparations in biotechnological reactions, involve homogenisation of the seedlings in Tricine buffer (pH 7.5) followed by centrifugation at $23\,000\,g$ (32). Such lipase preparations or acetone powders obtained from the seedlings have been used for the hydrolysis of sunflower oil, which resulted in selective cleavage of fatty acids esterified at the *sn*-1,3-positions of the triacylglycerols. In the hydrolysis reaction, the crude lipases from rape-seed and mustard seed have a pH optimum between 8 and 9. These lipase preparations were found to be relatively stable on storage at $-10\,°C$; however, a substantial proportion of the activity was lost by freeze-drying. More stable preparations of purified lipases have been isolated from cotyledons of germinating rape seedlings and immobilised on Celite (34, 37), and some unique substrate specificities of such lipase preparations in hydrolysis and esterification have been observed (110, 111).

In esterification and hydrolysis reactions, the ability of the rape-seed lipase to discriminate against fatty acids/acyl moieties having a *cis*-6 or a *cis*-4 double bond, such as all-*cis*-6,9,12-octadecatrieonic acid (γ-linolenic acid) or all-*cis*-4,7,10,13,16,19-docosahexaenoic acid, respectively (111) has been utilised in enzymic fractionation for enrichment of these fatty acids *via* kinetic resolution (112, 113).

For the enrichment of γ-linolenic acid, the fatty acids of evening primrose oil (125 mM) were allowed to react at 30 °C for various periods with *n*-butanol (250 mM) in hexane using the immobilised rape-seed lipase as biocatalyst (112, 113). The results are shown in Table 1. In accordance with the observed substrate specificities (111), as the degree of esterification of the fatty acids increases with

Table 1. Composition of fatty acids and butyl esters during esterification of fatty acids from evening primrose oil with n-butanol catalysed by immobilised rape-seed lipase (adapted from reference 113).

Reaction time (h)	Component	% of total	Composition (%) of acyl constituents*				
			16:0	18:0	18:1	18:2	γ-18:3
0	Fatty acids	100	9	2	6	73	10
48	Fatty acids	18	7	2	3	33	55
48	Butyl esters	82	9	2	6	83	Traces
72	Fatty acids	14	6	1	1	27	65
72	Butyl esters	86	9	2	6	83	<1

* 16:0, palmitic; 18:0, stearic; 18:1, oleic; 18:2, linoleic; γ-18:3, γ-linolenic.

time, there is a steep increase in the proportion of γ-linolenic acid in the non-esterified fatty acids and a concomitant decrease of this fatty acid in the butyl esters. At the same time, a steep decrease in the level of linoleic acid in the non-esterified fatty acids and concomitant increase of this fatty acid esterified in butyl esters was observed. Thus, after 72 h of reaction, resulting in about 86% esterification, the level of γ-linolenic acid in the non-esterified fatty acids rose from about 10% in the starting material to almost 65%, which corresponds to almost sevenfold enrichment and near quantitative (98%) yield.

In another experiment, the triacylglycerols of evening primrose oil (10 mM), emulsified with 5% (w/v) Gum Arabic, were subjected to selective hydrolysis in a medium containing Bis-tris-propane-HCl (50 mM, pH 7.5), dithiothreitol (2 mM), $CaCl_2$ (30 mM) and deoxycholic acid (0.1% w/v) using immobilised rape-seed lipase as biocatalyst (112). The results are given in Table 2. Again, in accordance with the observed substrate specificities (111), in the course of time the degree of hydrolysis increases and γ-linolenic acid is enriched in the acyl moieties of the non-hydrolysed acylglycerols. Thus, after 60 min of reaction, resulting in 83% hydrolysis of the oil, the γ-linolenic acid content of the non-hydrolysed acylglycerols increased from about 10% in the starting material to 28%, which corresponds to almost threefold enrichment of γ-linolenic acid. It is interesting to note that, among the individual classes of acylglycerols, the maximum enrichment of γ-linolenic acid was observed in the monoacylglycerol fraction.

For the enrichment of docosahexaenoic acid from cod liver oil *via* selective esterification, fatty acids of cod liver oil (125 mM) were treated at 30 °C for

Table 2. γ-Linolenic acid content of products formed upon hydrolysis of evening primrose oil catalysed by immobilised rape-seed lipase (adapted from reference 112).

Reaction time (min)	Proportions (%) of lipid classes in products of hydrolysis (% γ-linolenic acid content of lipid class)				
	Monoacyl glycerols	Diacyl glycerols	Triacyl glycerols	Total acyl glycerols	Fatty acids
0	Traces	6 (4)	87 (10)	93 (10)	7 (4)
30	5 (29)	10 (21)	23 (12)	38 (16)	62 (5)
60	5 (45)	5 (28)	7 (15)	17 (28)	83 (6)

various periods with n-butanol (125 mM) in hexane using the immobilised rape-seed lipase as biocatalyst (113). It was found, in accordance with the observed substrate specificities (111), that with increasing degree of esterification in the course of time the proportion of docosahexaenoic acid in non-esterified fatty acids increased with concomitant decrease in the level of this fatty acid in the butyl esters. After 48 h reaction, the level of docosahexaenoic acid in non-esterified fatty acids could be raised from about 9% in the starting material to about 17%, *i.e.*, about twofold enrichment was obtained.

The ability of the rape-seed lipase to esterify acids, including fatty acids, exclusively to the primary alcohols, but not to secondary or tertiary alcohols (114), could be utilised for the synthesis of 'designed' esters for a wide variety of speciality articles, such as personal-care and pharmaceutical products.

3.2 Triacylglycerol lipases from castor bean

It has been described in the foregoing section that castor bean, unlike most oilseeds, contains an active triacylglycerol lipase even in the dormant seed. As early as at the turn of this century, lipase preparations from castor bean were used to hydrolyse fats for the preparation of fatty acids (106) and synthesis of acylglycerols by esterification of fatty acids with glycerol (107).

Production of ricinoleic acid (12-hydroxy-9-*cis*-octadecenoic acid) – a valuable starting material for a variety of technical products (115) – using the conventional technology of 'steam splitting' at high temperatures and pressures (116) poses problems due to reactions of ricinoleic acid, such as dehydration, yielding conjugated fatty acids, and intermolecular esterification, yielding estolides (117).

Therefore, saponification of castor oil followed by acidification at relatively low temperature has so far been the method of choice for the production of ricinoleic acid.

Earlier, lipase preparations isolated from castor bean were used as powder or as protein–oil emulsions for the hydrolysis of the oil extracted from castor bean (39, 118, 119). Recently, lipolysis *in situ* of castor oil present in the bean without prior isolation of the oil or the lipase from castor bean has been reported (120). In order to carry out the lipolysis, ground castor beans were shaken with dilute acetic acid (*ca.* 0.1 M) at pH ranging from 4.0 to 5.5 at 25–37 °C for various periods. The extent of hydrolysis was found to be as much as 84% after only 1 h of reaction at 30 °C and pH 4.8. Scaling-up of the process led to about 90% lipolysis of the oil after 5 h and 97% lipolysis after 24 h without any formation of estolides (121).

3.3 Triacylglycerol lipases from oat

Oats have long been known to be a rich source of lipase (122) that is localised on the surface of oat caryopses (57).

A crude lipase preparation from oat seeds has been used for the lipolysis of olive oil and tallow in an emulsifier-free, two-phase system (108). Addition of calcium chloride stimulated lipolytic activity in the hydrolysis of both olive oil and tallow. Although the lipolysis of tallow by the oat lipase was incomplete after 18 h reaction at 48 °C, selective hydrolysis of the oleoyl moieties was observed, which is in agreement with the substrate selectivity of this lipase observed earlier (58, 123). Selective cleavage of the oleoyl moieties from the triacylglycerols of tallow led to enrichment of oleic acid in the fatty acids and enrichment of the saturated acyl moieties in the non-hydrolysed acylglycerols (108).

Ground caryopses of oat, de-fatted with an organic solvent such as diethyl ether, have been used as a lipase preparation for the hydrolysis of oils containing polyunsaturated acyl moieties (124). In the presence of 2,2,4-trimethylpentane as reaction medium and a definite amount of water, almost complete lipolysis of soybean oil has been observed at 35 °C in about 19 h. The oat preparation has been re-used in a second batch of lipolysis, which resulted in about 90% hydrolysis compared with the first batch.

By using a similar preparation of ground and de-fatted oat caryopses, lipolysis of castor oil has been carried out in the presence of 2,2,4-trimethylpentane (125). The reaction occurred in a three-phase mixture, *i.e.*, solid ground oats, a phase containing predominantly castor oil, and another phase containing predominantly 2,2,4-trimethylpentane. The rate of hydrolysis was found to be strongly dependent on the ratio of oil-dominant phase to the solvent-dominant phase. Thus, at low ratio of oil-dominant phase to solvent-dominant phase, as much as 90% lipolysis occurred after about 22 h, whereas at a high ratio between the two phases low rates of lipolysis were observed.

Instead of isolated oat lipase preparations or ground de-fatted caryopses of oat, whole caryopses of oat, moistened with water, have been used as lipase source for biotransformation of fats (126). The studies on the effects of various reaction parameters on the lipolysis of soybean oil showed that the addition of 20% water (by weight of the caryopses) and a temperature of 40 °C was optimal for lipolysis. Increased rates of lipolysis were obtained by gentle agitation and by reducing the viscosity of the oil phase by the addition of non-polar solvents, such as hexane. Hydrolysis was inhibited by both the reaction products, *i.e.*, fatty acids and glycerol. The lipase in oat caryopses was found to cleave the fatty acids from all the three positions of the glycerol backbone, and no accumulation of either mono- or diacylglycerols was observed. The overall rate of lipolysis using the oat caryopses was, however, low as compared with lipolysis catalysed by the ground and de-fatted oat caryopses (108, 125).

Moistened oat caryopses were also found to catalyse transesterification and esterification reactions (126). Thus, transesterification of oleic acid with triacylglycerols of soybean oil, esterification of oleic acid with glycerol and transesterification of triacylglycerols of soybean oil with primary alcohols were achieved using the moistened oat caryopses as biocatalyst. However, the overall reaction rates were rather low.

3.4 Other triacylglycerol lipases

There is another property of lipases from some other oilseeds that deserves closer examination with regard to their possible application in biotransformation of lipids. The ability of rice bran lipase to cleave preferentially the fatty acids esterified at the *sn*-1,3-positions of triacylglycerols (53) could be utilised for the enrichment of such fatty acids by kinetic resolution *via* selective hydrolysis. For example, erucic acid, a valuable starting material for oleochemicals and technical products, could be concentrated from seed oils of many cruciferae that contain large proportions of erucic acid almost exclusively esterified at the *sn*-1,3-positions. Similarly, polyunsaturated fatty acids, which are predominantly esterified at the *sn*-1,3-positions of triacylglycerols in many seed oils, could be concentrated by selective hydrolysis using rice bran lipase.

Another lipase, isolated from cotton plant, has been used for transesterification of mixtures of triacylglycerols (127). Moreover, the use of the lipase from *Vernonia galamansis* in the analysis of seed oils has been reported (128).

3.5 Phospholipase D from cabbage

It is theoretically possible to use phospholipases A_1 and A_2, isolated from plants, to modify the composition of the acyl moieties of phospholipids, such as commercial soya lecithin or egg lecithin, by transesterification or interesterification; similarly, it is possible to use phospholipase C from plants as a biocatalyst for the synthesis of glycerophospholipids by esterification of diacylglycerols to phosphocholine or phosphoethanolamine (42). However, the only phospholipase that has been used so far for the biotransformation of phospholipids is phospholipase D isolated from cabbage leaves, which has been employed successfully for transphosphatidylation reactions (109, 129–135).

Transphosphatidylation catalysed by phospholipase D involves interconversion of the polar head groups of phospholipids as follows (135):

$$R_2 \begin{bmatrix} R_1 \\ \\ (P) - X \end{bmatrix} + Y \quad \xrightarrow{\text{Phospholipase D}} \quad R_2 \begin{bmatrix} R_1 \\ \\ (P) - Y \end{bmatrix} + X$$

where (P) is the phosphate group, and X and Y are choline, ethanolamine, glycerol, serine *etc*. Typically, transphosphatidylation of phosphatidylcholines with ethanolamine yields phosphatidylethanolamines and choline. It should be noted, though, that the transphosphatidylation reaction is always accompanied by hydrolytic action of phospholipase D on the phospholipids yielding undesired phosphatidic acids and choline, ethanolamine, glycerol, serine *etc*. Therefore, the conditions of transphosphatidylation should be chosen such that the undesired hydrolytic cleavage of the terminal phosphate diester bond of glycerophospholipids is kept to a minimum.

Phospholipase D, partially purified from cabbage leaves, has been used for transphosphatidylation of phosphatidylcholines with glycerol to yield phosphatidylglycerols, which are potentially of use as artificial lung surfactants in infants suffering from respiratory distress syndrome (109). Transphosphatidylation has been carried out in a flow reactor consisting of two adjacent horizontal compartments separated by a microporous, hydrophobic polypropylene membrane. Phosphatidylcholine, dissolved in di-isopropyl ether, is pumped through the upper compartment in a countercurrent manner against a stream of a solution containing glycerol, phospholipase D and calcium chloride in an acetate buffer (pH 5.6) flowing through the lower compartment. Use of an ether as solvent stabilised the activity of the phospholipase D in the bioreactor. The optimal concentration of glycerol for the formation of phosphatidylglycerol was found to be between 10% and 20%. Low concentrations of phosphatidylcholines favoured the formation of phosphatidylglycerols. Only minor formation of phosphatidic acids was observed.

Transphosphatidylation of phosphatidylcholines with glycerol, catalysed by phospholipase D from cabbage, has also been carried out in a stirred reactor using micelles of phosphatidylcholines in glycerol buffer as well as emulsion of glycerol buffer in a solution of phosphatidylcholines in diethyl ether (130). Although the initial rate of formation of phosphatidylglycerols was found to be higher when micelles of phosphatidylcholines were used, the hydrolysis of the phospholipids producing phosphatidic acids was also found to be higher. In the emulsion system, on the other hand, almost quantitative transphosphatidylation with very little hydrolysis of phospholipids occurred.

The above transphosphatidylation reaction for the preparation of phosphatidylglycerol was also carried out using the cabbage phospholipase D, immobilised on octyl-Sepharose CL-4B, in repeated-batch and continuous operations (131). The reactions were carried out in a biphasic system, in which the aqueous phase was immobilised phospholipase D in acetate buffer (pH 5.6) containing calcium chloride and varying concentrations (30–50% v/v) of glycerol, while the substrate (phosphatidylcholines) and products (mainly phosphatidylglycerols) were in the organic phase (diethyl ether). In the repeated batch operations, carried out in a stirred reactor using a glycerol concentration of 30%, the yield of phosphatidylglycerols was almost quantitative, and essentially no formation of phosphatidic acids was observed. However, the activity of the enzyme preparation decreased substantially after several repetitions. In continuous operation, using a packed-bed reactor or a continuous stirred-tank reactor, a higher concentration of glycerol (50%) was found to be more favourable for obtaining higher yield and minimising enzyme deactivation.

Phospholipase D from cabbage has also been used for the transphosphatidylation of phosphatidylcholine with ethanolamine to yield phosphatidylethanolamine (132). Transphosphatidylation in a biphasic system was carried out in a stirred reactor using ethyl acetate as solvent for the substrate (phosphatidylcholines) and the product (phosphatidylethanolamines). The aqueous phase contained the enzyme preparation in a buffer to which calcium chloride and different amounts of ethanolamine were added. The phospholipase D from cabbage, as compared with other microbial phospholipase D preparations, gave almost quantitative conversion of phosphatidylcholines to phosphatidylethanolamines with essentially no formation of phosphatidic acids. Hydrolysis of phosphatidylethanolamines occurred only after the transphosphatidylation reaction was complete.

Phosphatidylserines, used extensively for the preparation of liposomes as drug delivery systems, have been prepared *via* transphosphatidylation of phosphatidylcholines with serine using phospholipase D from cabbage (133). The reactions were carried out in a manner similar to the one described above (132). In contrast to all phospholipase D preparations from microbial sources, the phospholipase D from cabbage was found to be highly stereoselective in the transphosphatidylation reactions. Thus, while transphosphatidylation of phosphatidylcholines with L-serine gave phosphatidylserine in up to about 40% yield, no transphosphatidylation occurred with D-serine; instead, the substrates (phosphatidylcholines) were hydrolysed completely to phosphatidic acids.

Transphosphatidylation of commercially available phospholipids, such as soybean lecithin and egg lecithin, with choline chloride has been carried out using various preparations of phospholipase D with the aim of enriching the phosphatidylcholine content of such products (134). As compared with several microbial phospholipase D preparations, the phospholipase D from cabbage was found to be a rather poor biocatalyst for this transphosphatidylation reaction.

Transphosphatidylation of radiolabelled ethanolamine plasmalogen, catalysed by phospholipase D from cabbage, has been used to prepare the dimethylethanolamine and choline analogues of these substances (129).

4. PERSPECTIVES

Recently, several physiological responses in plants, mediated by lipases, have received considerable attention. Thus, senescence, accompanied by degradation of phospholipids, has been shown to be caused by the action of phospholipase D (136, 137). Galactolipases, *i.e.*, acylhydrolases acting predominantly on galactolipids associated with the thylakoids of leaves (138), have been shown to be affected by environmental factors, such as cadmium ions (139) and chilling (140, 141).

Very recently, the identification of Ca^{2+}-stimulated polyphosphoinositide-specific phospholipase C (95, 96, 142) and that of phospholipases A and D in plant plasma membrane (143) and cell cultures (144) has lent support to the concept (145, 146) of the involvement of plant phospholipases in the signal-transduction mechanism coupled to hydrolysis of phosphoinositides. Extensive work in this area is to be expected in the future.

REFERENCES

1. Brockerhoff, H. & Jensen, R.G. (1974). *Lipolytic Enzymes.* New York: Academic Press.
2. Galliard, T. (1975). Degradation of plant lipids by hydrolytic and oxidative enzymes. In *Recent Advances in the Chemistry and Biochemistry of Plant Lipids*, ed. Galliard, T. & Mercer, E. I., pp. 319–337. New York: Academic Press.
3. Galliard, T. (1980). Degradation of acyl lipids: Hydrolytic and oxidative enzymes. In *The Biochemistry of Plants*, vol. 4, ed. Stumpf, P.K. & Conn, E., pp. 85–116. New York: Academic Press.
4. Huang, A.H.C. (1984). Plant lipases. In *Lipases*, ed. Borgström, B. & Brockman, H.L., pp. 419–442. Amsterdam: Elsevier.
5. Huang, A.H.C. (1987). Lipases. In *The Biochemistry of Plants*, vol 9, ed. Stumpf, P.K. & Conn, E., pp. 91–119. New York: Academic Press.
6. Huang, A.H.C., Lin, Y.-h. & Wang, S.-m. (1988). Characteristics and biosynthesis of seed lipases in maize and other plant species. *J. Am. Oil Chem. Soc.*, 65, 897–899.
7. Desassis, A. (1957). L'acidification de l'huille de palme. *Oléagineux*, 12, 525–534.

8. Hartley, C.W.S. (1977). *The Oil Palm*, 2nd edn. London: Longman.
9. Coursey, D. G. (1963). The deterioration of palm oil during storage. *J. West Afr. Sci. Assoc.*, 7, 101–114.
10. Oo, K. C. (1981). The absence of lipase activity in mesocarp of the palm fruit. *Oléagineux*, 36, 613–616.
11. Tombs, M.T. & Stubbs, J.M. (1982). The absence of endogenous lipase from oil palm mesocarp. *J. Sci. Food Agric.*, 33, 892–897.
12. Abigor, D.R., Opute, F.I., Opoku, A.R. & Osagie, A. U. (1985). Partial purification and some properties of the lipase present in oil palm (*Elaeis guineensis*) mesocarp. *J. Sci. Food Agric.*, 36, 599–606.
13. Henderson, J. & Osborne, D.J. (1991). Lipase activity in ripening and mature fruit of the oil palm. Stability *in vivo* and *in vitro*. *Phytochemistry*, 30, 1073–1078.
14. Mohankumar, C., Arumughan, C. & Kaleysa raj R. (1990). Histological localization of oil palm fruit lipase. *J. Am. Oil Chem. Soc.*, 67, 665–669.
15. Sambanthamurthi, R., Chong, C.L., Oo, K.C., Yeo, K.H. & Premavathy, R. (1991). Chill-induced lipid hydrolysis in the oil palm (*Elaeis guineensis*) mesocarp. *J. Exp. Bot.*, 42, 1199–1205.
16. Maycock, J.H. (1987). Extraction of crude palm oil. In *Palm Oil*, ed. Gunstone, F. D., pp. 29–38. Chichester: John Wiley.
17. Kiritsakis, A. (1984). Effect of olive collection regimen on olive oil quality. *J. Sci. Food Agric.*, 35, 677–678.
18. Hirayama, O. & Matsuda, H. (1975). Purification and characterisation of lipolytic acylhydrolases from rice bran. *Nippon Nogei Kagaku Kaishi*, 49, 569–576.
19. Aizono, Y., Funatsu, M., Fujiki, Y. & Watanabe, M. (1976). Biochemical studies on rice bran lipase. IV. Purification and characterisation of rice bran lipase. *Agric. Biol. Chem.*, 40, 317–324.
20. Shastry, B.S. & Raghavendra Rao, M.R. (1976). Chemical studies on rice bran lipase. *Cereal. Chem.* 53, 190–200.
21. Fujiki, Y., Aizono, Y. & Funatsu, M. (1978). Biochemical studies on rice bran lipase. V. Chemical properties of major subunit of rice bran lipase. *Agric. Biol. Chem.*, 42, 599–606.
22. Nasirullah, Krishnamurthy, M.N. & Nagaraja, K.V. (1989). Effect of stabilization on the quality characteristics of rice-bran oil. *J. Am. Oil Chem. Soc.*, 66, 661–663.
23. Thomas, P.P., Gopalkrishnan, N. & Damodaran, A.D. (1991). Rice bran lipase inactivation by microwave for low FFA oil. *Oléagineux*, 46, 245–246.
24. O'Connor, J., Perry, H.J. & Harwood, J.L. (1989). Solubilization and studies of cereal lipases. *Biochem. Soc. Trans.*, 17, 687–688.
25. List, G.R. (1980). Special processing for off-specification oil. In *Handbook of Soy Oil Processing and Utilization*, ed. Erickson, D.R., Pryde, E.H., Brekke, O.L., Mounts, T.L. & Falb, R.A., pp. 355–376. Champaign: American Oil Chemists' Society.
26. Simpson, T.A. (1991). Phospholipase activity in hexane. *J. Am. Oil Chem. Soc.*, 68, 176–178.
27. Lin, Y.-h. & Huang, A.H.C. (1984). Purification and initial characterization of lipase from the scutella of corn seedlings. *Plant Physiol.*, 76, 719–722.
28. Lin, Y.-h., Wimer, L.T. & Huang, A.H.C. (1983). Lipase in the lipid bodies of corn scutella during seedling growth. *Plant Physiol.*, 73, 460–463.
29. Wang, S. & Huang, A.H.C. (1987). Biosynthesis of lipase in the scutellum of maize kernel. *J. Biol. Chem.*, 262, 2270–2274.
30. Theimer, R.R. & Rosnitschek, I. (1978). Development and intracellular localization of lipase activity in rapeseed (*Brassica napus* L.) cotyledons. *Planta*, 139, 249–256.
31. Hills, M.J. & Murphy, D.J. (1988). Characterization of lipases from the lipid bodies and microsomal membranes of erucic acid-free oilseed-rape (*Brassica napus*) cotyledons.

    Biochem. J., 249, 687–693.
32. Hassanien, F.R. & Mukherjee, K.D. (1986). Isolation of lipase from germinating oilseeds for biotechnological processes. J. Am. Oil. Chem. Soc., 63, 893–897.
33. Lin, Y.-h. & Huang, A.H.C. (1983). Lipase in the lipid bodies of cotyledons of rape and mustard seedlings. Arch. Biochim. Biophys., 225, 360–369.
34. Hills, M.J. & Mukherjee, K.D. (1990). Triacylglycerol lipase from rape (Brassica napus L.) suitable for biotechnological purposes. Appl. Biochim. Biotechnol., 26, 1–10.
35. Weselake, R.J., Thomson, L.W., Tenaschuk, D. & MacKenzie, S.L. (1989). Properties of solubilized microsomal lipase from germinating Brassica napus. Plant Physiol., 91, 1303–1307.
36. O'Sullivan, J.N., Hills, M.J. & Murphy, D.J. (1990). Purification and properties of lipase from oilseed rape (Brassica napus L.). In Plant Lipid Biochemistry, Structure and Utilization, ed. Quinn, P.J. & Harwood, J.L., pp. 313–315. London: Portland Press.
37. Hills, M.J. & O'Sullivan, J.N. (1989). Partial purification of lipase from cotyledons of oilseed rape (Brassica napus L.). Biochem. Soc. Trans., 17, 480.
38. Maeshima, M. & Beevers, H. (1985). Purification and properties of glyoxysomal lipase from castor bean. Plant Physiol., 79, 489–493.
39. Ory, R.L., St. Angelo, A.J. & Altschul, A.M. (1962). The acid lipase of the castor bean: properties and substrate specificity. J. Lipid Res., 3, 99–105.
40. Ory, R.L. (1969). Acid lipase in the castor bean. Lipids, 4, 177–185.
41. Hills, M.J. & Beevers, H. (1987). Ca^{2+} stimulated neutral lipase activity in castor bean lipid bodies. Plant Physiol., 84, 671–674.
42. Mukherjee, K. D. (1990). Lipase-catalyzed reactions for modification of fats and other lipids. Biocatalysis, 3, 277–293.
43. Diez, T.A. & Mata-Segreda, J.F. (1985). Castor-bean acid lipase catalyzed hydrolysis of triacylglycerols does not involve C-2 to C1,3 transacylation. Phytochemistry, 24, 3047–3048.
44. Hills, M.J. & Beevers, H. (1987). An antibody to the castor bean glyoxysomal lipase (62 kD) also binds to a 62 kD protein in extracts from many young oilseed plants. Plant Physiol., 85, 1084–1088.
45. Sanders, T.H. & Pattee, H.E. (1975). Peanut alkaline lipase. Lipids, 10, 50–54.
46. Huang, A.H.C. & Moreau, R.A. (1978). Lipase in the storage tissues of peanut and other oilseeds during germination Planta, 141, 111–116.
47. Krewson, C.F., Ard, J.S. & Riemenschneider, R.W. (1962). Vernonia anthelmentica (L.) wild. Trivernolin, 1,3-divernolin and vernolic (epoxyoleic) acid from the seed oil. J. Am. Oil Chem. Soc., 39, 334–340.
48. Olney, C.E., Jensen, R.G., Sampugna, J. & Quinn, J.G. (1968). The purification and specificity of a lipases from Vernonia anthelmintica seed. Lipids, 3, 498–502.
49. Oo, K.C. & Stumpf, P.K. (1983). Some enzymic activities in the germinating oil palm (Elaeis guineensis) seedling. Plant Physiol., 73, 1028–1032.
50. Oo, K.C. & Stumpf, P.K. (1983). The metabolism of the germinating oil palm (Elaeis guineensis) seedling. Plant Physiol., 73, 1033–1037.
51. Tavener, R.J.A. & Laidman, D.L. (1972). The induction of triglyceride metabolism in the germinating wheat grain. Phytochemistry, 11, 981–987.
52. O'Conner, J. & Harwood, J.L. (1992). Solubilization and purification of membrane bound lipases from wheat flour. J. Cereal Sci., 16, 141–152.
53. Funatsu, M., Aizono, Y. Hayashi, K. Inmasu, M. & Yamaguchi, M. (1971). Biochemical studies on rice bran lipase: purification and physical properties. Agric. Biol. Chem., 35, 734–742.
54. Fujiki, Y., Aizono, Y. & Funatsu, M. (1978). Characterization of minor subunit of rice bran lipase. Agric. Biol. Chem., 42, 2401–2402.
55. Firn R.D. & Kende, H. (1974). Some effects of applied gibberellic acid on the synthesis

and degradation of lipids in isolated barley aleurone layers. *Plant Physiol.*, 54, 911–915.
56. Fernandez, D.E. & Staehlin, L.A. (1987). Does gibberellic acid induce the transfer of lipase from protein bodies to lipid bodies in barley aleurone cells? *Plant Physiol.*, 85, 487–496.
57. Martin, H.F. & Peers, F.G. (1953). Oat lipase. *Biochem. J.*, 55, 523–529.
58. Berner, D.L. & Hammond, E.G. (1970). Specificity of lipase from several seeds and *Leptospira pomona. Lipids*, 5, 572–573.
59. Hasson, E.P. & Laites, G.G. (1976). Separation and characterization of potato lipid acylhydrolases. *Plant Physiol.*, 57, 142–147.
60. Lin Y.-h., Yu, C. & Huang A.H.C. (1986). Substrate specificities of lipases from corn and other seeds. *Arch. Biochim. Biophys.*, 244, 346–356.
61. Lin, Y.-h , Moreau, R.A. & Huang, A.H.C. (1982). Involvement of the glyoxysomal lipase in the hydrolysis of storage triacylglycerols in the cotyledons of soybean seedlings. *Plant Physiol.*, 70, 108–112.
62. Somolenska, G. & Lewak, S. (1974). The role of lipases in the germination of dormant apple embryo. *Planta*, 116, 361–370.
63. Opute, F.I. (1975). Lipase activity in germinating seedlings of *Cucumeropsis edulis J. Exp. Bot.*, 26, 379–386.
64. Ching, T.M. (1968). Intracellular distribution of lipolytic activity in the female gametophyte of germinating Douglas fir seeds. *Lipids*, 3, 482–488.
65. Heimann-Matille, J. & Pilet, P.E. (1977). Lipase activity in growing roots of *Zea mays. Plant Sci. Letters*, 9, 247–252.
66. Aman, T., Waheed, M. & Jamshaid, F. (1989). Isolation and characterization of *Zea mays* root alkaline lipase. *Sci. Int. (Lahore)*, 1, 192–195.
67. Jacks, T.J., Yatsu, L.Y. & Altschul, A.M. (1967). Isolation and characterization of peanut spherosomes. *Plant Physiol.*, 42, 585–597.
68. St Angelo, A.J. & Altschul, A.M. (1964). Lipolysis and the free fatty acid pool in seedlings. *Plant Physiol.*, 39, 880–883.
69. Rosnitschek, R. & Theimer, R.R. (1980). Properties of a membrane bound triglyceride lipase of rapeseed (*Brassica napus* L.) cotyledons. *Planta*, 148, 193–198.
70. Hills, M.J., Murphy, D.J. & Beevers, H. (1989). Inhibition of neutral lipase from castor bean lipid bodies by CoA and oleoyl-CoA. *Plant Physiol.*, 89, 1006–1010.
71. Gargouri, Y., Julien, R., Pieroni, G., Verger, R. & Sarda, L. (1984). Studies on the inhibition of pancreatic and microbial lipases by soybean protein. *J. Lipid Res.*, 25, 1214–1221.
72. Chapman, G.W. (1987). A proteinaceous competitive inhibitor of lipase isolated from *Helianthus annuus* seeds. *Phytochemistry*, 26, 3127–3131.
73. Gargouri, Y., Pieroni, G., Riviäre, C., Sugihara, A., Sarda, L. & Verger, R. (1985). Inhibition of lipase by proteins. *J. Biol. Chem.*, 260, 2268–2273.
74. Murphy, D.J. (1990). Storage lipid bodies in plants and other organisms. *Prog. Lipid Res.*, 29, 299–324.
75. Tzen, J.T.C. & Huang, A.H.C. (1992). Surface structure and properties of plant seed oil bodies. *J. Cell Biol.*, 117, 327–335.
76. Wang, S. & Huang, A.H.C. (1984). Inhibitors of lipase activities in soybean and other oilseeds. *Plant Physiol.*, 76, 929–934.
77. Vance, V.B. & Huang, A.H.C. (1987). The major protein from lipid bodies of maize. *J. Biol. Chem.*, 262, 11275–11279.
78. Murphy, D.J., Keen, J.N., O'Sullivan, J.N., Au, D.M.Y., Edwards, E.-W., Jackson, P.J., Cummins, I., Gibbons, T., Shaw, C.H. & Ryan, A.J. (1991). A class of amphipathic proteins associated with lipid storage bodies in plants. *Biochim. Biophys. Acta*, 1088, 86–94.
79. Galliard, T. & Dennis, S. (1974). Phospholipase, galactolipase and acyltransferase

activities of a lipolytic enzyme from potato. *Phytochemistry*, 13, 1731–1735.
80. Andrews, D.L., Beames, B., Summers, M.D. & Park, W.D. (1988). Characterisation of the lipid acylhydrolase activity of the major potato (*Solanum tuberosum*) tuber protein, patatin, by cloning and abundant expression in a baculovirus vector. *Biochem. J.*, 252, 199–206.
81. Galliard, T. (1971). The enzymic deacylation of phospholipids and galactolipids in plants. *Biochem J.*, 121, 379–390.
82. Hirayama, O., Matsuda, H. Takeda, H. Maenaka, K. & Takatsuka, H. (1975). Purification and properties of a lipid acyl-hydrolase from potato tubers. *Biochim. Biophys. Acta*, 384, 127–137.
83. Racusen, D. (1984). Lipid acyl hydrolase of patatin. *Can. J. Bot.*, 62, 1640–1644.
84. Mignery, G.A., Pikaard, C.S., Hannapel, D.J. & Park, W.D. (1984). Isolation and sequence analysis of cDNAs for the major potato storage protein, patatin. *Nucl. Acids Res.*, 12, 7987–7999.
85. Helmsing, P.J. (1969). Purification and properties of galactolipase. *Biochim. Biophys. Acta*, 178, 519–533.
86. Matsuda, H., Tanaka, G. Morita, K. & Hirayama, O. (1979). Purification of a lipolytic acylhydrolase from *Phaseolus vulgaris* leaves by affinity chromatography on palmitoylated gauze and its properties. *Agric. Biol. Chem.*, 43, 563–570.
87. Burns, D.D., Galliard, T. & Harwood, J.L. (1979). Purification of acyl hydrolase enzymes from the leaves of *Phaseolus multiflorus*. *Phytochemistry*, 18, 1793–1797.
88. Burns, D.D., Galliard, T. & Harwood, J.L. (1980). Properties of acyl hydrolase enzymes from *Phaseolus multiflorus* leaves. *Phytochemistry*, 19, 2281–2285.
89. Matsuda, H. & Hirayama, O. (1979). Purification and properties of a lipolytic acyl-hydrolase from potato leaves. *Biochim. Biophys. Acta*, 573 155–165.
90. Moreau, R.A. & Isset, T.F. (1985). Autolysis of membrane lipids in potato leaf homogenates: effects of calmodulin and calmodulin antagonists. *Plant Sci.*, 40, 95–98.
91. Moreau, R.A. (1986). Regulation of phospholipase activity in potato leaves by calmodulin and protein phosphorylation/dephosphorylation. *Plant Sci.*, 47, 1–9.
92. Moreau, R.A. & Morgan C.P. (1988). Proteolytic activation of a lipolytic enzyme activity in potato leaves. *Plant Sci.*, 55, 205–211.
93. Einspahr, K.J. & Thompson, G.A. (1990). Transmembrane signaling *via* phosphatidylinositol 4,5-bisphosphate hydrolysis in plants. *Plant Physiol.*, 93, 361–366.
94. McMurray, W.C. & Irvine, R.F. (1988). Phosphatidylinositol 4,5 bisphosphate phosphodiesterase in higher plants. *Biochem. J.*, 249, 977–881.
95. Melin, P.-M., Pical, C., Jergil, B. & Sommarin, M. (1992). Polyphosphoinositide phospholipase C in wheat root plasma membranes: partial purification and characterization. *Biochim. Biophys. Acta*, 1123, 163–169.
96. Coté, G.G., Tate, B.T. & Crain, R.C. (1991). Purification of a polyphosphoinositide-specific phospholipase C from oat plasma membranes. *Plant Physiol.*, 96, 448.
97. Van den Bosch, H. (1982). Phospholipases. In *Phospholipids*, ed. Ansell, G.B. & Hawthorne, J.N., pp. 313–357. Amsterdam: Elsevier.
98. Yang, S.F., Freer, S. & Benson, A.A. (1967). Transphosphatidylation by phospholipase D. *J. Biol. Chem.*, 242, 477–484.
99. Heller, M., Mozes, N. & Maes, E. (1975). Phospholipase D from peanut seeds. In *Methods Enzymol.*, 35, 226–232.
100. Tsai, R. & Shapiro, B. (1972). Purification of phospholipase D from peanuts. *Biochim. Biophys. Acta*, 280, 290–296.
101. Heller, M., Mozes, N., Peri, I. & Maes, E. (1974). Phospholipase D from peanut seeds: final purification and some properties of the enzyme. *Biochim. Biophys. Acta*, 369, 397–410.
102. Heller, M. & Arad, R. (1970). Properties of the phospholipase D from peanut seeds.

Biochim. Biophys. Acta, 210, 276–286.
103. Allgyer, T.T. & Wells, M.A. (1979). Phospholipase D from Savoy cabbage: purification and preliminary kinetic characterization. *Biochemistry*, 18, 5348–5353.
104. Bruzik, K. & Tsai, M.D. (1984). Phospholipids chiral at phosphorus. Synthesis of chiral phosphatidylcholine and stereochemistry of phospholipase D. *Biochemistry*, 23, 1656–1662.
105. Fujikura, Y. & Baisted, D. (1985). Purification and characterization of a basic lysophospholipase in germinating barley. *Arch. Biochim. Biophys.*, 243, 570–578.
106. Connstein, W., Hoyer, E. & Wartenberg, H. (1902). Über fermentative Fettspaltung. *Ber. Dtsch. Chem. Ges.*, 35, 3988–4006.
107. Jalander, Y. W. (1911). Zur Kenntnis der Ricinuslipase. *Biochem. Z.*, 36, 435–476.
108. Piazza, G., Bilyk, A., Schwartz, D. & Haas, M. (1989). Lipolysis of olive oil and tallow in an emulsifier-free two-phase system by the lipase from oat seeds (*Avena sativa* L.). *Biotechnol. Letters*, 11, 487–492.
109. Lee, S.Y., Hibi, N., Yamané, T. & Shimizu, S. (1985). Phosphatidylglycerol synthesis by phospholipase D in a microporous membrane bioreactor. *J. Ferment Technol.*, 63, 37–44.
110. Hills, M.J., Kiewitt, I. & Mukherjee, K.D. (1989). Esterification reactions catalysed by immobilized lipase from oilseed rape (*Brassica napus* L.). *Biochem. Soc. Trans.*, 17, 478.
111. Hills, M.J., Kiewitt, I. & Mukherjee, K.D. (1990). Lipase from *Brassica napus* L. discriminates against *cis*-4 and *cis*-6 unsaturated fatty acids and secondary and tertiary alcohols. *Biochim. Biophys. Acta*, 1042, 237–240.
112. Hills, M.J., Kiewitt, I. & Mukherjee, K.D. (1989). Enzymatic fractionation of evening primrose oil by rape lipase: enrichment of γ-linolenic acid. *Biotechnol. Letters*, 11, 629–632.
113. Hills, M.J., Kiewitt, I. & Mukherjee, K.D. (1990). Enzymatic fractionation of fatty acids: enrichment of γ-linolenic acid and docosahexaenoic acid by selective esterification catalyzed by lipases. *J. Am. Oil Chem. Soc.*, 67, 561–564.
114. Hills, M. J., Kiewitt, I. & Mukherjee, K.D. (1991). Synthetic reactions catalyzed by immobilized lipase from oilseed rape (*Brassica napus* L.). *Appl. Biochim. Biotechnol.*, 27, 123–129.
115. Lakshminarayana, G. (1981). Castor oil derivatives. *J. Oil Technol. Assoc. India*, 13, 75–83.
116. Sonntag, N.O.V. (1989). Fat splitting and glycerol recovery. In *Fatty Acids in Industry*, ed. Johnson, R. W. & Fritz, E., pp. 23–72. New York: Marcel Dekker.
117. Lakshminarayana, G., Subbarao, R., Sastry, Y.S.R., Kale, V., Rao, T.C. & Gangadhar, A. (1984). High pressure splitting of castor oil. *J. Am. Oil Chem. Soc.*, 61, 1204–1206.
118. Longenecker, H.E. & Haley, D.E. (1935). Ricinus lipase, its nature and specificity. *J. Am. Chem. Soc.*, 57, 2019–2021.
119. Ory, R.L., St. Angelo, A.J. & Altschul, A.M. (1960). Castor bean lipase: action on its endogenous substrate. *J. Lipid Res.*, 1, 208–213.
120. Rao, K.V.S.A., Paulose, M.M. & Lakshminarayana, G. (1990). *In situ* lipolysis of castor oil in homogenised castor seeds. *Biotechnol. Letters*, 12, 377–380.
121. Rao, K.V.S.A. & Paulose, M.M. (1992). A process for splitting castor oil at ambient temperature using homogenised castor seed as lipase source. *Res. Ind.*, 37, 36–37.
122. Hammond, E.G. (1983). Oat lipids. In *Lipids in Cereal Technology*, ed. Barnes, P. J., pp. 331–352. New York: Academic Press.
123. Piazza, G.J. & Bilyk, A. (1989). Properties and substrate specificity of lipase activity from oat (*Avena sativa*). *J. Am. Oil Chem. Soc.*, 66, 489.
124. Piazza, G. J. (1991). Generation of polyunsaturated fatty acids from vegetable oils using the lipase from ground oat (*Avena sativa* L.) seeds as a catalyst. *Biotechnol. Letters*, 13, 173–178.

125. Piazza, G. J. & Farrell, Jr, H.M. (1991). Generation of ricinoleic acid from castor oil using the lipase from ground oat (*Avena sativa* L.) seeds as a catalyst. *Biotechnol. Letters,* 13, 179–184.

126. Lee, I. & Hammond, E.G. (1990). Oat (*Avena sativa*) caryopses as a natural bioreactor. *J. Am. Oil Chem. Soc.,* 67, 761–765.

127. Kadyrova, Z.Kh., Abdurakhimov, S.A. & Khalinyazov, K.K. (1983). Transesterification of mixtures of triglycerides in the presence of cotton plant lipase (*Gossypium*). *Chem. Nat. Comp.,* 19, 411–413.

128. Atolali, O.A., Ologunde, M.O., Anderson, W.A., Read, J.S., Dacosta, M.D., Epps, F.A. & Ayonide, F.O. (1991). The use of lipase (acetone powder) from *Vernonia galamanensis* in the fatty acid analysis of seed oils. *J. Chem. Technol. Biotechnol.,* 51, 41–46.

129. Achterberg, V., Fricke, H. & Gercken, G. (1986). Conversion of radiolabelled ethanolamine plasmalogen into the dimethylethanolamine and choline analogue *via* transphosphatidylation by phospholipase D from cabbage. *Chem. Phys. Lipids,* 41, 349–353.

130. Juneja, L.R., Hibi, N., Inagaki, R., Yamane, T. & Shimizu, S. (1987). Comparative study on conversion of phosphatidylcholine to phosphatidylglycerol by cabbage phospholipase D in micelle and emulsion systems. *Enzyme Microb. Technol.,* 9, 350–354.

131. Juneja, L.R., Hibi, N., Yamane, T. & Shimizu, S. (1987). Repeated batch and continuous operations for phosphatidylglycerol synthesis from phosphatidylcholine with immobilized phospholipase D. *Appl. Microbiol. Biotechnol.,* 27, 146–151.

132. Juneja, L.R., Kazuoka, T., Yamane, T. & Shimizu, S. (1988). Kinetic evaluation of conversion of phosphatidylcholine to phosphatidylethanolamine by phospholipase D from different sources. *Biochim. Biophys. Acta,* 960, 334–341.

133. Juneja, L.R., Kazuoka, T., Goto, N., Yamane, T. & Shimizu, S. (1988). Conversion of phosphatidylcholine to phosphatidylserine by various phospholipases D in the presence of L- or D-serine. *Biochim. Biophys. Acta,* 1003, 277–283.

134. Juneja, L.R., Yamane, T. & Shimizu, S. (1989). Enzymatic method for increasing phosphatidylcholine content of lecithin. *J. Am. Oil Chem. Soc.,* 66, 714–717.

135. Yamane, T., Juneja, L.R. & Shimizu, D. (1989). Conversion of phospholipids by phospholipase D. In *Fats for the Future,* ed. Cambie, R.C., pp. 129–138. Chichester: Ellis Horwood.

136. Brown, J.H., Paliyath, G. & Thompson, J.E. (1990). Influence of acyl chain composition on the degradation of phosphatidylcholine by phospholipase D in carnation microsomal membrane. *J. Exp. Bot.,* 41, 979–986.

137. Brown, J.H., Chambers J. A. & Thompson, J.E. (1991). Acyl chain and head group regulation of phospholipid catabolism in senescing carnation flowers. *Plant Physiol.,* 95, 909–916.

138. O'Sullivan, J.N., Warwick, N.W.M. & Dalling, M.J. (1987). A galactolipase activity associated with the thylakoids of wheat leaves (*Triticum aestivum* L.). *J. Plant Physiol.,* 131, 393–404.

139. Skórzynska, E., Urbanik-Sypniewska, T., Russa, R. & Baszynski, T. (1991). Galactolipase activity of chloroplasts in cadmium-treated runner bean plants. *J. Plant Physiol.,* 138, 454–459.

140. Gemel, J. & Kaniuga, Z. (1987). Comparison of galactolipase activity and free fatty acid levels in chloroplasts of chill-sensitive and chill-resistant plants. *Eur. J. Biochem.,* 166, 229–233.

141. Gemel, J., Saczynska, V. & Kaniuga, Z. (1988). Galactolipase activity and free fatty acid levels in chloroplasts of domestic and wild tomatoes with different chilling tolerance. *Physiol. Plant.,* 74, 509–514.

142. Kamada, Y. & Muto, S. (1991). Ca^{2+} regulation of phosphatidylinositol turnover in the

plasma membrane of tobacco suspension culture cells. *Biochim. Biophys. Acta*, 1093, 72–79.
143. Cho, M. H. & Boss, W.F. (1992). Phospholipase A (PLA) and D (PLD) activity are associated with plasma membranes. *Plant Physiol.*, 99, S 718.
144. Berglez, I., Kuralt, M.C. & Boss, W.F. (1992). Purification and properties of phospholipase A isolated from culture medium. *Plant Physiol.*, 99, S 717.
145. Boss, W. F. (1989). Phosphoinositide metabolism: Its relation to signal transduction in plants. In *Second Messengers in Plant Growth and Development*, ed. Boss, W. F. & Morré, J.D., pp. 29–56. New York: Alan R. Liss.
146. Scherer, G.F.E., André, B. & Martiny-Baron, G. (1990). Hormone-activated phospholipase A_2 and lysophospholipid-activated protein kinase: a new signal transduction chain and a new second messenger system in plants? *Curr. Topics Plant Biochem. Physiol.*, 9, 190–218.

4.

The three-dimensional structures of two lipases from filamentous fungi

DAVID M. LAWSON, ANDRZEJ M. BRZOZOWSKI, GUY G. DODSON, ROD E. HUBBARD, BIRGITTE HUGE-JENSEN, ESPER BOEL and ZYGMUNT S. DEREWENDA

Lipases cleave ester bonds in glyceride and related molecules using a nucleophilic mechanism that is effected by way of an activated serine (1). These enzymes are interesting to the biochemist both from a mechanistic viewpoint (this is true particularly of the serine proteases), and because of their inactivity under aqueous conditions and their catalytic activation by a lipid–water interface (2). These enzymes are also widely used in industry, and their use can potentially be extended to wider and more effective applications through alteration of their properties by protein engineering.

It is now known that a wide variety of fungi produce lipases, and since these are extracellular enzymes, they are relatively easy to isolate and tend to be very stable. This makes them particularly suitable for industrial use and academic research. The enzyme from *Rhizomucor miehei* (Rml) was selected for X-ray crystallographic study because it represented a typical fungal lipase and was not of immediate industrial significance (3). A second enzyme isolated from the fungus *Humicola lanuginosa* (Hll), which is widely used in industry, has also been studied. This enzyme exhibits about 32% sequence identity with the Rml (see Table 1), which suggests that there will be substantial structural similarities.

The kinetic behaviour, specificity and interfacial activation of Rml have been characterised. The magnitude of the activation is amongst the largest observed to date (R. Verger, personal communication). This property, by which lipases become catalytically competent at a lipid–water interface, is most striking, and it raises fundamental questions as to how the enzyme can be stable in both aqueous and non-polar environments, and also be able to undergo the specific structural changes required for it to become catalytically active.

It was hoped that X-ray analysis of the two fungal lipase crystal structures would make it possible to identify the nature of the structural changes associated with interfacial activation and to establish the three-dimensional organisation of the

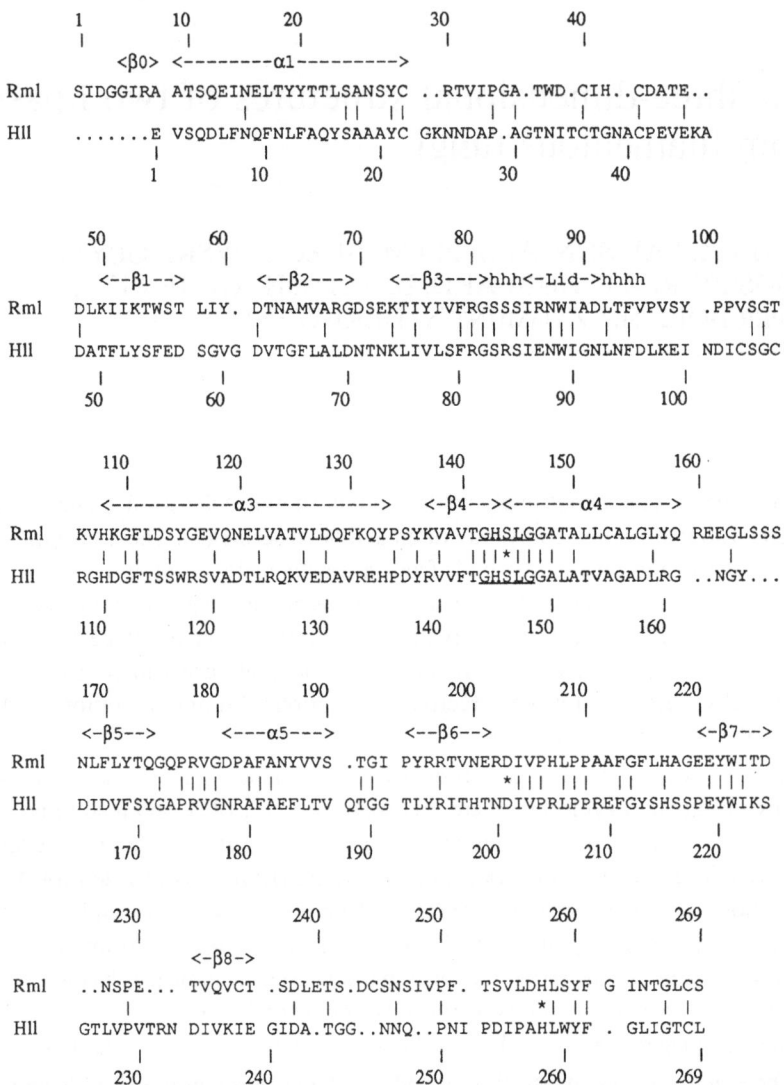

Table 1. Sequences of lipases from *Rhizomucor miehei* (Rml) and *Humicola lanuginosa* (Hll). The parts of the sequences that overlap closely in the two structures were aligned first (boxed). The intervening regions were then aligned with respect to sequence identity only. This alignment generates 87 absolute identities, which amounts to 32% overall sequence identity. The penta-peptide consensus sequences at the active serine have been underlined, and the catalytic residues are marked by asterisks. The secondary structure elements seen in Rml have been indicated β0–β8 for β strands and α1–α5 for α helices. The 'lid' helix has been indicated and corresponds to α2. Hinge regions are denoted by 'h'. The Rml sequence has been published previously (6); the Hll sequence, deduced from cDNA, is presented here for the first time. A detailed description of the purification, enzyme characterization and cDNA cloning of Hll is in preparation (15).

Table 2. Summary comparing the results of structural studies on Rml and Hll.

Lipase source	*Rhizomucor miehei*	*Humicola lanuginosa*
Crystal parameters		
Space group	$P2_12_12_1$	$P6_1$
Cell parameters (Å)	$a = 71.6, b = 75.0, c = 55.0$	$a = b = 143.0, c = 81.0$
Molecules in asymmetric unit	1	2
Solvent content (%)	51	69
Max. resolution of data (Å)	1.90	2.65
Components of model		
Total number of atoms	2289	4182
Protein residues present	5–269	A chain: 2–22, 29–269
		B chain: 1–22, 29–269
Number of water molecules	231	102 [a]
Model geometry		
Bond RMS deviation (Å)	0.015 (0.020)	0.019 (0.020)
Angle RMS deviation (Å)	0.077 (0.040)	0.059 (0.040)
RMS deviation from planes (Å)	0.013 (0.020)	0.015 (0.020)
Overall temperature factor (Å^2)	20.39	25.42
Crystallographic R-factor [b]	19.19%	24.16% [c]

The figures in brackets under the heading 'model geometry' refer to ideal values.
[a] This figure will undoubtedly increase with further refinement and model building.
[b] Calculated on all data.
[c] This value is relatively high; it reflects poor agreement between observed and calculated data at low resolution resulting from the high proportion of solvent in the crystal.

catalytic residues and the stereochemistry of their specificity and binding pockets. The analyses would equally highlight how the differences in sequence between the two enzymes is related to their different chemical and physical behaviour.

1. CRYSTALLOGRAPHIC ANALYSIS

The crystals of the lipase from *Rhizomucor miehei* (Rml) belong to the ortho-rhombic space group $P2_12_12_1$ and have unit cell dimensions of $a = 71.6$ Å, $b = 75.0$ Å and $c = 55.0$ Å (Table 2). Native data extending to 1.9 Å resolution were collected on a Xentronics area detector. A total of 21 165 independent reflections (87.8% of the available data at this resolution) were measured with an average redundancy of 5.5, and processed to give a merging R-factor of 8.1%. The essentials of the crystal structure determination by multiple isomorphous replacement have been described elsewhere (4).

The crystallographic refinement of the Rml has been carried out by least-squares minimisation methods in which the calculated shifts are restrained to conform to realistic peptide geometry (5). During the refinement, simulated annealing techniques were applied to improve the modelling of some of the more poorly defined peptide loops within the structure. At completion the protein atomic parameters were accurately defined at 1.9 Å resolution (4). Details of the fully refined model are given in Table 2.

The *Humicola lanuginosa* lipase (Hll) crystal structure was solved at 3.25 Å resolution, also by the multiple isomorphous replacement method, and is described here for the first time. Hexagonal crystals of space group $P6_1$ were grown from KH_2PO_4 in Tris.HCl buffer at pH 8.0 and had the cell parameters $a = b = 143.0$ Å, $c = 81.0$ Å. Native data to 3.25 Å resolution were collected on a Xentronics area detector. A total of 15 070 independent reflections (99.8% of the available data at this resolution) were measured, with an average redundancy of 5.4, and processed to give a merging R-factor of 9.7%. Three heavy-atom-derivative data sets (2 gold, 1 mercury) were used to solve the structure, although the high solvent content (69%) made subsequent analysis and refinement very difficult. Refinement is currently in progress against further native data to 2.65 Å resolution, collected by the Laue method on a synchrotron source. Consequently, the atomic co-ordinates are not nearly so well defined as those of the *Rml* model (the current model still contains a number of disordered residues that cannot be confidently placed in the electron density maps). A more detailed description of the structure solution will be published elsewhere. The parameters defining the current status of the model are shown in Table 2.

2. MAIN-CHAIN ORGANISATION

Figure 1 illustrates schematically the main-chain structure in the Rml enzyme. It is in some ways very simple and generates a more or less spherical molecule. There is a central β-sheet system of nine strands, eight of which are ordered with respect to the sequence (4). Only the N-terminal fragment, residues 5–8, which forms a β-sheet interaction with the C-terminal strand (residues 231–236), breaks this pattern. The central five strands (strands 3–7) are all parallel, and the two terminal pairs are antiparallel to the others. All the α–β connections are right-handed, which generates on one side of the central β sheet a series of linking helices and segments which form a complex structure. Generally, the helices are constrained by their connections to the β strands and consequently they run approximately parallel to the strand directions.

On the other side, the nine strands present a much simpler concave surface, across which runs the N-terminal helix, roughly perpendicular to the strand

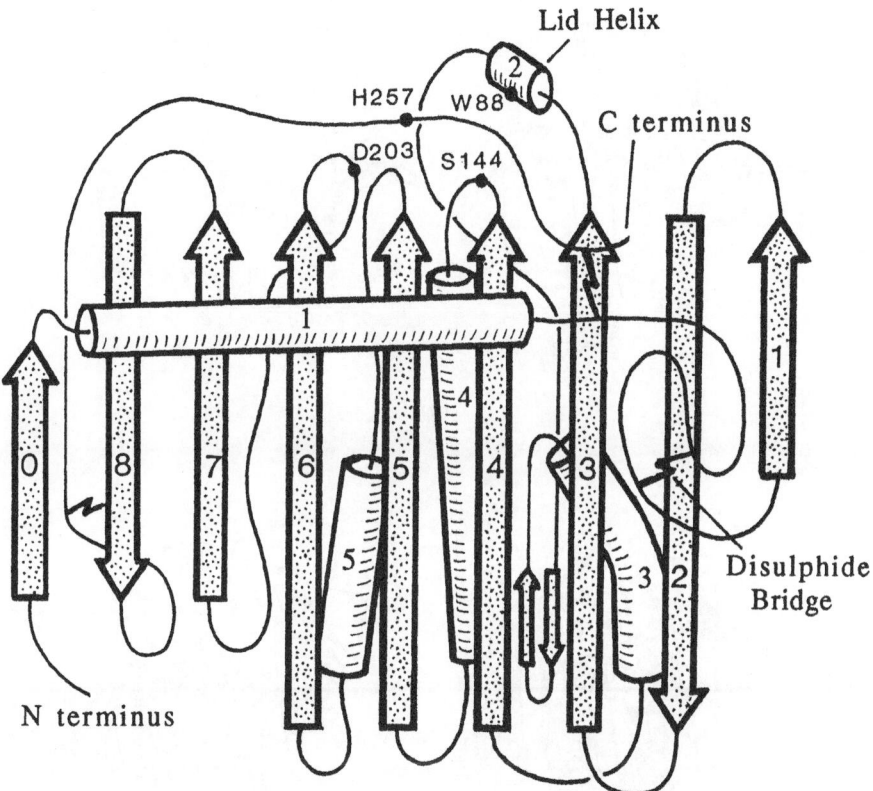

Figure 1. Schematic diagram showing the secondary structure of Rml. The β strands are depicted as arrows and labelled 0–8, and the helices are shown as cylinders, labelled 1–5. The positions of the catalytic triad residues and the tryptophan of the 'lid' helix have been indicated. It should be noted that the β sheet is not flat, as suggested by this diagram, but twisted along its length. See Figure 2 for a three-dimensional representation.

directions. There is a striking structural relationship between the N-terminal helix and the β sheet. The side-chains of this extensive helix make a series of contacts along the β sheet and serve to fill the shallow cavity generated by its curvature. A number of these contacts in the Rml enzyme are polar (see later).

The three-dimensional structure of the Hll enzyme proves to be essentially identical with that of Rml, as can be seen in Figure 2, which illustrates the backbone folding of the two molecules. Despite being comprised of the same number of amino-acid-residues (269, reference 6), Hll lacks the short N-terminal β strand seen in Rml (strand O), such that it begins with the long N-terminal

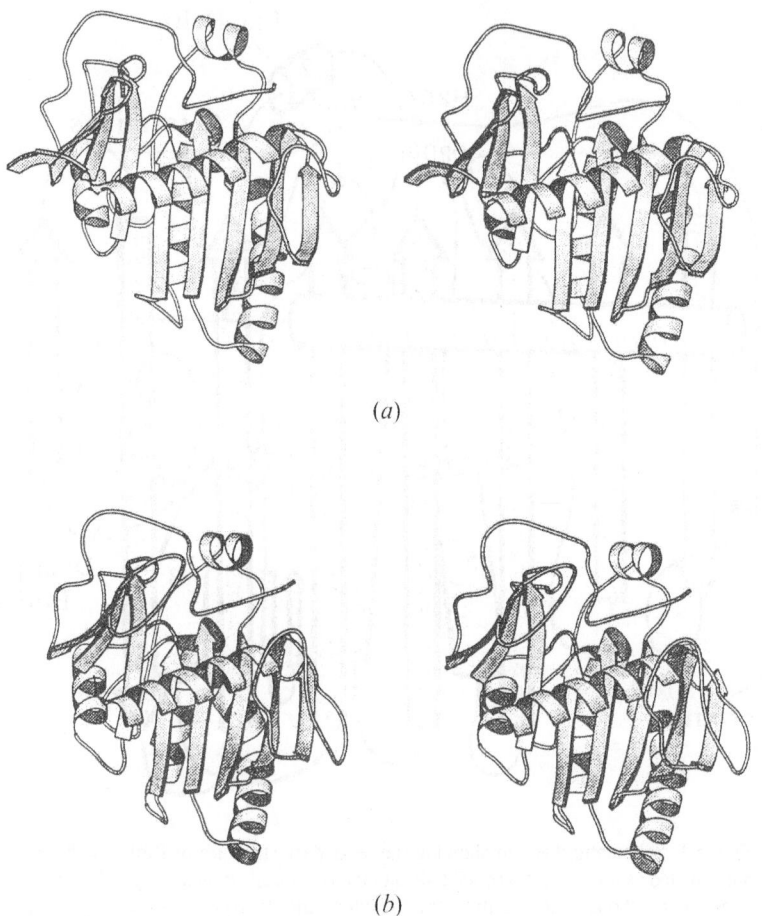

(a)

(b)

Figure 2. Stereo figures showing the main secondary structure elements of the Rml (a) and Hll (b) models. It should be noted that the Hll structure lacks an N-terminal β strand. However, this is compensated for by a much longer external loop between the N-terminal helix and β strand 1.

helix. The extra residues are found in the external loop regions, particularly the one that occurs between this N-terminal helix and β strand 1. Indeed, the main elements of secondary structure are essentially conserved between the two structures, and it is in these connecting regions that the majority of the structural differences occur (see Table 1).

Generally, the disulphide bridges are not conserved in the lipase family. However, both these fungal structures contain three disulphide bridges. One is

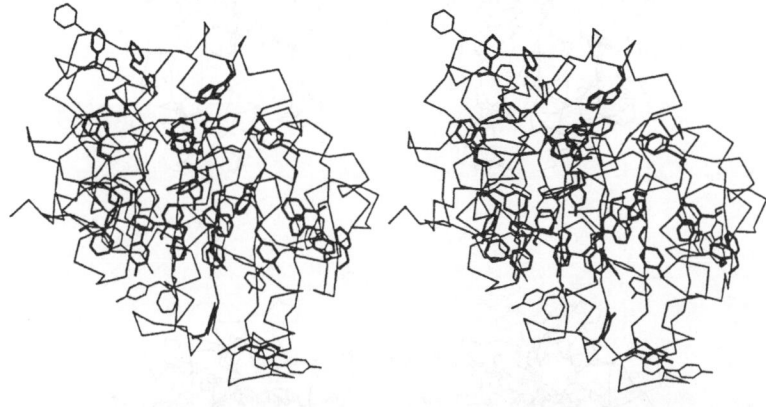

Figure 3. Stereo figure showing the distribution of hydrophobic residues in the two fungal lipases. The side-chains of Rml (thin lines) and Hll (bold lines) are superimposed on the α-carbon backbone of Rml only.

found in the loop which occurs between the N-terminal helix and β strand 1 in both models. The bridge in Rml is notable in that the two cysteines are only two residues apart, and therefore this bridge generates an extremely tight turn. The second cystine bridge is also conserved, and effectively secures the extended C terminus to the main body of the molecule by linking the penultimate residue to the C terminus of helix 1. The remaining bridge is not conserved. In Rml it links β strand 8 to the start of the extended C terminus, while in Hll it stabilises the loop that arises between the lid helix (see section 6) and helix 3.

3. THE INTERNAL STRUCTURE

As a general rule, globular proteins contain a hydrophobic core made up of non-polar side-chains, which provide an entropic element important for the folding and stability of the molecule (7). However, it is notable that in both Rml and Hll there are several groupings of polar residues and associated molecules. One of these of course is the active-site triad of Asp, His and Ser residues, which is essential for catalysis.

From a comparison of the two fungal enzymes, it can be seen that a considerable proportion of the buried side-chains are not strictly preserved, which of course is to be expected given the limited similarity of around 32% between their sequences. The variation in the aromatic side-chains is seen in Figure 3.

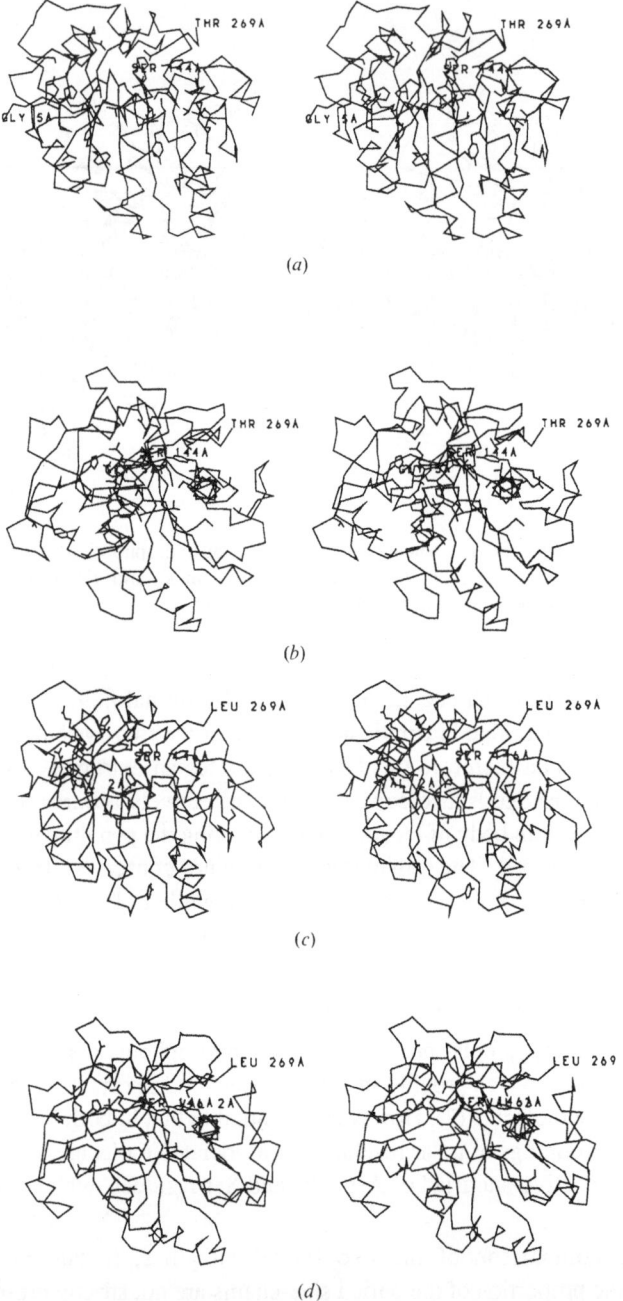

Figure 4. Stereo figures showing orthogonal views of the buried hydrophilic residues in Rml (*a* and *b*) and Hll (*c* and *d*). Only the side-chains and α-carbon backbones are shown.

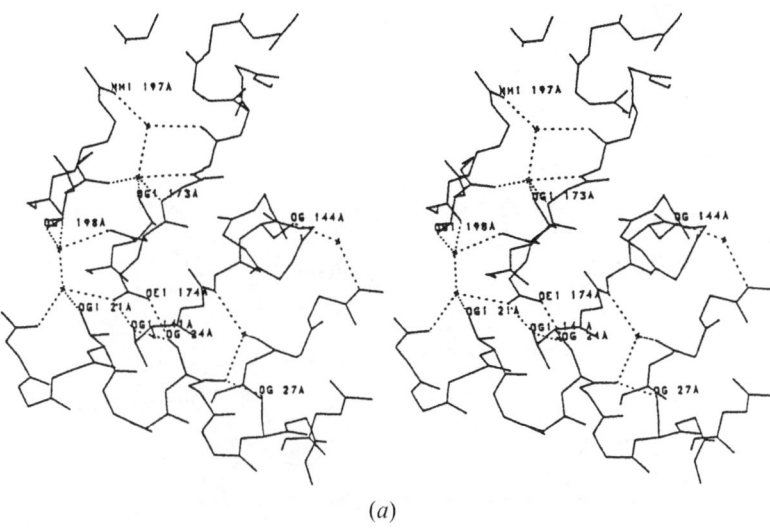

(a)

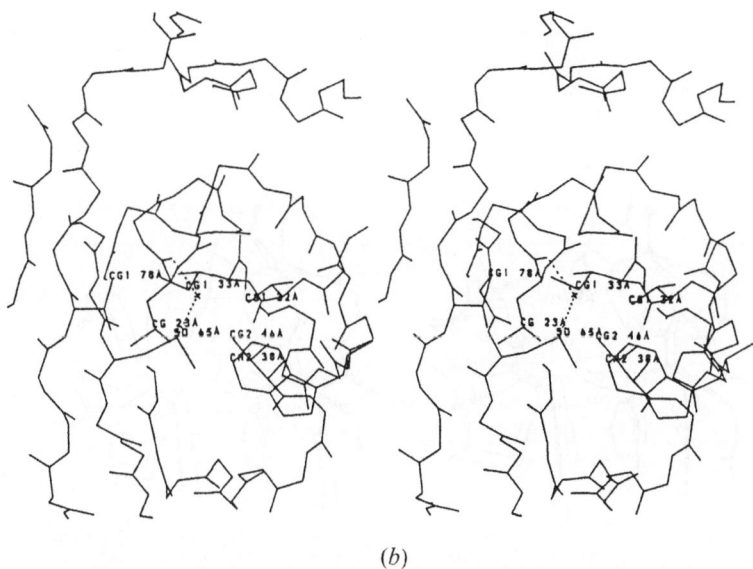

(b)

Figure 5. Stereo figures showing buried water molecules in Rml. (*a*) Six water molecules and associated hydrophilic side-chains are shown (all hydrophobic side-chains have been omitted). (*b*) A single water molecule sitting in a hydrophobic pocket. Apart from Met 65, which hydrogen-bonds to the water, only hydrophobic side-chains are shown.

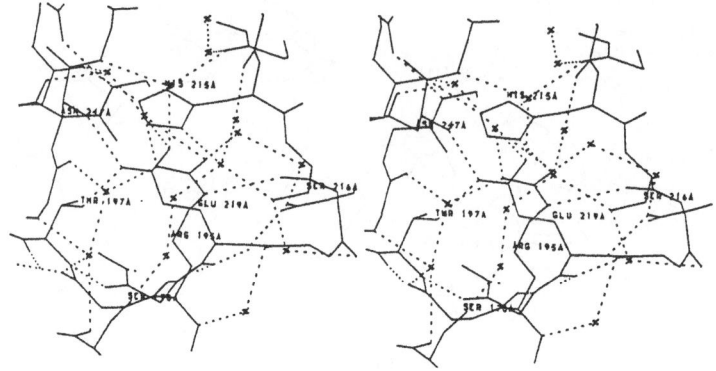

Figure 6. Stereo figure showing one of the four buried arginine residues in Hll. This one sits in a particularly hydrophobic pocket, being surrounded by a cluster of water molecules and several hydrophilic side-chains (hydrophobic side-chains have been omitted for clarity).

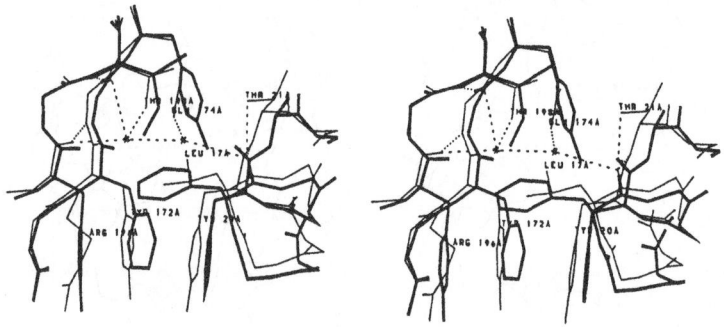

Figure 7. Stereo overlay showing how two water molecules in the Rml structure (thin lines) become replaced by side-chains in the Hll model (bold lines). See main text for a fuller description.

There are marked changes in their positions and in their conformations. The distribution of buried hydrophilic side-chains is also quite different. This variation indicates that the buried polar clusters seen in Rml are not critical for lipase function. The positions and interactions of those that occur in both Rml and Hll are seen in Figure 4.

4. BURIED WATER STRUCTURE

There are some 22 buried water molecules in the Rml molecule, an unusually large number. These waters are almost always associated with the buried polar side-chains. Their contacts largely complete the hydrogen-bonding potentials of the buried polar atoms which cannot otherwise be satisfied, because of the generally non-polar environment and the specific stereochemical requirements of hydrogen bond formation. In addition, by filling space, the buried water molecules increase the Van der Waals' contacts and dispersion forces in the core, thus increasing the overall stability of the folded molecule. Figure 5(a) illustrates some of these buried water containing clusters. There is one example in Rml of a buried water molecule that is making weak hydrogen-bonding contacts, one to a bifurcated main-chain carbonyl oxygen and the other to a methionyl sulphur atom. This water is seen in Figure 5(b), and is clearly filling space.

The situation in Hll is somewhat different. However, there also appears to be a significant number of buried polar residues and associated buried water molecules, although this latter observation can only be confirmed by a more precise description of the structure (which is not possible with the resolution and quality of the X-ray data currently available). The Hll model is unusual in that it contains four buried arginine residues (see Figure 6 for an example). In Figure 7, the polar cluster Thr21, Thr198, and Gln174 in Rml is seen to be replaced in Hll by Ala, Ile, and Tyr respectively. These changes in Hll displace the two water molecules in Rml. Interestingly, one of the waters is essentially replaced by the tyrosyl hydroxyl, which both fills space and makes a favourable hydrogen bond to the adjacent main-chain carbonyl oxygen atom of residue 17. The space occupied by the other water in Rml is filled by the C^δ atom of the Ile side-chain in Hll.

The great majority of the buried water molecules in Rml are limited to the half of the molecule that contains the active site (see Figure 8). It is this portion of the molecule that is most likely to be enveloped by the lipid at the water–lipid interface (8). The existence of this extensively hydrated and polar volume about the Rml active site raises the possibility that these waters assist the enzyme to function within the lipid environment. A polar core in the enzyme would confer extra stability against unfolding in non-polar conditions.

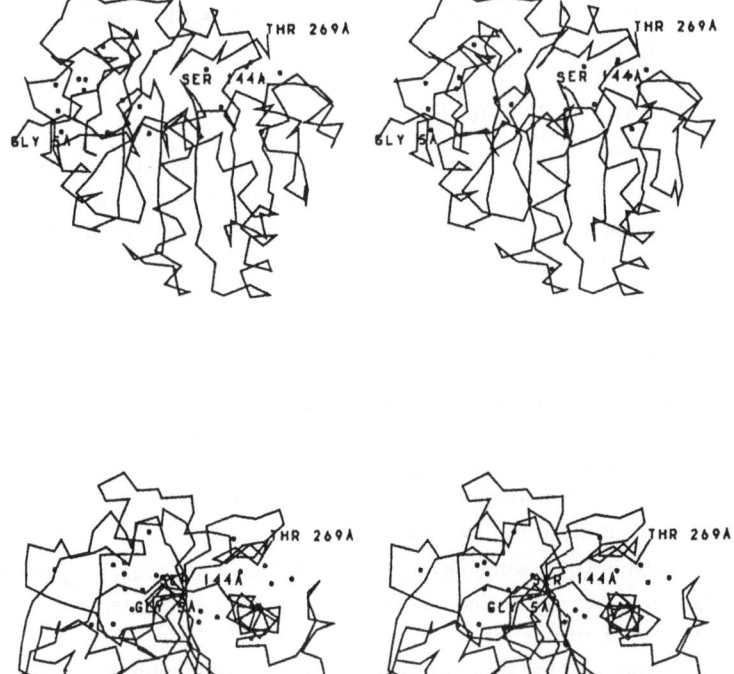

Figure 8. Stereo figure showing the distribution of buried water molecules in orthogonal views of the Rml structure. It should be noted that most of these waters are clustered around the top half of the molecule, which contains the active site, and is thought to be engulfed by lipid upon activation.

5. ACTIVE SITE

The importance of the activated serine in lipases has been established clearly by chemical modification experiments. From subsequent sequencing, the active serine was seen to occur in the pentapeptide consensus sequence Gly-X_1-Ser-X_2-Gly similar to that reported for serine proteases (6). This enabled the location of the active serine to be determined unambiguously in electron-density maps of the two fungal enzymes (4). Analysis of the immediate vicinity revealed a hydrogen-

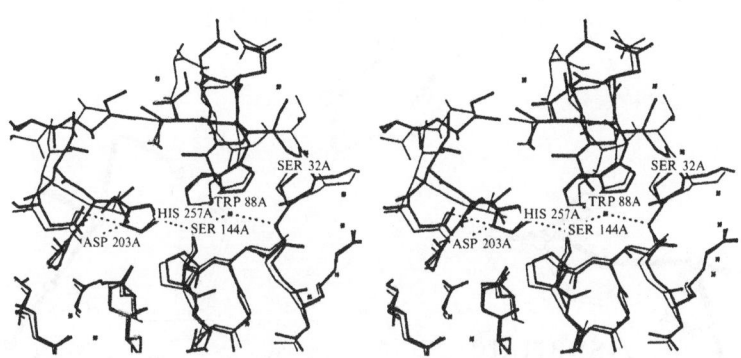

Figure 9. Stereo overlay of active site regions of Rml (thin lines) and Hll (bold lines). Only water molecules for the Rml structure are shown. The labels refer to the Rml structure.

bonding network linking the serine to His and Asp side-chains in a manner closely reminiscent of the catalytic triad seen in serine proteases (see Figure 9). However, although all the active atoms lie roughly in the plane of the imidazole ring, in both enzyme families, in lipases, the Ser O^γ is presented from the opposite side of the plane (see Figure 10). Such a triad was also reported for human pancreatic lipase (9) and more recently in another fungal lipase from *Geotrichum candidum* (10), although the latter differs in that the carboxyl function is provided by a glutamate side-chain.

6. ACTIVATION

A cursory examination of the two fungal enzyme structures is all that is needed to conclude that they do not represent the active conformation of the enzyme. These structures undoubtedly represent the appearance of the molecule in an aqueous medium, where lipases are found to be poorly active (2). The catalytic sites are clearly buried beneath a short stretch of helix, referred to as the 'lid' (see Figures 1 and 2), which arises from a long external loop between β strands 3 and 4. In both structures, the lids present a Trp side-chain into the catalytic site region, which is essentially hydrophobic except for the catalytic triad (see Figure 9). The lid is amphiphilic in character – the surface that bears the conserved Trp is completely apolar and interacts favourably with the hydrophobic residues surrounding the catalytic centre. The exposed surface is hydrophilic and faces

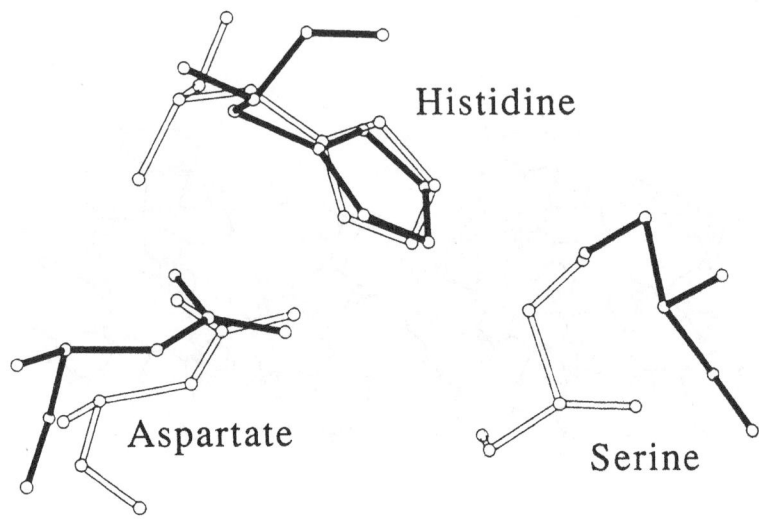

Figure 10. Comparison of catalytic triads of Rml (filled bonds) and *Streptomyces griseus* trypsin (21) (open bonds). It should be noted that, although all the active atoms are superposed and lie roughly in the plane of the imidazole ring, in Rml the seryl O^γ is presented from the opposite side of this plane to that in trypsin. See also Figure 11.

outwards, where it hydrogen-bonds to a number of well-ordered water molecules.

In an aqueous environment, exposure of the active centre by a conformational change of the lid would presumably be thermodynamically unfavourable, because of the large hydrophobic patch that would necessarily be exposed as a result. Conversely, in the presence of hydrophobic interfaces such as substrate lipid micelles, this non-polar surface would be favoured, and hence the displacement of the lid would be stabilised. This provides a simple explanation for the phenomenon of 'interfacial activation' seen in lipases, whereby increasing the level of water-insoluble substrates to beyond their critical micelle concentration causes a dramatic rise in lipolytic activity.

A number of crystallographic studies using serine protease inhibitors or specific lipase substrate analogues have confirmed the hypothesis that the lid is displaced during activation (8, 11, 12). In all cases, the lid has rolled back as a rigid body into a hydrophilic trench previously filled with water molecules, exposing both the active site and the hydrophobic surface of the amphiphilic lid helix. This is achieved by specific changes in the dihedral angles of residues located at each end of the helix ('hinge' regions, see Table 1). The large hydrophobic patch created by this movement, as discussed above, will no doubt be stabilised by a non-polar interface during normal activation. In the crystal structure, this surface is stabilised by interactions with the equivalent surface of a symmetry-related molecule. Part of this hydrophobic patch forms along narrow groove extending from the active centre to the edge of the molecule. This is thought to be the binding pocket for the acyl moiety of the triglyceride substrate during catalysis (8). It has been estimated that this groove is long enough to accommodate a 12-carbon fatty-acid chain. It should be noted that the lid movement observed is the same for small inhibitors such as diethyl-*p*-nitrophenyl phosphate (11, 12) as it is for larger inhibitors such as *n*-hexyl chlorophosphonate ethyl ester (8). This suggests that these lipases have only one activated conformation.

Recent simulations of Rml by molecular dynamics (K. Hult *et al.*, personal communication) have indicated that during activation, it is Arg86 in the lid that moves first and appears to draw the lid after it. In fact, the rôle of Arg86 is not clear from crystal structures of the activated complex, since it does not form strong hydrogen bonds with the main body of the molecule. Indeed, the residue appears to be disordered and may interact with a symmetry-related molecule – its conformation may therefore be a crystallographic artefact. However, its importance has been demonstrated experimentally (13). It is observed that guanidine will inhibit competitively the activation of Rml. This is thought to be because guanidine is chemically analogous to the end of an Arg side-chain, and therefore competes with Arg86 for hydrogen-bonding partners. When assayed on tributyrin substrate at pH 7.0 and 25 °C, the Rml showed approximately 26% of its normal activity in the presence of guanidine. Conversely the Hll, which has an Arg in the hinge region but not in the lid helix itself, was much more active, exhibiting 88% of its normal activity. Furthermore, if guanidine is added after the addition of substrate, the inhibition is not observed in Rml. This suggests that the Arg in Rml is only important during the activation process, and may not be necessary to maintain the protein in the activated conformation, which would agree with crystallographic evidence.

From a close examination of the stereochemistry of the transition states inferred from crystallographic studies of enzyme-inhibitor complexes (8, 14), it is apparent that the 'handedness' of the tetrahedral carbon with respect to the seryl OY, the oxyanion hole, the leaving group and the acyl group is reversed in lipases relative to that seen in proteases (see Figure 11). However, as a direct result of the hydrogen-bonding network, all the active atoms of the triad lie roughly in the same plane in both cases. Therefore it must be the immediate environment of the active centre that determines the chirality of this carbon, rather than the active atoms themselves.

7. CONCLUSION

The use of enzymes in industry is a relatively new development, and the potential of lipases, in particular, is only just being realised. Their unique characteristics of

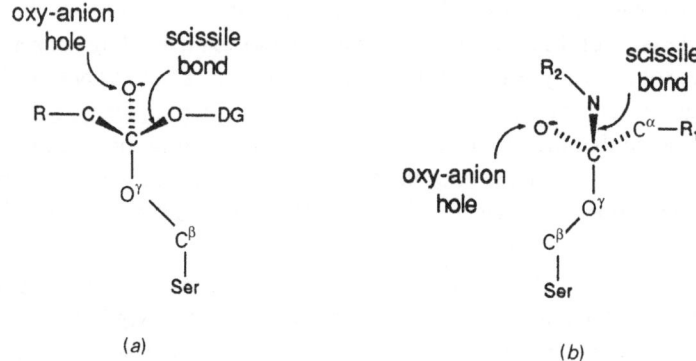

Figure 11. Diagram showing the arrangement of the tetrahedral intermediates of lipases (*a*) and serine proteases (*b*) inferred from crystal structures of enzyme–inhibitor complexes. These are viewed from the direction of the catalytic histidine. The imidazole ring is parallel to the O^γ–C bond and perpendicular to the plane of the paper. R, fatty acyl group; DG, diglyceride; R_1, carboxyl moiety; R_2, amino moiety.

being stable in both aqueous and organic media makes them particularly suitable catalysts for a number of synthetic processes, which would otherwise require very harsh conditions to proceed (*e.g.*, high temperatures/pressures, extremes of pH).

Until comparatively recently, the location and nature of the catalytic centre and substrate-binding pocket, and the structural basis for interfacial activation and the catalytic mechanism were all unknown. Now that a handful of X-ray structures are available, we are a much better position to explain the molecular basis of lipase properties. Armed with this knowledge, we can now in principle make rational changes to these enzymes, using the techniques of protein engineering, in order to improve catalytic efficiency, alter substrate specificity, increase temperature or pH stability *etc.* These experiments are in progress.

The origin of lipases and related enzymes is of interest to the geneticist. In all the lipase structures available to date, there is a common central β sheet, which supports the catalytic residues. These motifs are also seen in apparently unrelated enzymes, such as wheat serine carboxypeptidase (16), acetylcholine esterase (17) from *Torpedo californica*, dienelactone hydrolase (18) from *Pseudomonas* species, and haloalkane dehalogenase (19) from *Xanthobacter autotrophicus*. These observations suggest that all these enzymes share a common evolutionary origin, and therefore exhibit divergent evolution. A recently solved structure of the enzyme cutinase isolated from the fungus *Fusarium solani pisi* (20), also has these motifs, although in this case, its relationship to lipases is much more obvious, since it is active against similar substrates.

The link between lipases and serine proteases is much more tenuous, because there are no apparent similarities in secondary structure. The only common element is the catalytic triad. Since both enzyme families have evolved to catalyse a fundamentally similar hydrolytic reaction, that of cleaving an ester bond, it seems quite reasonable that they should both acquire the same catalytic apparatus through evolutionary selection. In other words, the relationships seen between lipases and serine proteases appear to be the result of convergent evolution.

The observation that the two fungal lipases studied in this monograph have a relatively high proportion of buried hydrophilic residues and solvent molecules, is worthy of note. It appears to contradict most of the theories concerning protein folding, whereby the most stable conformation of a protein is generated by burying the non-polar residues and exposing the polar residues to solvent. This partially polar core may confer on lipases their unique properties, which enable them to remain intact in a hydrophobic environment. It will be interesting to see if these findings can be extended to other lipase structures as and when they become available.

REFERENCES

1. Blow, D. (1990). More of the catalytic triad. *Nature*, 343, 694–695.
2. Desnuelle, P., Sarda, L. & Ailhard, G. (1960). Inhibition de la lipase pancreatique par le diethyl-*p*-nitrophenyl phosphate en emulsion. *Biochim. Biophys. Acta*, 37, 570–571.
3. Huge-Jensen, B., Gailuzzo Rubano, D. & Jensen, R.G. (1987). Partial purification and characterization of free and immobilized lipases from *Mucor miehei*. *Lipids*, 22, 559–565.
4. Brady, L., Brzozowski, A.M., Derewenda, Z.S., Dodson, E., Dodson, G.G., Tolley, S., Turkenburg, J.P., Christiansen, L., Huge-Jensen, B., Norskov, L., Thim, L. & Menge, U. (1990). A serine protease triad forms the catalytic centre of a triacylglycerol lipase. *Nature*, 343, 767–770.
5. Konnert, J.H. & Hendrickson, W.A. (1980). A restrained-parameter thermal-factor refinement procedure. *Acta Cryst.*, A36, 344–250.
6. Boel, E., Huge-Jensen, B., Christensen, M., Thim, L. & Fiil, N. (1988). *Rhizomucor miehei* triglyceride lipase is synthesized as a precursor. *Lipids*, 23, 701–706.
7. Kauzmann, W. (1959). Some factors in the interpretation of protein denaturation. *Advan. Prot. Chem.*, 14, 1–63.
8. Brzozowski, A.M., Derewenda, U., Derewenda, Z.S., Dodson, G.G., Lawson, D.M., Turkenburg, J.P., Björkling, F., Huge-Jensen, B., Patkar, S.A. & Thim, L. (1991). A model for interfacial activation in lipases from the structure of a fungal lipase-inhibitor complex. *Nature*, 351, 491–497.
9. Winkler, F.K., D'Arcy, A. & Hunziker, W. (1990). Structure of human pancreatic lipase. *Nature*, 343, 771–774.
10. Schrag, J.D., Li, Y., Wu, S. & Cygler, M. (1991). Ser-His-Glu triad forms the catalytic site of the lipase from *Geotrichum candidum* lipase. *Nature*, 351, 761–764.
11. Derewenda, U., Brzozowski, A.M., Lawson, D.M. & Derewenda, Z.S. (1992). Catalysis at the interface: the anatomy of a conformational change in a triglyceride lipase. *Biochemistry*, 31, 1532–1541.
12. Lawson, D.M., Brzozowski, A.M. & Derewenda, Z.S., unpublished results.

13. Holmquist, M., Norin, M. & Hult, K. Stabilization of the active open-conformation of *Rhizomucor miehei* lipase: Role of Arg 86 in the lid. To be published.
14. Chambers, J.L. & Stroud, R.M. (1977). Difference Fourier refinement of the structure of DIP-trypsin at 1.5 Å resolution with a minicomputer refinement technique. *Acta Cryst.*, B33, 1824–1837.
15. Huge-Jensen, B. & Boel, E., to be published.
16. Liao, D.I. & Remington, S.J. (1990). Structure of wheat germ serine carboxypeptidase II at 3.5 Å resolution: a new class of serine proteinase. *J. Biol. Chem.*, 265, 6528–6531.
17. Sussman, J.L., Harel, M., Frolow, F., Oefner, C., Goldman, A., Toker, L. & Silman, I. (1991). Atomic structure of acetylcholinesterase from *Torpedo Californica*, a prototypic acetylcholine-binding protein. *Science*, 253, 872–879.
18. Pathak, D. & Ollis, D. (1990). Refined structure of dienelactone hydrolase at 1.8 Å. *J. Mol. Biol.*, 214, 497–525.
19. Franken, S.M., Rozeboom, H.J., Kalk, K.H. & Dijkstra, B.W. (1991). Crystal structure of haloalkane dehalogenase: an enzyme to detoxify halogenated alkanes. *EMBO J.*, 10, 1297–1302.
20. Martinez, C., De Geus, P., Lauwereys, M., Matthyssens, G. & Cambillau, C. (1992). *Fusarium solani* cutinase is a lipolytic enzyme with a catalytic serine accessible to solvent. *Nature*, 356, 615–618.
21. Read, R.J. & James, M.N.J. (1988). Refined crystal structure of *Streptomyces griseus* trypsin at 1.7 Å resolution. *J. Mol. Biol.*, 200, 523–551.

5.

Structural aspects of phospholipase C from *Bacillus cereus* and its reaction mechanism

EDWARD HOUGH and SISSEL HANSEN

1. PHOSPHOLIPASES C

Phospholipases C cleave membrane phospholipids at the phosphate moiety, liberating the polar head group and leaving behind diacylglycerol or ceramide. The substrate specificities of these enzymes vary widely, ranging from the highly specific phosphatidylinositol- and sphingomyelin-hydrolysing enzymes (PI-PLC and SMaseC) to the less specific phosphatidylcholine-hydrolysing enzymes (PC-PLC), which usually also accept phosphatidylserine (PS) and phosphatidyl-ethanolamine (PE) as substrates. Bacterial phospholipases C are usually small Ca^{2+} or Zn^{2+} metalloenzymes ($M_r = 20$–30 kDa).

The involvement of phospholipases C in the agonist-induced generation of second messengers has been proposed as a critical event in signal transduction pathways in mammalian cells (1, 2). Over 40 hormones, growth factors and other agents have been shown to stimulate the hydrolysis of phosphatidylinositol-4,5-bisphosphate (PIP_2), catalysed by phospholipase C, to yield inositol-1,4,5-trisphosphate (IP_3) and diacylglycerol (DAG). IP_3, together with its metabolite 1,3,4,5-inositol-tetraphosphate, is involved in the regulation of intracellular calcium levels (3). DAG is an important activator of protein kinase C and thus functions as a second messenger in mitogenic signalling (4, 5). There is now increasing evidence that signal transduction also occurs *via* agonist-induced breakdown of phosphatidylcholine (PC) to produce DAG. This can occur in a single step catalysed by phospholipase C, or else *via* phospholipase-D-catalysed hydrolysis of the phospholipid to phosphatidic acid (PA) with subsequent cleavage of PA by phosphatidic acid phosphohydrolase to yield DAG. Recently, it has been shown that activation of PC-PLC by platelet-derived growth factor is an important, relatively late step in the mitogenic response to this growth factor (6). These results have been confirmed by studies in *Xenopus laevis* oocytes, where it has been shown that the PC-PLC-catalysed hydrolysis of PC is required for maturation of the oocytes in response to insulin and *ras* p21 (7). Furthermore, studies on NIH 3T3 fibroblasts suggest that DAG derived from PC, rather than PIP$_2$ turnover, is responsible for the atypical regulation of protein kinase C in cell lines transformed by *ras* and *src* oncogenes (8). PC has also been shown to be a major source of the eicosanoid precursor arachidonic acid, either *via* a PLA2-catalysed hydrolysis or *via* several two-stage processes (2).

Several recent experiments have shown that phospholipase C from *Bacillus cereus* (hereinafter referred to as PLC$_{Bc}$) is a valuable reagent for mimicking the action of mammalian PC-PLC. Levine *et al.* (9) demonstrated in 1988 that incubation of several mammalian cell types with PLC results in increased prostaglandin formation. In a number of recent reports, some of them mentioned above, PLC$_{Bc}$ has been used as a mimicking agent to study the rôle of phospholipase C-mediated hydrolysis of PC in growth-factor-induced mitogenic signalling in fibroblasts, keratinocytes and *Xenopus* oocytes (6, 7, 10, 11). Such studies have also implicated PC-PLC in the tumorigenic process initiated by *ras* and *src* oncogenes (see above) and in the inhibition of adenylate cyclase following activation of muscarinic cholinergic receptors (8).

Mammalian PC-PLCs have not been purified, and genes encoding such enzymes have not been cloned. This is in marked contrast to mammalian PI-specific phospholipases C, where several isozymes and their genes have been isolated and carefully studied (for a recent review, see reference 12). However, since antibodies to PLC$_{Bc}$ cross-react with at least some carrier of the mammalian activity (13), and also neutralise the PC-PLC activity in *Xenopus* oocytes (7, 10), it seems likely that the bacterial enzyme may be structurally similar to some mammalian PC-PLCs.

2. PHOSPHOLIPASE C FROM *Bacillus cereus*

PLC$_{Bc}$ is a monomeric, exocellular Zn^{2+} metalloenzyme. Its gene has been cloned and sequenced, showing that the enzyme is synthesised as a 283-residue precursor with a 24-residue signal peptide and a 14-residue propeptide (14). PLC$_{Bc}$ is highly stable, even in the presence of 8 M urea at 40 °C, but its stability is closely linked to the presence of Zn^{2+}. The enzyme may be unfolded and refolded reversibly in guanidinium hydrochloride by the removal or addition, respectively, of this ion (15). The Zn^{2+} ion may be exchanged with other metals although in some cases this causes loss of activity (16, 17) or changes in the substrate specificity (18). PLC$_{Bc}$ has been shown to inhibit blood coagulation *via* inactivation of thromboplastin (19), and is widely used in the study of lipids and cell membranes.

PLC$_{Bc}$ has extensive sequence similarity to the recently reported 29-kDa phospholipase C from *Listeria monocytogenes* (20), a PC-PLC from *Clostridium bifermentans* (21) and with the first two-thirds of the α-toxin from *Clostridium perfringens* (21). The latter enzyme plays a key rôle in the pathogenesis of gas gangrene (caused by *C. perfringens*) by promoting local cell disruption. In addition to the hydrolysis of membrane phospholipids, the α-toxin is also active against sphingomyelin and is hæmolytic. Early reports on PLC_{Bc} claimed that it was hæmolytic, but this was later shown to be due to the presence of a sphingomyelinase C, the two enzymes acting in combination to produce an effect similar to that of the clostridial enzyme. Snyder (22) has also shown that the enzyme is active against both enantiomers of phosphorylcholine, although V_{max} for the *S*-form is only 40 times lower than that for the *R*-form.

Gene-sequencing has shown that the phospholipase C gene in *B. cereus* is followed by that for sphingomyelinase C (14), and Gilmore *et al.* (23) have proposed that the two enzymes function together to form a heat-stable cytolytic unit, which they have termed 'cereolysin AB'. Since the phospholipases C from *B. cereus*, *C. perfringens* and others are relatively insensitive to the nature of the polar head group, they are often described as 'non-specific'.

3. MOLECULAR STRUCTURE OF PLC_{Bc}

The crystal structure of PLC_{Bc} has been solved by the method of multiple isomorphous replacement, with $PtCl_4{}^{2-}$, $PtI_6{}^{2-}$, Cd^{2+} and I^- derivatives and solvent flattening of the final multiple isomorphous replacement map (24). The structure has now been refined to $R = 15\%$ at a resolution of $1.5\,\text{Å}$ by the Konnert–Hendrickson restrained-refinement technique with standard restraints (25). The final model includes 243 water molecules.

3.1 General features

The PLC_{Bc} molecule (Figure 1) is roughly ellipsoidal in shape, with overall dimensions $40\times30\times20\,\text{Å}$. Apart from a wide cleft, which is $8\,\text{Å}$ deep by $5\,\text{Å}$ wide and lies approximately perpendicular to the major axis, the surface of the molecule is smooth. Three zinc ions lie on the inner surface of this cleft and clearly play a rôle in the stabilisation of the enzyme, since they are co-ordinated to widely separated parts of the amino-acid chain.

In spite of the presence of both α-helix and β-sheet bands in the far-ultra-violet circular dichroism spectrum of PLC_{Bc} (26), the crystal structure shows that the enzyme is in fact an all-helix protein. The overall tertiary structure consists of ten α-helical regions folded into a single, tightly packed domain. These helices contain 66% of the residues, the remainder forming the loops between them.

The greater part of the molecular surface shows a uniform distribution of acidic, basic and neutral hydrophilic residues, but there are two regions that deviate from this. Both are adjacent to the active site and may influence substrate–enzyme interactions prior to the formation of the active complex. The first region involves

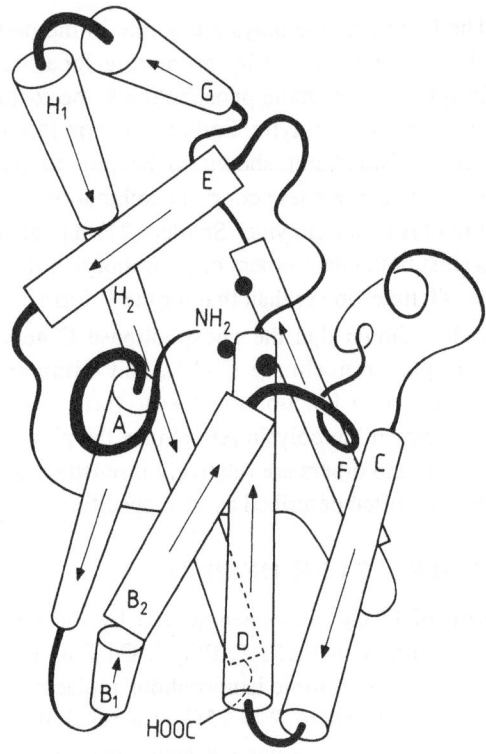

Figure 1. Schematic drawing of the PLC$_{Bc}$ structure with α helices represented as cylinders and metal ions as filled circles.

residues in the loops between helices B$_2$ and C and helices D and E (see Figure 1). These form a non-polar surface that extends over part of helix E to the highly flexible loop between helices G and H$_1$. The second region, which contains predominantly acidic residues, involves part of the N-terminal loop and the first part of the loop between helices B$_2$ and C.

With the exception of residues 4–11 in the N-terminal loop, residues 81 and 82 in the B$_2$–C loop, a region from residue 184 in helix F through to residue 210 in helix H$_1$ and the last four residues in the protein chain, temperature factors for the protein backbone lie under 15 Å^2. This suggests that the molecule contains a highly stable inner core and probably explains its high thermal stability and its resistance to denaturing agents. Structural studies of complexes between PLC$_{Bc}$ and various inhibitors, substrate analogues and reaction products (discussed in sections 4 and 5) support this, since the binding of these ligands causes very little change in the enzyme's structure.

At the time of first publication, the PLC_{Bc} fold was unique, but the subsequent determination of the crystal structure of P_1 nuclease from *Penicillium citrinum* (27) shows that, in spite of its different biological function, this enzyme is structurally very similar to PLC_{Bc}. Surprisingly, PLC_{Bc} and P_1 nuclease show no activity toward each other's substrates (28). Sequence comparisons suggest that the lecithinase from *Listeria monocytogenes* (20) has a structure similar to that of PLC_{Bc} and that the *Clostridium perfringens* α-toxin contains a PLC-like domain (21).

3.2 Topology

Helices A, B, C, D, F and H_2 form a compact bundle of mixed parallel and antiparallel helices with a predominantly right-handed twist. Helix E lies approximately perpendicular to the direction of this bundle and forms one side of the active-site cleft. Residues 206–243 (helices H_1 and H_2) are folded into a ten-turn helix with a 45° bend at residue 216, which is caused by a proline at position 218. This proline, together with the flexible loop between helices F and G, allows helices G and H_1 to fold around helix E to form the 'upper' end of the molecule. Helix E seems to function as a hinge for this region, since thermal motion increases as one moves away from helix E to reach a maximum in the G–H_1 loop.

Although they do not overlap completely and are not consecutive helices in the PLC_{Bc} sequence, helices C, D, F and H_2 can be regarded as forming a four-helix bundle similar to that found in a number of proteins including myohæmerythrin (29) and several cytochromes. Helices F and H_2 are closely parallel (divergence = 2.5°) whereas the angle between the principle axes of helices C and D is somewhat larger (22°). The C/D pair crosses the F/H_2 pair at an angle of 34°. Helix B_2 crosses helix D at an angle of 45°, and the crossing angle between helix A and helix H_2 is 62°. The angles between the principal axis of helix B_1 and the principal axes of helices A and B_2 are 26° and 67°, respectively, so that the B_2 helix effectively closes the six-helix bundle that forms the lower part of the PLC_{Bc} molecule.

3.3 The metal cluster

The three zinc ions in PLC_{Bc} form the metal cluster, which is shown in Figure 2. An almost identical cluster is present in P_1 nuclease (27) and probably in the α-toxin (21). A similar cluster also occurs in the active site of alkaline phosphatase of *Escherichia coli* (30), although PLC_{Bc} has no sequence homology with this enzyme. Among other substrates, alkaline phosphatase binds and hydrolyses the reaction products from PLC_{Bc}, so that the similarity between the active sites is probably more than coincidental. In soil, which is the natural habitat for *Bacillus cereus*, scarcity of inorganic phosphate (P_i) is the major growth-limiting factor.

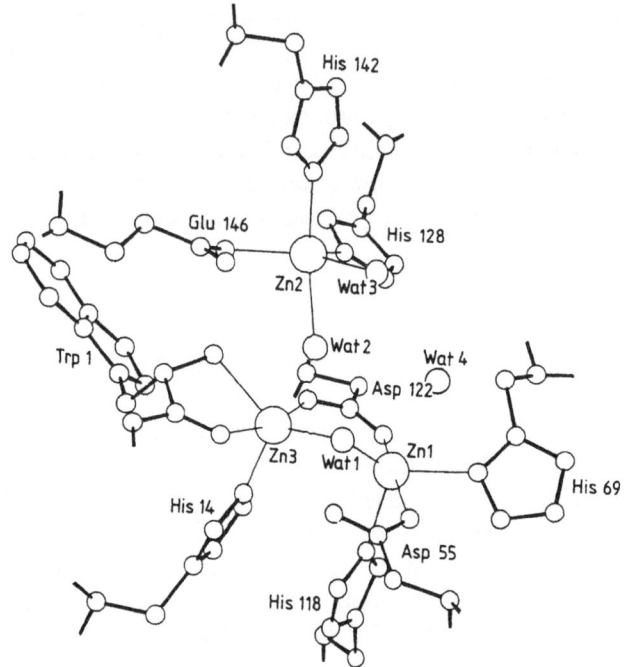

Figure 2. Metal-ion co-ordination at the active site of PLC$_{Bc}$, including water molecules.

Since the production of PLC$_{Bc}$ and an associated alkaline phosphatase is repressed by P$_i$, it seems likely that these two enzymes form a P$_i$-scavenging system for the bacterium (31).

Zn3 is co-ordinated to the N-terminal amino group in the protein and to the carbonyl group from the first peptide bond, to form a five-membered chelate ring similar to that found in a series of zinc–aminoacid complexes, but not in the active sites of enzymes (the chelate ring present in avian pancreatic polypeptide (32) is probably an artifact of the crystallisation procedure). Zn3 is also co-ordinated to His14, so that it secures the N-terminal end of helix A, and is linked to Zn1 via the carboxyl group of Asp122 and a shared water molecule. All three zinc ions are five-co-ordinate with approximately trigonal-bipyramidal symmetry.

Binuclear metal clusters in which carboxyl or other groups form a bridge between two metal ions have been reported in leucine aminopeptidase (33), hæmerythrin (29), ribonucleotide reductase (34), DNA polymerase (35), glutamine synthetase (36), xylose isomerase (37), concanavalin A (38), Cu,Zn superoxidase dismutase (39), alkaline phosphatase (28) and P$_1$ nuclease (27). The clusters are usually involved in catalysis. Metal–metal distances range from

2.88 Å in leucine aminopeptidase to 6.8 Å in Cu-Zn superoxide dismutase, where a 'bridging' histidine actually lies between the metal ions. The metal–metal distance is, as expected, shortest when there is more than one bridging ligand.

Apart from concanavalin A, the metal ions in all the above enzymes are involved in the catalytic function. Furthermore, for all the four enzymes that are phosphohydrolases (DNA polymerase, P_1 nuclease, alkaline phosphatase and PLC_{Bc}), studies of the binding of substrate analogues or inhibitors (27, 35, 40) have shown that the phosphate group in the substrate is co-ordinated directly to one or more of the metal ions and may well be involved in the stabilisation of a trigonal-bipyramidal transition state.

4. THE METAL ION IN SITE 2

Since there is no biochemical evidence to suggest the involvement of metals other than zinc, the structure has been refined using this metal. However, there is an element of uncertainty regarding the metal ion in site 2. Conventional restrained refinement of the final structure at 1.5 Å resolution resulted in temperature factors of 12.2, 16.6 and 12.8 Å^2 for Zn1, Zn2 and Zn3 respectively. Temperature factors for all of the ligands and for the surrounding protein structure, however, were approximately 12.5 Å^2, implying that Zn2 is subject to greater thermal movement than the ligands that bind it to the enzyme. This increased temperature factor implies that the effective scattering power of Zn2 is lower than that of the other Zn ions, a result which can also be explained in two other ways. The first of these is that the Zn2 site is not fully occupied in the crystal, that is, the crystal contains a mixture of two-zinc and three-zinc phospholipase C molecules. Since Zn2 cross-links the D–E loop to helix E, it is possible that its absence in some of the molecules could lead to a partial disordering of the crystal structure which, at least, would be reflected in higher temperature factors for the disordered regions. This is not observed. The second explanation is that the Zn2 site is fully occupied, but by a metal lighter than zinc. The two most obvious candidates for this are Ca^{2+} or Mg^{2+}. These alternatives will be discussed in the light of biochemical and structural evidence and by comparison with a number of relevant metalloenzymes.

4.1 Biochemical evidence

In 1965 Ottolenghi (41) published the results of an extensive study of the effect of EDTA and the Zn^{2+}-specific reagent 1,10-phenanthroline on the activity of PLC_{Bc}, concluding that the enzyme was a zinc metalloenzyme whose activity could also partially be restored by Mn^{2+}, while Ca^{2+} 'has no function when the enzyme is tested in the present system' (egg yolk lecithin as substrate). The Zn-dependence of PLC_{Bc} was confirmed by Otnæss *et al.* in 1971 (19) and, in a later study using highly purified PLC_{Bc}, Little and Otnæss (16) showed that the enzyme 'has two metal-binding sites, one of which is specific for zinc and binds the metal very tightly '. The other binds zinc less tightly and will probably accept Co^{2+}, Mn^{2+} and possibly Mg^{2+}. The range of possible metal ions was later extended (42) to include Ni^{2+}, but Cu^{2+} was shown to destroy enzymic activity (17). The exchangeable metal site was termed the 'catalytic site' as opposed to the more tightly binding 'structural site'.

4.2 Analytical evidence

Atomic-absorption analyses of the metal content in PLC_{Bc} (16, 17, 43) have consistently indicated the presence of 2 moles of metal per mole PLC_{Bc}. However, these analyses were carried out using a standard procedure that included a preliminary treatment of the enzyme solution with an ion-exchange resin to remove 'free' divalent metal ions – a technique that could also remove loosely bound co-factor metal ions, so that the possibility of there being three metal-binding sites is not excluded. Furthermore, results from the atomic-absorption analyses were calculated using a molecular weight of 23.0 kDa. Re-calculation using the gene-derived molecular weight (28 500 Da; reference 14) leads to $Zn:PLC_{Bc}$ ratios in the range 2.3–2.6, which implies the presence of more than two Zn-binding sites.

4.3 Crystallographic evidence

The crystal structure of PLC_{Bc} was solved by the isomorphous replacement method using, among others, a derivative in which the Zn^{2+} ions in the native enzyme had been replaced by Cd^{2+}. Crystallographic difference calculations at 3.2 Å resolution indicated the presence of two metal ions, 6 Å apart, which lay in a cleft in the surface of the molecule (44). However, interpretation of electron-density maps in what later turned out to be the N-terminal region of the enzyme was complicated by the presence of high density that could only be explained satisfactorily by the presence of a third metal ion. Subsequent difference calculations have shown that the Cd^{2+} derivative in fact contains *three* metal ions. In two of the metal-binding sites (1 and 2), Zn^{2+} has been at least partly replaced by Cd^{2+}, but the third contains a Zn^{2+} ion, which is bound so tightly that it was not exchanged by equilibrium dialysis against a Cd^{2+} solution.

Difference calculations showed a significant increase in electron density at the Zn2 site when PLC_{Bc} crystals had been equilibrated with 10 mM Zn^{2+}. This implies an increase in occupancy of this site, and suggests that the site binds the metal ion relatively weakly. Phase refinement of the metal occupancies, assuming that the correct temperature factor for all three metal ions is 12.5 $Å^2$, gave the numbers of electrons at the Zn sites listed in the following Table.

Zinc site	Native PLC$_{Bc}$	PLC$_{Bc}$ + extra Zn *	Cd-PLC$_{Bc}$
1	28	27	46
2	18	26	40
3	30	30	30

* PLC$_{Bc}$ after equilibration with 10 mM Zn^{2+}

The occupancy of site 2 in the native enzyme, 18, is close to the atomic number of calcium, which is 20. The values obtained for the Cd derivative, 46 and 40 electrons in sites 1 and 2, respectively, correspond to approximately 90% and 56% replacement of Zn^{2+} in a fully occupied Zn site by Cd^{2+}.

The widely used buffer compound tris(hydroxymethyl)methylamine (Tris) inhibits the binding of PLC$_{Bc}$ to the affinity columns used for the purification of the enzyme. Tris is known to form stable complexes with Cu^{2+} and Zn^{2+} (45). The study of PLC$_{Bc}$ crystals equilibrated with 10 mM Tris (secton 5.1) has shown that it binds to the metal in site 2, at the active site. This suggests strongly that this metal is zinc.

4.4 Structural evidence

The metal ion at site 2 is co-ordinated to one of the carbonyl oxygen atoms of Glu146, N$^\varepsilon$ in His128 and N$^\varepsilon$ in His142. Two water molecules complete the co-ordination shell. Although many examples where Ca^{2+} is co-ordinated to nitrogen have been found in the study of small molecules, surveys of the interaction of metal ions with histidine residues (46) and with carboxylic and carboxamide side-chains (47) reveal no known cases where this occurs in proteins. On the other hand, histidine is the most common ligand for Zn^{2+}, and co-ordination to two such residues is common (46). In a recent review (48), Higaki *et al.* have suggested criteria for the design of an 'ideal binding site for Zn^{2+}', which contains two histidine ligands. The Zn2 site in PLC$_{Bc}$ fits these criteria admirably.

4.5 Evidence from related enzymes

α-Toxin from Clostridium perfringens

This enzyme has an exceptionally high turnover rate for the hydrolysis of sphingomyelin and phosphatidylcholine in eucaryotic cell membranes, and it promotes the cell disruption associated with *Clostridium perfringens* gas gangrene. The α-toxin has extensive amino-acid identity (21) with PLC$_{Bc}$, this including all of the ligands involved in zinc-binding in PLC$_{Bc}$. Although activity after treatment with phenanthroline is almost completely restored by addition of Mn^{2+} or Co^{2+} (49), the α-toxin is considered to be a Zn-metalloenzyme. Atomic-absorption analysis has shown that activity is greatest when the enzyme contains 2.3 moles of zinc per mole of protein (50).

P₁-nuclease from Penicillium citrinum

This enzyme, which cleaves the P–O3' bond in single stranded RNA or DNA (27), has extensive homology with PLC$_{Bc}$ again including the Zn ligands. In spite of its different biological function, P₁ nuclease is structurally very similar to PLC$_{Bc}$. Fujimoto *et al.* (51) have reported that the active form of this enzyme contains three zinc ions in the active site. The crystal structure shows that these are bound in a cluster which is almost the same as that in PLC$_{Bc}$.

PC-PLC from Listeria monocytogenes

Geoffroy *et al.* (52) have shown that this enzyme, which has extensive amino-acid identity with PLC$_{Bc}$ and complete conservation of the metal-binding ligands (20), is zinc-dependent.

4.6 Zinc occupancy of catalytically active PLC$_{Bc}$

On the basis of the biochemical, analytical, crystallographic and structural evidence summarised in this section, and from comparison with homologous or otherwise relevant enzymes, it seems likely that PLC$_{Bc}$ is indeed a three-zinc metalloenzyme. The Zn2 binding site seems to be only partially occupied in our crystals of the native enzyme, and it may have a lower affinity for Zn^{2+} than the other sites. In spite of the fact that Zn^{2+} was not included in the soaking solutions used to prepare PLC$_{Bc}$ complexes with several inhibitors (section 5.5), the higher temperature factor and lower occupancy for Zn2 found in the native enzyme have not been found in the refined structures of these PLC$_{Bc}$ complexes. Since all three zinc ions are involved in inhibitor-binding, it seems likely that the catalytically active form of the enzyme contains three zinc ions.

5. MECHANISM OF ACTION OF PLC$_{Bc}$

5.1 The active site

PLC$_{Bc}$ is inhibited by several univalent anions (44, 53) especially I^-, CNO^-, NO_3^- and Cl^-. The rate of degradation of phospholipids is also lowered markedly in the presence of inorganic phosphate (50% inhibition at 50 mM P$_i$) and IO_3^- (60% inhibition at 50 mM IO_3^-; reference 54) and binding to substrate-based affinity columns is inhibited by the zinc-co-ordinating buffer compound Tris (55). Crystals of PLC$_{Bc}$ were therefore treated with P$_i$, I^-, IO_3^- and Tris in the hope of clarifying their mode of inhibition. Difference electron-density maps for PLC$_{Bc}$ treated with P$_i$ and Tris are shown in Figure 3 (*a*)–(*d*) along with the structures of the active-site regions with these inhibitors in their refined positions. Comparison with Figure 2 shows that P$_i$ binds to all three metal ions by replacing two of the ligated water molecules (Wat1 and Wat2). The phosphate oxygen replacing Wat2 is co-ordinated to Zn2, while that replacing Wat1 is bonded asymmetrically to Zn1 and Zn3, with respective Zn–O bond distances of 1.96 and 2.23 Å (the Zn1–O–Zn3 bridge is symmetric in the native enzyme). In the [PLC$_{Bc}$.Tris] complex the amine nitrogen in the Tris molecule is co-ordinated to Zn2, although the Zn–N bond is rather long (2.43 Å); all three hydroxyl groups of the Tris are involved in water-mediated hydrogen bonds to the protein, while the Tris methylene groups lie parallel to the plane of the aromatic ring in Phe66 at a distance of 3.8 Å from it. Iodate was found to bind to the metal ions in a manner very similar to that of P$_i$, whereas in I$^-$-inhibited PLC$_{Bc}$ an I$^-$ ion lies 5 Å from Zn2, blocking access to the active site. In all these inhibitor complexes, the structure of the protein itself is almost completely unchanged. These results confirm the location of the active site and that the metal ions are involved in the catalytic process.

5.2 Manual substrate-docking

The metal cluster at the active site carries a nett positive charge, which will clearly be attractive to the phosphate group in a substrate molecule. If one assumes that the phosphate group in a substrate molecule binds to these ions in the same manner as that found in the complex with P$_i$, then there are two possible orientations for the phospholipid molecule. The difference between these lies in the amphiphilic nature of both substrate and active site. In the substrate, the head group and the phosphate group are hydrophilic, the ester groups are of intermediate nature and the fatty-acid side-chains are hydrophobic. There is an equally clear segregation of polarity at the active site, with polar residues concentrated around the metal ions at one end of the cleft, the remaining residues being hydrophobic or neutral. Manual docking studies show that a phospholipid molecule can be placed in both possible orientations at the active site, with its phosphate group in the position found for Pi and without undue strain. However, matching of the hydrophobic and hydrophilic regions in substrate and active site suggests that the orientation with the choline head group at the polar end of the active site is the more likely one.

5.3 Crystallographic studies

A possible scheme for the reaction pathway for the hydrolysis of phospholipids is as follows, where E, H, P and L represent the enzyme, the polar head group, the phosphate group and diacylglycerol, respectively.

$$E + H.P.L \rightleftharpoons E.H.P.L \begin{array}{c} \nearrow \quad E.H.P \rightleftharpoons E + H.P + L \\ \\ \searrow \quad E.L + H.P \rightleftharpoons E + L + H.P \end{array}$$

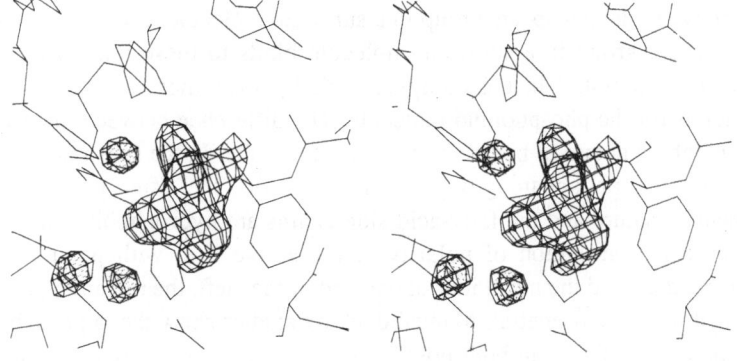

Figure 3. Difference electron density at the active site of PLC$_{Bc}$ for (a) Pi and (b) Tris. Refined positions of these inhibitors at the active site are shown in (c) and (d) on the opposite page.

In an attempt to clarify experimentally the question of substrate orientation, crystals of PLC$_{Bc}$ were treated with a number of reaction products and a substrate analogue. Studies of this type are restricted by factors such as the availability of suitable ligands, the solubility of these ligands in the high-salt mother liquor (45% saturated ammonium sulphate), the stability of the crystals after treatment with the ligand, access to the active site *via* channels through the crystal lattice and an unhindered active site in the crystalline enzyme. Fortunately, the active site in PLC$_{Bc}$ is exposed and is adjacent to large channels running through the lattice.

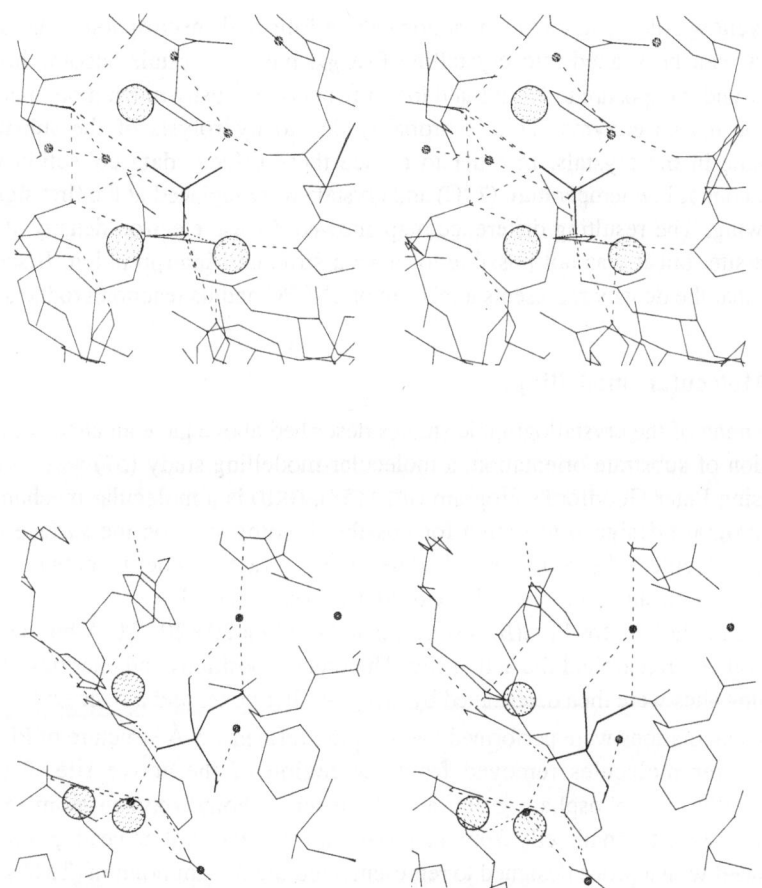

Crystals were treated with the reaction products phosphorylcholine, phosphorylserine and phosphorylethanolamine at the highest concentrations (up to 100 mM; reference 53) that could be used without the crystals being destroyed. Although these compounds are at best very poor inhibitors, it was hoped that this soaking might result in a ligand occupancy at the active sites of the crystalline enzyme that was high enough to give interpretable difference electron-density maps. Unfortunately, this did not turn out to be the case. A similar treatment with 1,2-diheptanoylglycerol, which causes product inhibition (56) was also unsuccessful, although it does appear that this compound diffuses into the crystals, since they decompose extremely rapidly on exposure to X-rays.

PLC$_{Bc}$ is often assayed using *p*-nitrophenyl phosphorylcholine (pNPPC) as a substrate analogue (49). This compound is a poor substrate ($K_m = 0.2$ M), possibly owing to the lack of the diacylglycerol moiety, but *p*-nitrophenolate has a deep yellow colour, so that the reaction can be followed spectrophotometrically. pNPPC can be soaked into crystalline PLC_{Bc}, but the crystals become bright yellow under exposure to X-rays and lose diffraction intensity much more quickly that the native enzyme. This is probably due to hydrolysis of the substrate analogue in the crystals. In order to reduce these effects, data collection was carried out at low temperature (7 °C) and crystals were replaced at the first sign of yellowing. The resulting difference map showed diffuse electron density at the active site, but it was not possible to make a structural interpretation. It seems likely that the density represents a mixture of pNPPC and its reaction products.

5.4 Molecular modelling

Since none of the crystallographic studies described above gave an answer to the question of substrate orientation, a molecular-modelling study (57) was carried out using Peter Goodford's program GRID (58). GRID is a molecular mechanics-type program designed to search for possible binding sites on the surface of a macromolecule. The program calculates the interaction energy between the macromolecule and probes which are designed to simulate all or part of the ligand. Initial calculations for PLC(*Bc*) were carried out in a $20 \times 20 \times 20$-Å box with a 1-Å grid centred around the active site. The precise positions and energies of the resulting sites were then determined by using smaller boxes and a finer grid.

The calculations were performed by using the refined 1.5-Å structure of PLC_{Bc} with water molecules removed from the region of the active site. Probes representing the phosphate group and the ester carbonyl oxygen atom in the substrate were taken directly from the program, while the choline head group was simulated with a probe designed to represent a tetramethylammonium (TMA) ion. The calculations were able to predict both the binding site and the orientation of the phosphate group in P_i-inhibited PLC_{Bc} very well, and it also gave single minima for TMA and the ester carbonyl group. The identification of single preferred positions for the TMA and ester carbonyl probes may have solved the question of substrate orientation in the active site of PLC_{Bc}.

A short-chain phospholipid molecule was then constructed stepwise from a phosphate group placed in the predicted position in such a way that the choline head group and an ester carbonyl group occupied the appropriate energy minima from the GRID calculations. An energy minimisation was carried out at each stage. The resulting orientation of the substrate molecule and energy minima for the probes are shown in Figure 4. It can be seen from this Figure that the choline head group lies in the more polar part of the active site, with its positively charged nitrogen atom lying close to the phenolic oxygen atom of Tyr56. The *sn*-2 carbonyl oxygen atom is co-ordinated to Zn2, replacing a water molecule in the native enzyme. This proposed substrate docking resembles that observed in

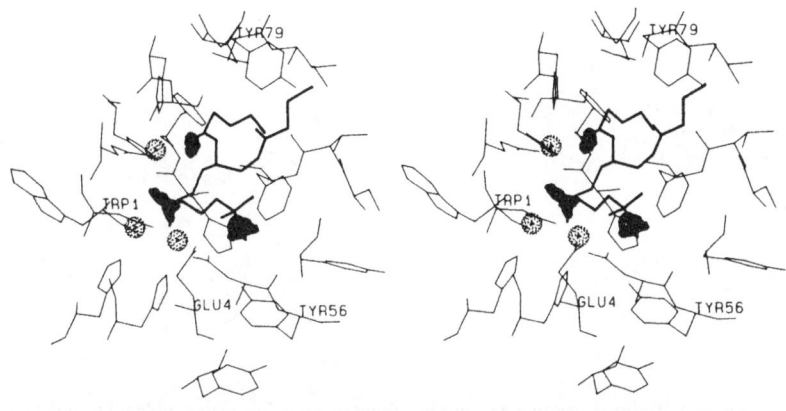

Figure 4. Stereo drawing of the active site of PLC_{Bc} showing energy contours for the TMA, phosphate and ester carbonyl probes as determined by GRID together with the final orientation of dipentanoyl phosphatidylcholine. The zinc ions are shown as filled circles.

complexes between phospholipase A_2 and a substrate-based inhibitor (59) and two transition-state analogues (60, 61).

5.5 Inhibitor complexes

A series of substrate-based specific inhibitors for PLC_{Bc} have been synthesised by Martin *et al.* (62). In these compounds, the PLC_{Bc}-labile P–O–C bond has been replaced by a phosphonate (P–C–C), phosphoramidate (P–N–C) or thiophosphate (P–S–C) bond. X-ray diffraction data from PLC_{Bc} crystals soaked for 20 h in 0.5 mM 3(S),4-dihexanoyl-oxybutyl-1-phosphonylcholine has been collected to a resolution of 1.9 Å by using a Siemens imaging proportional area detector and a Rigaku RU-200 BEH rotating-anode X-ray generator at the University of Lund, Sweden. Difference electron density for the complex is shown in Plate 3, and the refined position of the inhibitor molecule is shown in Figure 5.

The inhibitor binds to the metal ions in the active site of PLC_{Bc} with the phosphonyl group co-ordinated to the zinc ions in a manner very similar to that found for P_i, confirming the underlying assumption behind all our attempts to model the docking of substrate. This is similar to the situation observed in Cd^{2+}-substituted alkaline phosphatase (40) and at the 3'–5' exonuclease site in DNA polymerase I (35). The conformation of the inhibitor is very similar to those of

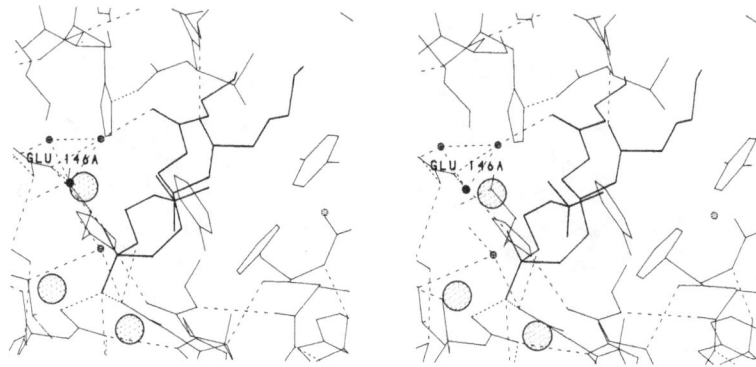

Figure 5. Refined position of the phosphonate, showing water molecules and metal ions. Hydrogen bonds are represented as broken lines. The water molecule most suitably placed for nucleophilic attack on the phosphorus atom is emphasized.

crystalline phospholipids (63), with the acyl side-chains roughly parallel and a sharp bend at the α-carbon atom in the sn-2 acyl chain. The major difference is that in the [PLC$_{Bc}$.phosphonate] complex the choline head group has been forced by the docking process to fold back into a strained conformation almost parallel to the acyl chains. The inhibitor displaces eight water molecules from the native structure. The carbonyl oxygen atom of the sn-2 acyl residue is hydrogen-bonded to the peptide nitrogen atom of Asn134 and that from the sn-1 chain forms part of a hydrogen-bonded network that includes several water molecules and side-chains in the interior of the active-site cleft. Again, inhibitor binding causes no significant movement in the enzyme or the metal ions. The absence of electron density for parts of the sn-1 acyl chain indicates that it is less firmly bound than the sn-2 chain. For both acyl side-chains, thermal motion increases along the chain as the number of enzyme–inhibitor contacts decreases.

The location and conformation of the bound inhibitor molecule resembles one of the two suggested substrate orientations in our manual docking studies (64). More importantly, the choline head group and phosphonyl group are located less than 1 Å from the positions that were predicted for them by GRID (57). Although the diacylglycerol moiety is located, as predicted, in the hydrophobic part of the active site with its roughly parallel acyl chains pointing out of the cleft, the sn-2 carbonyl oxygen atom is not co-ordinated to Zn2.

Roberts and her co-workers have investigated the effect of structural modifications of phosphatidylcholines on their suitability as substrates for PLC$_{Bc}$ (56, 65,

66, 67). It was shown that hydrolysis, in addition to requiring the phosphoryl head group, is influenced by the following factors. (a) The fatty-acid side-chains must be long enough to contain hydrophobic region which can bind to the enzyme. (b) Introduction of bulky side-chains on the first two methylene groups in the fatty acids reduces activity. (c) Replacement of the ester linkages by ether linkages results in loss of activity. (d) Ether lipids do not inhibit the enzyme. In a related study, Snyder (22) have shown that, at least, an ester group in the *sn*-2 position is essential for effective binding.

From Plate 4, it is clear that the crystal structure supports these conclusions very well, with the head group, ester groups and the first three methylene groups in the fatty-acid side-chains buried in the active site cleft. It is readily apparent that substitution of large side-chains on the first three methylene groups will hinder entry into the active site, and that these side-chains are involved in hydrophobic interactions with the enzyme (points (a) and (b) above). Although both ester carbonyl groups are involved in hydrogen-bonding (Figure 5), hydrogen-bonds involving the *sn*-1 acyl group are mediated by water, whereas the *sn*-2 carbonyl group is hydrogen-bonded directly to a peptide NH group (points (c) and (d) above).

5.7 Common structural fragments

With the exception of Tris and the inorganic anions (which function by blocking the active site), hexanol (56) and pNPPC, all compounds that bind to PLC_{Bc} either as substrates or as inhibitors possess the phosphate group, which must be present in the substrate for a phospholipase, and an *sn*-2 carbonyl group. This is emphasized in Figure 6. The importance of the *sn*-2 carbonyl group is made clear by the fact that neither glycerophosphorylcholine nor the reaction products phosphorylcholine, phosphorylserine and phosphorylethanolamine bind to the enzyme, even though they contain the phosphate group; in contrast, the activity of PLC_{Bc} toward phosphatidic acid, which contains both common features but no head group, seems to indicate that the latter plays a relatively minor rôle in substrate recognition and binding. The other reaction product, diacylglycerol (DAG), causes product inhibition which is probably associated with its release from the active site. This species does contain the *sn*-2 carbonyl group, and a phosphate moiety was present during binding of the intact substrate to the active site before hydrolysis. There is also some crystallographic evidence suggesting that DAG may bind to PLC alone (section 1).

5.8 Surface activation

PLC_{Bc} exhibits biphasic activity/substrate concentration curves (68) which are typical for enzymes where activity is influenced by the state of aggregation of the substrate. k_{cat}/K_m increases by a factor of approximately 30 as the substrate concentration passes the critical micelle concentration (cmc). However, the effect is much less than that reported for phospholipases A_2, where it is attributed to 'surface activation' of the enzyme. It has been suggested that PLA2 undergoes a conformational charge on binding to the aggregated substrate and that this results in increased activity, but there is no firm structural evidence for such a change (69). The crystallographic studies of inhibitor and substrate–analogue complexes with PLC_{Bc}, described above, show that this enzyme is very rigid. It thus seems likely that the increase in activity on passing the cmc is associated with the hydrophobic nature of diacylglycerols rather than with a conformational change. No detailed kinetic studies of the reaction mechanism for PLC_{Bc} have been carried out, so it is uncertain which step is rate-determining. However, diacylglycerol is known to cause product inhibition (56), and activity is increased in the presence of anionic detergents that solubilise these compounds (70), so it is possible that release of this product is rate-determining. The hydrolysis of a monodisperse phospholipid generates a water-soluble phosphorylated base and a predominantly hydrophobic diacylglycerol. Both of these must be released into an aqueous medium. This is not necessary when the lipid is in an aggregated state, since the fatty-acid side-chains will remain in the lipid bilayer when the enzyme is bound to the micelle. In phospholipases A_2, substrate-binding requires partial withdrawal of the substrate from the lipid bilayer into a long hydrophobic channel in the surface of the enzyme. On the other hand, binding of the phospholipid to PLC_{Bc} requires only a small withdrawal from lipid bilayer, with the first two to three methylene groups in the fatty-acid side-chains in contact with the enzyme. This probably explains the much lower 'surface activation' effect.

5.9 Hydrolysis of phosphoesters

Most enzymes that catalyse reactions at the phosphorus of phosphate esters require at least one divalent metal ion, more or less tightly bound. The phosphate group is usually co-ordinated to the metal ion(s), and its susceptibility to nucleophilic attack is thus markedly increased. The metal ion(s) may also serve to stabilise the transition state. Phosphoryl transfer reactions can proceed by either of two general mechanisms which are analogous to the SN1 and SN2 mechanisms in carbon chemistry. The SN1 mechanism is dissociative, commencing with departure of the leaving group to form an unstable planar metaphosphate. This is followed by rapid addition of a nucleophile to form the final product. This can occur with retention or inversion of configuration at the phosphorus. The associative SN2 mechanism is initiated by a nucleophilic attack on the phosphorus to form a five-co-ordinate trigonal-bipyramidal transition state. The attacking nucleophile may enter opposite to the leaving group, in a so-called 'in-line' mechanism, or on the same side as the leaving group in an 'adjacent' mechanism.

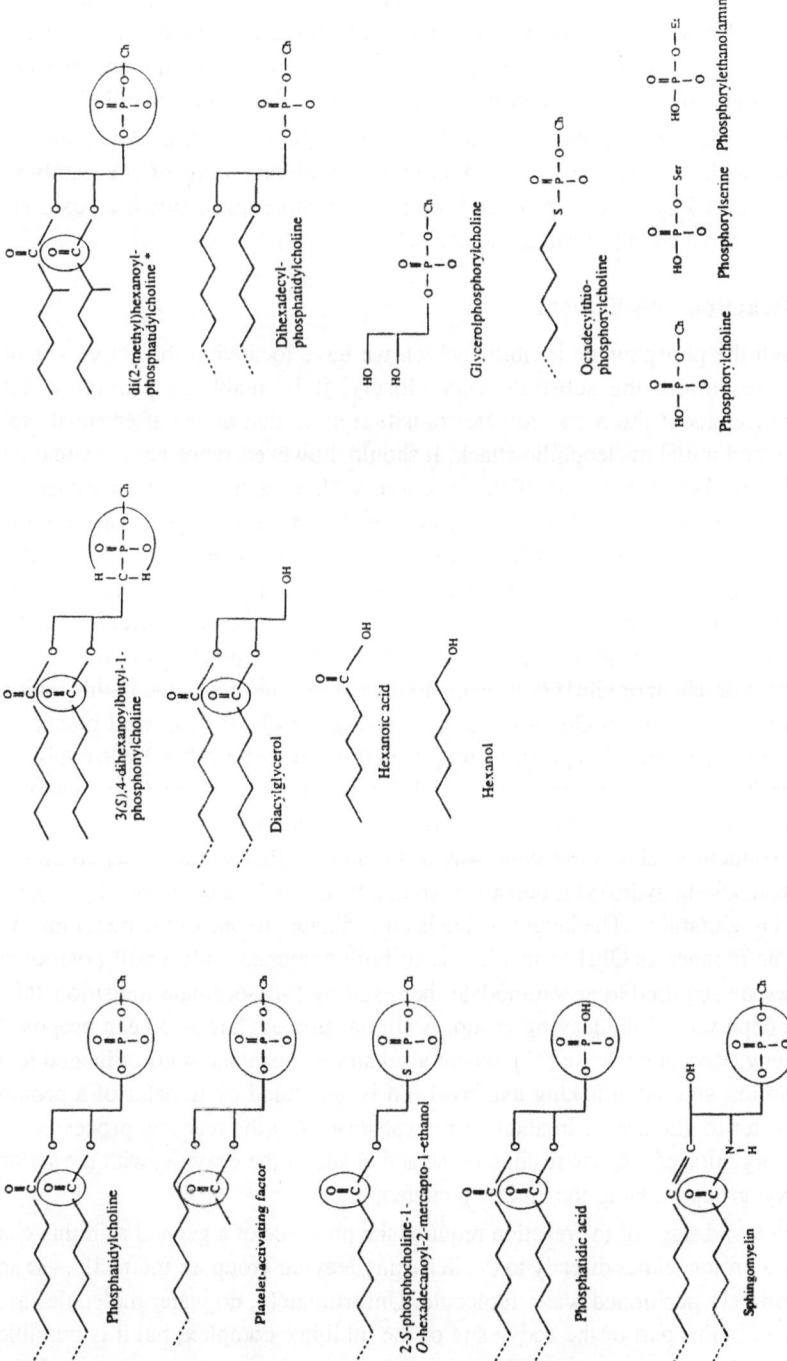

Figure 6. Survey of substrates, inhibitors and related compounds that do not bind to PLC$_{Bc}$, emphasizing features that seem to be necessary for binding to the enzyme.

The first of these results in inversion of configuration, while the latter involves pseudorotation and results in retention. However, the active site of an enzyme is relatively restricted, with fixed locations for the attacking, leaving and charge-stabilising groups, so that a mechanism involving pseudorotation is unlikely.

Both mechanisms require an attacking nucleophile and a proton that can combine with the leaving group. A complete understanding of the catalytic mechanism of PLC_{Bc} must provide both of these moieties in suitable positions with respect to a bound substrate molecule.

5.10 Reaction mechanism

Although the phosphonate inhibitor which we have located at the active site of PLC_{Bc} resembles the substrate very closely, it is unable to provide exact information about the 5-co-ordinate transition state that arises after substrate-docking and initial nucleophilic attack. It should, however, represent the situation immediately before the start of the reaction, with an activated water molecule suitably positioned for such an attack and in an apical position with respect to the DAG leaving group. It is clear from Figure 5 that, of the several water molecules in the relevant region of the active site in the inhibitor complex, the water molecule marked, although not ideally placed, is the most obvious candidate. In the inhibitor complex this water molecule is within hydrogen-bonding distance of the carboxyl side-chain of Glu146. It seems likely that a water molecule in this region will donate a proton to Glu146, thus generating a hydroxyl ion well placed to attack the substrate phosphorus atom and thus to initiate the hydrolysis. A comparable process has been proposed for DNA polymerase (34), where a phosphate oxygen atom in the substrate is co-ordinated to the two metal ions in the $3'$–$5'$ exonuclease site in the same way as O3 in our [PLC_{Bc}.inhibitor] complex, and the attacking hydroxyl group arises from a water molecule which is hydrogen-bonded to glutamine. The latter residue is co-ordinated to one of the metal ions in the same manner as Glu146 in PLC_{Bc}. In both enzymes, only small positional changes are required to accommodate the resulting 5-co-ordinate transition state before departure of the leaving group. A similar process has also been proposed for staphylococcal nuclease (71), where a substrate phosphate is co-ordinated to a calcium ion and the attacking hydroxyl ion is generated by transfer of a proton from water to glutamine. In alkaline phosphatase (40) the reaction proceeds *via* phosphorylation of a serine residue in the active site of the enzyme, with the serine hydroxyl group acting as the initiating nucleophile.

The second stage of the reaction requires the presence of a general acid that can donate a proton, either directly to the departing leaving group, or indirectly, *via* an appropriately positioned water molecule. Unfortunately, no water molecules are observed in this part of the active site of the inhibitor complex, but it is possible that the absence of water in this region may be due to the presence of the phosphonyl methylene group in a site which is occupied by the labile phosphate oxygen in the natural substrate – a methylene group is bulkier than oxygen and is also hydrophobic. Neither is there an obvious amino-acid side-chain that could act as a general acid.

It is also possible that the proton source is one of the water molecules that are ligated to Zn2 (Wat2 and Wat3 in Figure 2) and are displaced by the substrate as it enters the active site. This could occur by a mechanism where the substrate is bound and oriented by Zn1 and Zn3 in a manner similar to that observed for alkaline phosphatase (40). The proton required by the leaving group may then be donated by a water molecule that is co-ordinated to Zn2. Structural studies of other inhibitor complexes, including, it is hoped, a ligand that resembles the penta-co-ordinate transition state, are under way.

REFERENCES

1. Berridge, M.J. (1987). Inositol trisphosphate and diacylglycerol: two interacting second messengers. *Annu. Rev. Biochem.*, 56, 159–194.
2. Exton, J.H. (1990). Signalling through phosphatidylcholine breakdown. *J. Biol. Chem.*, 265, 1–4.
3. Berridge, M.J. & Irvine, R.F. (1989). Inositol phosphates and cell signalling. *Nature*, 341, 197–205.
4. Nishizuka, Y. (1986). The role of protein kinase C in cell surface signalling transduction and tumour promotion. *Nature*, 308, 693–698.
5. Nishizuka, Y. (1986). Studies and perspectives of protein kinase C. *Science*, 233, 305–312.
6. Larodera, P., Cornet, M.E., Diaz-Meco, M.T., Lopez-Barahona, M., Diaz-Laviada, I., Guddal, P.H., Johansen, T. & Moscat, J. (1990). Phospholipase C mediated hydrolysis of phosphatidylcholine is an important step in PDGF-stimulated DNA synthesis. *Cell*, 6, 1113–1120.
7. Garcia de Herreros, A., Dominguez, I., Diaz-Meco, M.T., Graziani, G., Cornet, M.E., Guddal, P.H., Johanessen, T. & Moscat, J. (1991). Requirement of phospholipase C catalysed hydrolysis of phosphorylcholine for maturation of *Xenopus laevis* oocytes in response to insulin and *ras* P21. *J. Biol Chem.* 266, 6825–6829.
8. Diaz-Laviada, I., Larrodera, P., Diaz-Meco, M.T., Cornet, M.E., Guddal, P.H., Johansen, T. & Moscat, J. (1990). Evidence for a role of phosphatidylcholine- hydrolysing phospholipase C in the regulation of protein kinase C by *ras* and *src* oncogenes. *EMBO J.*, 9, 3907–3912.
9. Levine, L., Xiaou, D.-M. & Little, C. (1988). Increased arachidonic metabolites from cells in culture after treatment with PLC$_{Bc}$. *Prostaglandins*, 34, 633–642.
10. Dominguez, I., Marshall, M.S., Gibbs, J.B., Garcia de Herreros, A., Cornet, M.E., Graziani, G., Diaz-Meco, M.T., Johansen, T., McCormick, S. & Moscat, J. (1991). Role of GTPase activating protein in mitogenic signalling through phosphatidylcholine hydrolysing phospholipase C. *EMBO. J.*, 10, 3215–3220.
11. Diaz-Meco, M.T., Dominguez, I., Sanz, L., Municio, M.M., Berra, E., Cornet, M.E., Garcia de Herreros, A, Johansen, T. & Moscat, J. (1992). Phospholipase C mediated hydrolysis of phosphatidylcholine is a target of transforming growth factor β1 inhibitory signals. *Mol. Cell Biol.*, 12, 302–308.

12. Rhee, S.G. (1991). Inositol phospholipid-specific Phospholipase C: interactions of the γ_1 iso form with tyrosine kinase. *Trends Biochem. Sci.*, 17, 297–301.
13. Clark, M.A., Shorr, R.G.L. & Bomalski, J.S. (1986). Antibodies prepared to PLC_{Bc} crossreact with a phosphatidylcholine preferring phospholipase C in mammalian cells. *Biochem. Biophys. Res. Commun.*, 140, 114–119.
14. Johansen, T., Holm, T., Guddal, P.H., Sletten, K., Haugli, F.B. & Little, C. (1988). Cloning and sequencing of the gene encoding the phosphatidylcholine-preferring phospholipase C of *Bacillus cereus*. *Gene*, 65, 293–304.
15. Little, C. & Johansen, S. (1979). Unfolding and refolding of PLC_{Bc} in solutions of guanidinium chloride. *Biochem. J.*, 179, 509–514.
16. Little, C. & Otnæss, A.-B. (1975). The metal ion dependence of PLC_{Bc}. *Biochim. Biophys. Acta*, 391, 326–323.
17. Little, C. (1981). Effect of some divalent cations on PLC_{Bc}. *Acta Chem. Scand.*, B35, 39–44.
18. Little, C. (1982). Effect of Co^{2+} substitution on the substrate specificity of PLC_{Bc} during attack on two membrane systems. *Biochem. J.*, 207, 117–121.
19. Otnaess, A.-B., Prytz, H., Bjørklid, E. & Berre, Å. (1972). PLC_{Bc} and its use in studies of tissue thromboplastin. *Eur. J. Biochem.*, 27, 238–243.
20. Vazquez-Boland, J-A., Kocks, C., Dramsi, S., Ohayon, H., Geoffroy, C., Mengaud, J. & Cossart, P. (1992). Nucleotide sequence of the lecithinase operon of *Listeria monocytogenes* and possible role of lecithinase in cell-to-cell spread. *Infect. Immun.*, 60, 219–230.
21. Yun Tso, J. & Siebel, C. (1989). Cloning and expression of the phospholipase C gene from *Clostridium perfringens* and *Clostridium bifermentans*. *Infect. Immun.*, 57, 468–476.
22. Snyder, W.R. (1987). PLC_{Bc}: carboxylic acid ester specificity and stereoselectivity. *Biochim. Biophys. Acta*, 920, 155–160.
23. Gilmore, M.S., Cruz-Rodz, A.L., Leimeister-Waechter, M., Kreft, J. & Goebel, W. (1989). A *Bacillus cereus* cytolytic determinant, Cereolysin AB, which comprises the phospholipase C and sphingomyelinase genes: nucleotide sequence and genetic linkage. *J. Bacteriol.* 171, 744–753.
24. Hough, E., Hansen, L.K., Birknes, B., Jynge, K., Hansen, S., Hordvik, A., Little, C., Dodson, E. & Derewenda, Z. (1989). High resolution (1.5 Å) crystal structure of PLC_{Bc}. *Nature*, 338, 357–360.
25. Hendrickson, W.A. (1985). Stereochemically restrained refinement of macromolecular structures. *Methods Enzymol.*, 115, 252–270.
26. Little, C. (1978). Conformational studies of PLC_{Bc}. *Biochem. J.*, 175, 977–986.
27. Volbeda, A., Lahm, A., Sakiyama, F. & Suck, D. (1991). Crystal structure of *Penicillium citrinum* P_1 nuclease at 2.8 Å resolution. *EMBO. J.*, 10, 1607–1618.
28. Guddal, P.H. personal communication.
29. Sheriff, S., Hendrickson, W. A. & Smith, J.L. (1987). Structure of myohemerythrin in the azidomet state at 1.7/1.3 Å resolution. *J. Mol. Biol.*, 197, 273–296.
30. Sowadski, J.M., Handschumaker, M.D., Krishna Murthy, H.M., Foster, B.A. & Wyckoff, H.W. (1985). Refined structure of alkaline phosphatase from *Escherichia coli* at 2.8 Å resolution. *J. Mol. Biol.*, 186, 417–433.
31. Guddal, P.H., Johansen, T., Schulstad, K. & Little, C. (1989). Apparent phosphate retrieval system in *Bacillus cereus*. *J. Bacteriol.*, 171, 5702–5706.
32. Blundell, T.L., Pitts, J.E., Tickle, I.J., Wood, S.P. & Wu, C.-W. (1981). X-ray analysis (1.4 Å) of avian pancreatic polypeptide: small globular protein. *Proc. Natl. Acad. Sci. USA*, 78, 4175–4179.
33. Burley, S.K., David, P.R., Taylor, A. & Lipscomb, W.N. (1990). Molecular structure of leucine aminopeptidase at 2.7 Å resolution. *Proc. Natl. Acad. Sci. USA*, 87, 6878–6882.
34. Nordlund, P., Sjöberg, B.-M. & Eklund, H. (1990). Three dimensional structure of the free

radical protein of ribonucleotide reductase. *Nature*, 345, 593–598.

35. Beese, L.S. & Steitz, T.A. (1991). Structural basis for the 3′–5′ exonuclease activity of Escherichia coli DNA polymerase 1: a two metal ion mechanism. *EMBO. J.*, 10, 25–33.

36. Yamashita, M.M., Almassy, R.J., Janson, C.A., Cascio, D. & Eisenberg, D. (1989). Refined atomic model of glutamine synthetase at 3.5Å resolution. *J. Biol. Chem.*, 264, 17681–17690.

37. Collier, C.A., Henrick, K. & Blow, D.M. (1990). Mechanism for the aldose–ketose interconversion by D-xylose isomerase involving ring opening followed by a 1,2-hydride shift. *J. Mol. Biol.*, 212, 211–235.

38. Hardman, K.D., Agarwal, R.C. & Freiser, M.J. (1982). Magnesium and calcium binding sites of conconavalin A. *J. Mol. Biol.*, 157, 69–86.

39. Tainer, J.A., Getzoff, E.D., Richardson, J.S. & Richardson, D.C. (1983). Structure and mechanism of copper, zinc superoxide dismutase. *Nature*, 306, 284–287.

40. Kim, E.E. & Wyckoff, H.W. (1991). Reaction mechanism of alkaline phosphatase based on crystal structures – two metal ion catalysis. *J. Mol. Biol.*, 218, 449–464.

41. Ottolenghi, A.C. (1965). PLC$_{Bc}$, a zinc requiring metalloenzyme. *Biochim. Biophys. Acta*, 106, 510–518.

42. Otnaess, A.-B., Little, C., Sletten, K., Wallin, R., Flengsrud, R. & Prydtz, H. (1977). Some characteristics of PLC$_{Bc}$. *Eur. J. Biochem.*, 79, 459–468.

43. Bicknell, R., Hanson, G.R. & Holmquist, B. (1986). A spectral study of cobalt(II)-substituted PLC$_{Bc}$. *Biochemistry*, 25, 4219–4223.

44. Aalmo, K., Hansen, L., Hough, E., Jynge, K., Krane, J., Little, C. & Storm, C.B. (1984). An anion binding site in the active site of PLC$_{Bc}$. *Biochem. Internat.*, 8, 27–33.

45. Dawson, R.M.C., Elliott, D.C., Elliott, W.H. & Jones, K.M. (1989). *Data for Biochemical Research*. Oxford: Oxford Science Publications.

46. Chakrabarti, P. (1990). Geometry of interactions of metal ions with histidine residues in protein structures. *Prot. Engin.*, 4, 57–63.

47. Chakrabarti, P. (1990). Interaction of metal ions with carboxylic and carboxamide groups in protein structures. *Prot. Engin.*, 4, 49–56.

48. Higaki, J.N., Fletterick, R.J. & Craik, C.S. (1992). Engineered metalloregulation in enzymes. *Trends Biochem. Sci.*, 17, 100–104.

49. Kurioka, S. & Matsuda, M. (1976). Phospholipase C assay using *p*-nitrophenyl-phosphorylcholine together with sorbitol and its application to studying the metal and detergent requirement of the enzyme. *Anal. Biochem.*, 75, 281–289.

50. Krug, E.L. & Kent, C. (1984). Phospholipase C from *Clostridium perfringens*: preparation and characterisation of the homogeneous enzyme. *Arch. Biochem. Biophys.*, 231, 400–410.

51. Fujimoto, M., Kuninaka, A, & Yoshino, H. (1974). Identity of phosphodiesterase and phosphomonoesterase activities with nuclease P$_1$. *Agric. Biol. Chem.*, 38, 785–790.

52. Geoffroy, C, Raveneau, J, Beretti, J.-L., Lecroisey, A., Vazquez-Boland, J.-O., Alouf, J.E. & Berche, P. (1991). Purification and characterisation of an extracellular 29-kilodalton phospholipase C from *Listeria monocytogenes*. *Infect. Immun.*, 59, 2382–2388.

53. Gerasimene, G.B., Glemzha, A.A, Kulene, V.V., Kulis, Y.Y. & Makaryunaite, Y.P. (1977). Chromatographic purification of PLC$_{Bc}$ on a monoalkylpolysaccharide adsorbent. *Biokhimya*, 42, 919–925.

54. Hansen, S. Hansen, L.K. & Hough, E. (1992). Crystal structures of phosphate, iodide and iodate inhibited PLC$_{Bc}$ and structural investigations of the binding of reaction products and a substrate analogue. *J. Mol. Biol.*, 225, 543–549.

55. Aakre, S.E. & Little, C. (1982). Inhibition of PLC$_{Bc}$ by univalent anions. *Biochem. J.*, 203, 799–801.

56. Burns, R.A., Friedman, J.M. & Roberts, M.F. (1981). Characterisation of short-chain alkyl ether lecithin analogues: ^{13}C NMR and Phospholipase studies. *Biochemistry*, 20, 5945–5950.

57. Byberg, J.R., Jørgensen, F.S., Hansen, S. & Hough, E. (1992). Substrate–enzyme interactions and catalytic mechanism in phospholipase C: a molecular modelling study using the GRID program. *Proteins*, 12, 331–338.
58. Goodford, P.J. (1985). A computational procedure for determining energetically favourable binding sites on biologically important macromolecules. *J. Med. Chem.*, 28, 849–857.
59. Thunnissen, M.M.G.M., Eiso, A.B., Kalk, K.H., Drenth, J., Dijkstra, B.W., Kuipers, O.P., Dijkman, R., de Haas, G.H. & Verheij, H.M. (1990). X-ray structure of a phospholipase A_2 complexed with a substrate derived inhibitor. *Nature*, 347, 689–691.
60. White, S.P., Scot, D.L., Otwinowski, Z., Gelb, M.H. & Sigler, P.B. (1990). Crystal structure of cobra-venom phospholipase A_2 in a complex with a transition-state analogue. *Science*, 250, 1560–1563.
61. Scott, D.L., Otwinowski, Z., Gelb, M.H. & Sigler, P.B. (1990). Crystal structure of bee-venom phospholipase A_2 in a complex with a transition state analogue. *Science*, 250, 1563–1566.
62. Wong, Y-L. & Martin, S.F. (1992). Synthesis and evaluation of phospholipid analogues as potential inhibitors for PLC_{Bc}. *J. org. chem.* (submitted)
63. Hauser, H., Pascher, I., Pearson, R.H. & Sundell, S. (1981). Preferred conformation of phosphatidylethanolamine and -choline. *Biochim. Biophys. Acta*, 650, 21–51.
64. Hough, E., Hansen, L.K., Hansen, S., Hordvik, A, Jynge, K. & Little, C. (1990). The active site in PLC_{Bc}. In *Frontiers in Drug Research*, ed B. Jensen, F.S. Jørgensen, & H. Kofoed, pp. 237–239 (Alfred Benzon Symposium 28). Copenhagen: Munksgaard.
65. Burns, R.A. & Roberts, M.F. (1989). Carbon-13 NMR studies of short chain lecithins. Motional and conformational characteristics of micellar and monomeric phospholipid. *Biochemistry*, 19, 3100–3106.
66. DeBose, C.D., Burns, R.A., Donovan, J.M. & Roberts, M.F. (1985). Methyl branching in short-chain lecithins: are both chains important for effective phospholipase A_2 activity? *Biochemistry*, 24, 1298–1306.
67. El-Sayed, M.Y., DeBose, C.D., Coury, L.A. & Roberts, M.F. (1985). Sensitivity of PLC_{Bc} to phosphatidylcholine structural modifications. *Biochim. Biophys. Acta*, 837, 325–335.
68. Little, C. (1977). PLC_{Bc}. Action on some artificial lecithins. *Acta Chem. Scand.*, B31, 267–272.
69. Gelb, M.H., Berg, O. & Jain, M.K. (1991). Qualitative and structural analysis of inhibitors of phospholipase A_2. *Curr. Opinion Struct. Biol.*, 1, 836–843.
70. El-Sayed, M.H. & Roberts, M.F. (1985). Charged detergents enhance the activity of PLC_{Bc} towards micellar short-chain phosphatidylcholine. *Biochim. Biophys Acta*, 831, 133–141.
71. Cotton, F.A., Hazen, E.E. Jr, & Legg, M.J. (1979). Staphylococcal nuclease: proposed mechanism of action based on the structure of enzyme-thymidine 3'-5'-bisphosphate-calcium ion complex at 1.5Å resolution. *Proc. Natl. Acad. Sci. USA*, 76, 2551–2555.

6.

Phospholipase A$_2$: mechanism and structure

HUBERTUS M. VERHEIJ and BAUKE W. DIJKSTRA

The lipolytic enzymes belong to a large family of enzymes that facilitate the degradation of lipids. Phospholipases and lipases are members of this family that have been investigated extensively (1). Phospholipases constitute a diverse group of enzymes that attack phospholipids; this group can be divided into two broad categories, acylhydrolases and phosphodiesterases. The nomenclature of phospholipases follows their functional specificity, as indicated in Figure 1. Lipolytic enzymes are characterised by their ability to hydrolyse aggregated (phospho)lipids with a much higher velocity than the same (phospho)lipid in its monomolecular form. This rate enhancement at lipid–water interfaces is the central

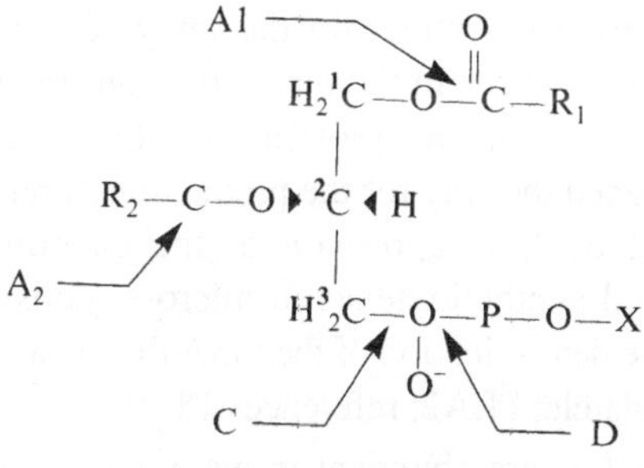

Figure 1. Cleavage sites of different phospholipases. R_1 and R_2 are alkyl chains, X represents any of the moieties (such as choline, ethanolamine *etc.*) found in 3-*sn*-phosphoglycerides. The nomenclature of phospholipases follows their functional specificity. The acyl ester bond at position 1 of 3-*sn*-phosphoglycerides is attacked by phospholipase A$_1$ (EC 3.1.1.32) and the 2-*sn* acyl ester bond by phospholipase A$_2$ (EC 3.1.1.4). Phospholipase B (EC 3.1.1.5) displays both activities of type A$_1$ and A$_2$, and it also hydrolyses lysophospholipids. Two phospholipase species attack the phosphodiester linkage, and this results either in the release of diacylglycerol by the action of phospholipase C (EC 3.1.4.3) or phosphatidic acid by phospholipase D (EC 3.1.4.4).

theme in the study of lipolysis and distinguishes (phospho)lipases from all other kinds of water-soluble enzymes, which invariably act on monomolecularly dispersed substrates.

Phospholipase A_2 (PLA2) is probably the most thoroughly characterised species of all lipolytic enzymes. The specificity of PLA2 had already been determined by 1960 (2) and, from then on, a continuing interest in this enzyme has led to a detailed knowledge of its mode of action (1, 3–6). The enzyme is strictly dependant on calcium ions and is also highly stereospecific, since only the naturally occurring L-isomers of phospholipid molecules (3-sn-phosphoglycerides) can be degraded.

Phospholipases A_2 can be encountered both inside and outside the cell. The intracellular enzymes are found in low concentrations in almost every mammalian cell (1, 7), where they may play an important rôle in membrane metabolism and in the release of arachidonic acid, serving as a precursor for the biosynthesis of molecules such as prostaglandins, leucotrienes and thromboxanes, which in their turn play a crucial part in inflammatory and allergic reactions (8, 9). Certain intracellular PLA2s also have a digestive rôle in the breakdown of phagocitosed materials (10). Currently, a keen pharmacological interest in the design and the use of specific inhibitors for these enzymes has arisen, because of their possible use in the treatment of tissue damage and inflammation. Initially, the study of intracellular PLA2s was hampered by the low abundance of the enzyme in cells and tissues, and by the lack of structural knowledge of these enzymes. However, in the last few years new perspectives for the study of this class of PLA2s have arisen, owing to recent advances in the microscale purification, characterisation and sequence-determination at the protein and cDNA levels (11–17). These developments have opened the way for the production of relatively large amounts of intracellular PLA2 by the expression of the encoding cDNA or by the expression of optimised synthetic genes in micro-organisms or eucaryotic cell lines, culminating in the determination of the three-dimensional structure of human synovial fluid PLA2 (platelet PLA2; references 18, 19).

The extracellular PLA2s are abundant in mammalian pancreatic juice and in snake and bee venom. The pancreatic enzymes are synthesized in the form of inactive precursor molecules (proPLA2s), which are stored in secretory granules. Following transport to the intestine, limited trypsinolysis occurs, which produces the active enzyme and a small activation peptide. For the snake-venom enzymes, precursor molecules have never been found. The pancreatic enzymes clearly serve a digestive function. For the PLA2s from snake and bee venom, a whole variety of functions has been described, ranging from a neurotoxic, myotoxic, anti-coagulant or hæmolytic to a digestive function (1, 20). Irrespective of the source, all extracellular PLA2s are small proteins (13–14 kDa) containing many disulphide bridges (6 or 7). A notable exception is the enzyme from honey-bee venom,

```
                    10                      20
1   A L W Q F R S M I K C A I P G S H P L M D F N N Y G C Y C
2   * * * * * N G * * * K * * S * E * * L * * * * * * * * * *
3   N * Y * * K N * * Q * T V * - * R S W W * * A D * * * * *
4   N * F * * E K L * * - K M T * K S G M L W Y S A * * * * *
5   S V L E L G K * * L - Q E T * K N A I T S Y G S * * * N *
6   S * V * * E T L * M - K * A * R S G * L W Y S A * * * * *
7   N * V N * H R * * - L T T * K E A A L S Y G F * * * H *
8   H G * * I * D R * G D N E L E E R - - - I I Y P G T L W *
                                                    * * *   *

    30                      40                          50
1   G L G G S G T P V D E L - - - - - - D R C C E T H D N C Y
2   * * * * * * * * * * D * - - - - - - * * * * Q * * * * * *
3   * R * * * * * * * D A * - - - - - - * * * * Q V * * * * *
4   * W * * Q * R * K * A T - - - - - - * * * * F V * * C * *
5   * W * H R * Q * K * A T - - - - - - * * * * F V * * C * *
6   * W * * H * L * Q * A T - - - - - - * * * * F V * * C * *
7   * V * * R * S * K * A T - - - - - - * * * * V * * * C * *
8   * H * N K S S G P N E * G R F K H T * A * * R * * * M * P
    *     *   *     *                   *     *   *     *   *

                60                      70
1   R D A K N L D S C - - - K F L V D N P Y T E S Y S Y S C S
2   K Q * * K * * * * - - - * V * * * * * * * N N * * * * * *
3   N E * E K I S G * - - - - - - W * * F K T * * * E * * *
4   G - - - K V T G * - - - - - - * * K M D I * T * * V E
5   K - - - K * T D * - - - - - - * H K * D R * * * W K
6   G - - - K A T D * - - - - - - * K * V * * T * * E E
7   K R L E K R - G * - - - - - - G T K F L * * K F * N *
8   D V M S A G E S K H G L T N T A S H T R - - - - - - - -
                    *                       *

    80                      90                      100
1   N T E I T C N S K N N A C E A F I C N C D R N A A I C F S
2   * N * * * * S * E * * * * * * * * * * * * * * * * * * * *
3   Q G T L * K G G * - * A * A V * D * * L * * * * * R
4   * G N * V * G G T * - P * K K Q * * E * * A * * * * L R R
5   * K A * I * E E * - P * L K Q M * E * * K A V * * * L R R
6   * G * * I * G G D D - P * G T Q * * E * * K A * * * * R R
7   G S R * * * A K - Q D S * R S Q L * E * * K A * * T * * A
8   - - - - - - - - - - - - - L S * D * * D K F Y D * L K
                    *                       *     *   *     *

    110                     120                     130
1   K A - P - Y N K E H K N L D T K K Y C
2   * V - * - * * * * * * * * * * * * N *
3   G * - * - * * D N D Y * I N L * A R * Q E
4   D N L L T * D - S K T Y W K Y P * N * T K E E S E P C
5   E N L D T * * - K K Y K A Y F * L K * K K - - P D T C
6   D N I P S * D - N K Y W * F P P * D * R E E - P E P C
7   R N K T T * * - K K Y Q Y Y S N * H * R G S - T P R C
8   N S - A - D T - I S S Y F V G * M * F N L I - D T K C Y K
                *                           *

          140                 150                     160
8   L E H P V T G C G E R T E G R C L H Y T V D K S K P K V Y

          170
8   Q W F D L R K Y
```

Figure 2. Sequence comparison of PLA2s. The sequences listed are: 1, pig; 2, cow; 3, *Naja naja atra*; 4, *Agkistrodon piscivorus piscivorus*, Asp49 enzyme; 5, *Agkistrodon piscivorus piscivorus* Lys49 PLA2 homologue; 6, *Crotalus atrox*; 7, human synovial fluid (platelet) PLA2; 8, bee venom. Sequences 1–6 are taken from reference 22, and sequence 7 is from references 14 and 15. The bee-venom sequence is taken from the cDNA sequence (21); it should be noted that the protein sequence starts at Ile (here number 21). Asterisks indicate identity with the sequence of the porcine pancreatic PLA2. In order to obtain a best alignment, deletions (indicated by dashes) were introduced. The asterisks below the sequences show residues that are conserved in most (95%) of the active phospholipases. The numbering used throughout this chapter is the numbering indicated in this Figure and may differ from the numbering used in the original references.

which has five disulphides and a molecular weight of 15 kDa (21). The many disulphide bridges in PLA2 are probably the reason for the great stability of these enzymes under denaturing conditions.

At the time of writing, over 80 PLA2s, comprising pancreatic, snake venom and intracellular species, have been isolated and sequenced. Their amino-acid sequences (some of them derived from cDNA sequencing) have recently been reviewed (22). All these PLA2s, with the exception of the bee-venom enzyme, show a high degree of sequence similarity, which makes it plausible that these enzymes have evolved from a common ancestor (23). In Figure 2 a limited sequence comparison is made of those PLA2s for which the three-dimensional structure has been solved. Although the bee-venom PLA2 sequence shows less amino-acid identity, its three-dimensional structure shows an active-site topology surprisingly similar to that of other PLA2s (*vide infra*). The most conspicuous difference in primary structure between the pancreatic and the snake-venom enzymes is seen in a stretch comprising residues 62–66 in the pancreatic enzymes, which is invariably absent from sequences of the snake-venom PLA2s.

The kinetic behaviour of human synovial fluid PLA2 (24) and of *Naja melanoleuca* venom PLA2 is illustrated in Figure 3, where their mode of action on a short-chain lecithin molecule, 1,2-diheptanoyl-*sn*-glycero-3-phosphocholine, is depicted. This substrate is present in its monomolecular form below the critical micelle concentration (CMC), but readily forms phospholipid aggregates, *i.e.* micelles, when it exceeds the CMC. The synovial fluid enzyme, like the porcine pancreatic PLA2 (25), only displays full activity when the substrate is presented in micellar form. In contrast the *Naja melanoleuca* venom PLA2 does not show such a biphasic curve, probably because of a lipid-induced aggregation of the enzyme (26). Similar substrate-induced aggregation processes have been reported for other PLA2s as well (27–29).

Several explanations have been proposed for the activation of the enzyme by a lipid–water interface, and these proposals have been reviewed elsewhere (4). In this chapter, the different hypotheses have been divided into two categories: substrate models and enzyme models. In the substrate models, the interfacial activation is considered to be a result of changes in the substrate itself, caused by the aggregation process. Essential features of the different substrate models are discussed in terms of: (a) the high local concentration of substrate near the active site of the enzyme, once it is adsorbed to the interface (30); (b) the restricted conformation and orientation of the substrate molecules in the interface, resulting in a lower entropy of activation for aggregated substrates relative to monodisperse substrates (5, 31, 32); (c) the dehydration of the ester bonds of lipid molecules in an interface (33–36); (d) the influence of surface effects, caused for example by the products of hydrolysis, on the adsorption/desorption rates of the enzyme, thereby regulating its activity (37).

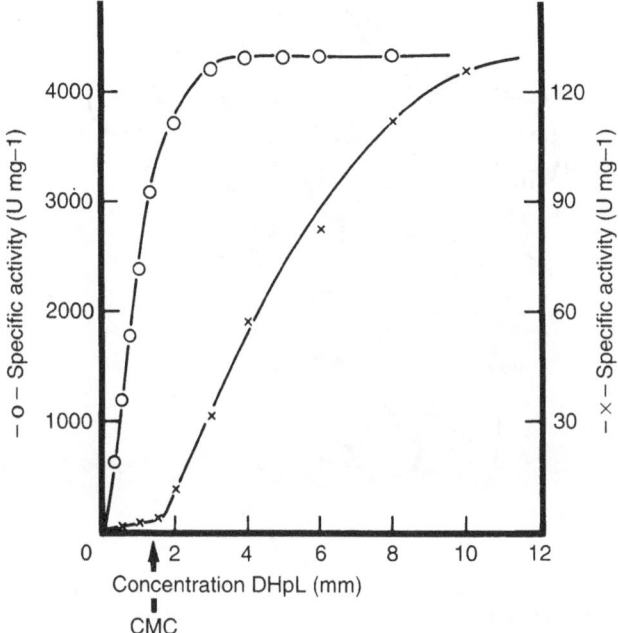

Figure 3. Hydrolysis of 1,2-diheptanoyl-*sn*-glycerophosphocholine by a recombinant human platelet (synovial fluid) PLA2 (—×—) and by PLA2 DE III from *Naja melanoleuca* venom (—○—).

Two different enzyme models have been proposed. In both, the interfacial activation is regarded as a direct consequence of the influence of the lipid–water interface upon the conformation of the enzyme. (a) Surface-induced dimerisation may result in a more favourable conformation of the enzyme (38–40). (b) In the so-called interface-recognition site (IRS) model (41, 42) it is postulated that a surface region of the PLA2 molecule, topographically distinct from the active site, is responsible for the interaction of the enzyme with organised lipid–water interfaces and that this interaction enhances the catalysis by the active site.

In this chapter, these different theories will be considered in the light of the results obtained by the combined use of recombinant DNA techniques, competitive inhibitors and X-ray crystallography.

1. THE THREE-DIMENSIONAL STRUCTURES OF SECRETED PHOSPHOLIPASES A₂

Table 1 gives an overview of the three-dimensional structures, as determined by X-ray crystallography, published up to now (5, 18, 19, 43–56). The first

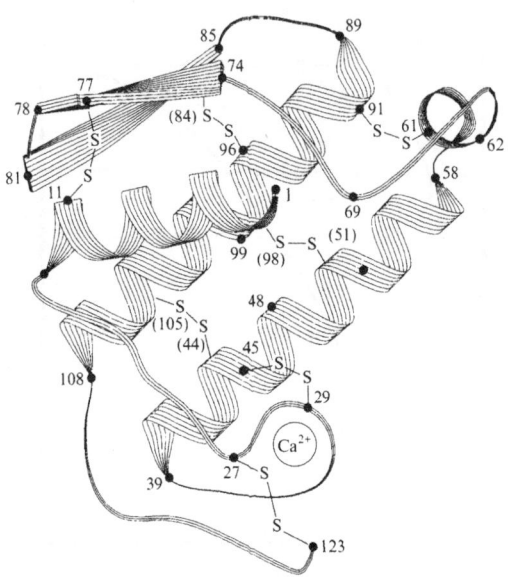

Figure 4. Schematic representation of the three-dimensional structure of bovine phospholipase A$_2$. See also Plate 5.

structure determined was that of the mature phospholipase from bovine pancreas, to a resolution of 1.7 Å with an estimated root-mean-square (r.m.s.) error in the co-ordinates of 0.12 Å. The molecule is kidney-shaped and has dimensions of approximately 22×30×42 Å. It has an appreciable content of regular secondary structure: 50% of the amino-acid residues are in an α-helical conformation and 10% in an anti-parallel, two-stranded element of β structure (see Figure 4). All the elements of secondary structure are knitted together by seven disulphide bonds. The active site is a hydrophobic pocket near His48. This histidine is hydrogen-bonded to the side-chain of Asp99. The wall of the active site is covered by residues Phe5, Ile9, Phe22, Ala102, Ala103, Phe106 and the disulphide bridge between Cys29 and Cys45. All these residues are invariant in the homologous sequences of active PLA2 species, except for an occasional replacement of Phe22 by a Tyr. The essential calcium is also bound in this pocket. It is ligated octahedrally by the carboxyl group of Asp49 and the backbone carbonyl oxygen atoms of Tyr28, Gly30 and Gly32, which form the calcium-binding loop. In the native structure, two water molecules are co-ordinated to the Ca^{2+}.

In accordance with the 85% amino-acid sequence identity, the three-dimensional structure of porcine pancreatic PLA2 is very similar to that of the bovine enzyme. The active sites are identical within the limit of error. However, a large conformational difference exists in the surface loop from residue 59 to residue 72,

Table 1. X-ray structures of phospholipases A$_2$.

Phospholipase source	Space group	Resolution	R-factor	PDB entry[†]	Crystallisation medium	Reference
Bovine pancreas, mature enzyme	$P2_12_12_1$	1.7 Å	0.171	1BP2	50% 2-methyl-2,4-pentanediol (MPD)	43
Bovine pancreas, pro-enzyme	$P3_121$	3.0 Å	0.22	2BP2	50% 2-methyl-2,4-pentanediol	44
Bovine pancreas, pro-enzyme	$P3_121$	1.6 Å	0.194	4BP2	37.5% 2-methyl-2,4-pentanediol	45
Bovine pancreas, transaminated	$P2_12_12_1$	2.1 Å	0.173	3BP2	50% 2-methyl-2,4-pentanediol	46
Bovine pancreas, *p*-bromo-phenacyl-	$P3_121$	2.5 Å	0.197	–	50% 2-methyl-2,4-pentanediol	47
Porcine pancreas, mature enzyme	$P3_121$	2.6 Å	0.241	1P2P	20–30% methanol	48
Porcine pancreas, mature enzyme	$P3_121$	2.4 Å	0.210	4P2P	50% methanol or 50% MPD	49
Porcine pancreas, Δ62–66 mutant	$P2_1$	2.1 Å	0.186	3P2P	50% methanol	50
Porcine pancreas, mutant, inhibited	$P2_12_12_1$	2.4 Å	0.189	5P2P	40% 2-methyl-2,4-pentanediol	51
Crotalus atrox venom (calcium-free)	$P2_12_12_1$	2.5 Å	0.178	1PP2	Distilled water	52,53
Naja naja atra venom	$I4$	1.5 Å	0.143	–	50% 1,4-butanediol/10% methanol	5
Naja naja atra venom, inhibited	$C222_1$	2.0 Å	0.179	–	1.4 M (NH$_4$)$_2$SO$_4$	54
Agkistrodon piscivorus piscivorus (D49K)	$P4_12_12$	2.0 Å	0.158	–	2.0–2.3 M (NH$_4$)$_2$SO$_4$	55
Bee venom, inhibited	$I4_122$	2.0 Å	0.198	–	60% saturated NaCl	56
Synovial fluid	$P2_1$	2.2 Å	0.178	–	4 M NH$_4$Cl	18
Synovial fluid	$P6_122$	2.2 Å	0.195	–	4.9 M NaCl	19
Synovial fluid, inhibited	$P4_32_12$	2.1 Å	0.204	–	4.5 M NaCl	19

[†]Identification code for co-ordinate data sets deposited with the Brookhaven Protein Data Bank.

remote from the active site. In the bovine enzyme, the first part of this loop is a stretch of α helix (residues 59–66) and the second part forms a surface loop in random-coil conformation. In the porcine enzyme, one substitution has occurred in this region: Val63 in the bovine enzyme has been replaced by a Phe. Whereas in the bovine enzyme Val63 is at the surface of the molecule, Phe63 is in the molecule's interior, probably because of its more hydrophobic character. The short α helix has disappeared in the porcine enzyme, and, instead, residues 67–71 form now a stretch of 3_{10}-helix.

The structures of pancreatic pro-phospholipase A_2 and transaminated phospholipase (with the α-amino group replaced by carbonyl), which also has zymogen-like catalytic properties, differ from the structures of the mature enzymes only in that residues 1–3 and 62–72 are disordered, and not visible in the electron density maps. The seven additional N-terminal residues in the pro-enzyme are also not visible. NMR experiments on the transaminated enzyme (46) indicate that the disorder seen in the crystal structure is due to flexibility (dynamic disorder). In the crystal structures of the mature bovine and porcine enzymes, the N terminus is hidden in the protein's interior, making hydrogen bonds with the $O^{\varepsilon 1}$ atom of Gln4, the carbonyl oxygen of residue 71 and an internal water molecule. This water molecule is, in turn, hydrogen-bonded to the side-chains of Tyr52 and Asp 99 in the central core of the enzyme molecule. If the α-NH$_3^+$ group is not present, as is the case in both the transaminated enzyme and the pro-phospholipase, these hydrogen bonds can not be made any more. As a result, the N-terminal part and the loop 62–72 lose their link to the central part of the molecule, and become flexible.

The conformational freedom of the 62–72 loop of the molecule is also apparent from the fact that 5–8 amino-acid residues are deleted in snake-venom and in synovial-fluid PLA2s (Figure 2). From the three-dimensional structure of the *C. atrox*, *N. naja atra* and synovial-fluid PLA2s, it is clear that such a deletion has no influence on the conformation of the rest of the molecule (18, 54, 57). In particular, the residues in the active site have identical configurations. Furthermore, a deletion of five residues in the porcine PLA2 appears to have only a local effect (59); the effect of this deletion on the catalytic properties will be discussed later.

The amino-acid sequence of bee-venom PLA2 differs considerably from those of the pancreatic, snake-venom and synovial-fluid PLA2s. Nevertheless, there are three stretches that show clear homology. The conserved sequence segments preserve the functionality of the active site and the calcium-binding loop. This is corroborated by the three-dimensional structure of the enzyme, from which it appears that the monomeric substrates are bound in a manner resembling their binding in the snake-venom, pancreatic and synovial-fluid enzymes.

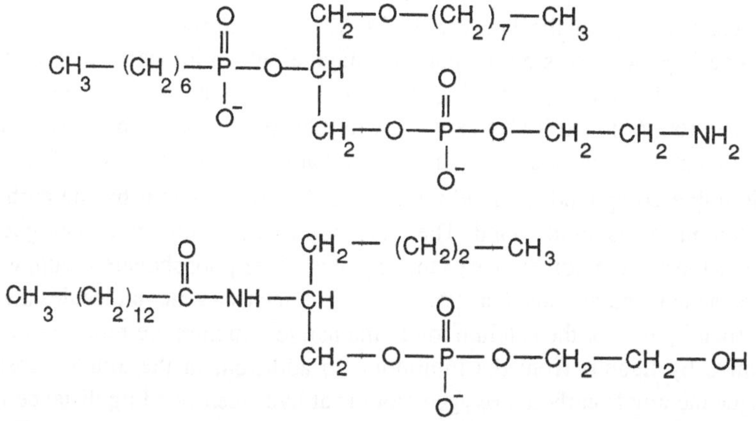

$$CH_3-(CH_2)_6-\overset{\overset{\displaystyle O}{\|}}{\underset{\underset{\displaystyle O^-}{|}}{P}}-O-\underset{\underset{\displaystyle CH_2-O-\overset{\overset{\displaystyle O}{\|}}{\underset{\underset{\displaystyle O^-}{|}}{P}}-O-CH_2-CH_2-NH_2}{\overset{\overset{\displaystyle CH_2-O-(CH_2)_7-CH_3}{|}}{CH}}}$$

$$CH_3-(CH_2)_{12}-\overset{\overset{\displaystyle O}{\|}}{C}-NH-\underset{\underset{\displaystyle CH_2-O-\overset{\overset{\displaystyle O}{\|}}{\underset{\underset{\displaystyle O^-}{|}}{P}}-O-CH_2-CH_2-OH}{\overset{\overset{\displaystyle CH_2-(CH_2)_2-CH_3}{|}}{CH}}}$$

Figure 5. The structures of two competitive inhibitors that were co-crystallised with various PLA2s.

2. THE CATALYTIC MECHANISM OF THE HYDROLYSIS OF MONOMERIC SUBSTRATES

As in the serine proteases, an Asp–His couple occurs in the active site of PLA2. This suggests that the catalytic mechanism of phospholipase might be similar to that of the serine proteases, where there are three main stages in the splitting of a peptide bond: (1) nucleophilic attack by the serine OH group; (2) proton transfer by the histidine's imidazole ring; and (3) fixation and stabilisation of the substrate's carbonyl oxygen atom by slightly positively charged NH groups. In the deacylation step, a water molecule replaces the serine OH group as nucleophile. Phospholipase A_2 has a similar configuration in its active site, but has a water molecule instead of a serine OH. In the mature bovine pancreatic PLA2, this water molecule is in the plane of the imidazole ring, 3.1 Å from the His48 $N^{\delta 1}$ atom. Thus, the ester hydrolysis by phospholipase A_2 seems to be similar to the deacylation step, which is also ester hydrolysis, in the serine proteases, the only difference being that in the latter the substrate is covalently bound to the enzyme (58).

To test this hypothesis, crystal structures of phospholipases complexed with a substrate analogue or a transition state analogue have been determined (51, 54, 56, 19). The inhibitor used in these studies was either (*R*)-2-dodecanoyl-amino-1-hexanol-phosphoglycol ('amide') or (*R*)-1-O-octyl-2-heptylphosphonyl-glycero-3-phospho-ethanolamine ('phosphonate'); see Figure 5 for structural formulae. Both inhibitors were found to be bound to the calcium ion in the active site of the phospholipase A_2 molecule by one of the free oxygen atoms of their phosphate groups, with the other free oxygen atom hydrogen-bonded to the side-chain of residue 69 (Tyr in the pancreatic enzymes, Tyr or Lys in the snake-venom enzymes, and Arg in the bee-venom PLA2). This agrees with site-directed mutagenesis experiments on the porcine enzyme, which indicated that Tyr69 is important for the precise positioning of the phospholipid substrate in the enzyme's active site (60, 61), and it also explains why phospholipase A_2 hydrolyses only one enantiomer of phospholipids that are chiral at the phosphorus atom (62).

The amide compound is, in addition, bound to the calcium by the carbonyl oxygen atom of its amide bond. The phosphonate transition-state analogue co-ordinates the calcium ion by one of the oxygens of the phosphonate group, which mimics the tetrahedral transition state. As a result, the two water molecules that are the normal ligands of the calcium ion in the native structure are replaced in both structures by ligands from the inhibitors. In addition, in the amide substrate analogue the amide carbonyl oxygen atom is at hydrogen-bonding distance from the NH group of Gly30. In the phosphonate compound, the second free oxygen atom of the phosphonate group is also at hydrogen-bonding distance from the NH group of Gly30.

The 2-*sn*-acyl chain of either inhibitor makes extensive hydrophobic contacts with the disulphide bond between residues 29 and 45, and with the side-chains of residues Leu2, Phe5, Ile9, Leu19, Phe22 and Tyr52. Figure 6 shows a schematic overview of the interactions of the 'amide' inhibitor with the enzyme. The nitrogen of this inhibitor forms a strong hydrogen bond with the $N^{\delta 1}$ atom of His48. The alkyl chains at the C1 atom of the glycerol have much less extensive interactions with the protein. All residues involved in binding are either identical with, or replaced by, an equivalent functional group in the pancreatic and snake-venom PLA2s. The conformation of the inhibitor in the active site is similar to that of natural phospholipids in aggregates as determined by X-ray crystallography and NMR in solution (63, 64). As was also found in those studies, the 2-acyl fatty-acid chain has a kink of about 90° at the α-carbon atom of the fatty acid. The active-site residues of PLA2 do not, within experimental error, change their position when the inhibitor is bound. Thus the active site has a conformation that is pre-formed to bind optimally a substrate molecule in the conformation imposed upon the phospholipid by the aggregate. In all active phospholipases known today, the active-site conformation is the most conserved part of the structure.

From these studies a picture of the catalytic events can be obtained (5, 58). The enzyme binds a substrate molecule at its active site, in which both polar interactions (phosphate–calcium, phosphate–side-chain of residue 69) and hydrophobic interactions play a rôle. After binding of the substrate, a water molecule, which is hydrogen-bonded to His48, attacks the carbonyl carbon atom of the substrate's scissile bond. A tetrahedral intermediate is formed, which is stabilised by both the calcium ion and the peptide NH group of residue 30. The tetrahedral intermediate breaks down into a fatty acid and a lysophospholipid, with His48 donating a proton.

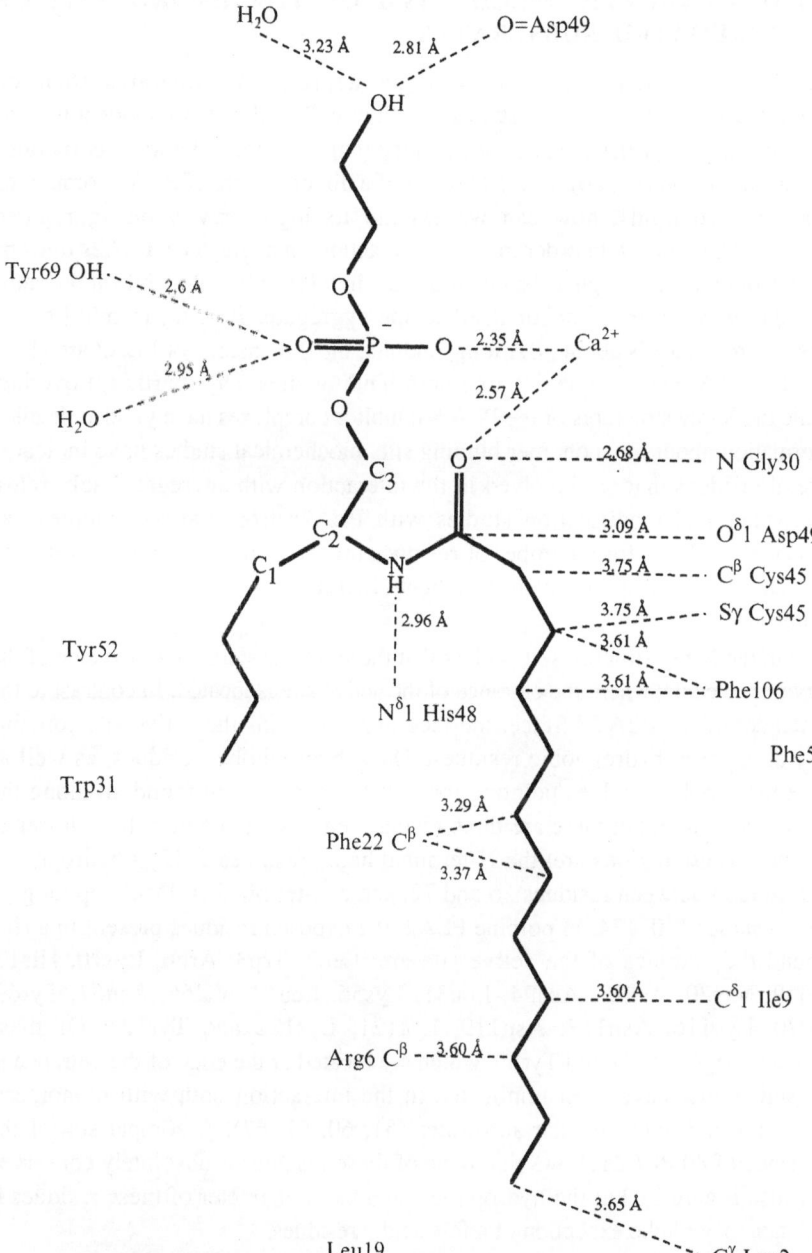

Figure 6. Schematic representation of the interaction of (*R*)-2-dodecanoyl-amino-1-hexanol-phosphoglycol with the active site of a mutant porcine pancreatic phospholipase A₂.

3. THE CATALYTIC MECHANISM OF THE HYDROLYSIS OF AGGREGATED SUBSTRATES

PLA2 has a much higher activity on aggregated substrates than on monomolecularly dispersed substrates (see Figure 3), and it is also evident that the enzyme is inhibited efficiently only when the inhibitor molecules are incorporated into the aggregate (65, 66). In the absence of a direct picture of PLA2 complexed to aggregated lipids, how can we explain its high activity on aggregated substrates? Obviously, in order to exert their action on aggregates, PLA2s must be capable of binding a single substrate molecule in their active site while at the same time the enzyme must anchor itself to the aggregate. Bearing in mind that a substrate molecule is about 30 Å long, and that the dimensions of PLA2 are about $22 \times 30 \times 42$ Å, we can imagine that both binding sites may (partially) overlap. While the X-ray structures of the PLA2–inhibitor complexes have yielded detailed information about the monomer binding site, biochemical studies have indicated several residues that are involved in the interaction with aggregated substrates. Thus, chemical modification studies with PLA2s from various sources has presented evidence for a number of residues involved in aggregate binding, and the list of residues involved has been increased further by site-directed mutagenesis.

From the X-ray structures, it is clear that these residues are all at one face of the enzyme molecule, where the entrance of the active site is located. In contrast to the remainder of the PLA2 surface, the face that surrounds the active site contains several exposed hydrophobic residues. These hydrophobic residues, as well as some more polar residues pointing towards the surface, are found all along the primary structure, but there are three distinct regions that contribute a cluster of residues. These regions are: the N-terminal helix, residues 1–12; a hydrophobic surface loop between residues 56 and 72; and a C-terminal surface loop ranging from residues 110–124. In porcine PLA2, the exposed residues present in a ring around the entrance of the active site are: Leu2, Trp3, Arg6, Lys10, His17, Leu19, Met20, Asn23, Asn24, Leu31, Lys56, Leu64, Val65, Asn67, Tyr69, Thr70, Lys116, Asn117, Asp119, Lys121, Lys122 and Tyr123. Of these residues, Arg6, Leu31 and Tyr69, which are located at the edge of the entrance to the active site, have been implicated in the interaction both with monomeric substrates and with micellar substrates (51, 60, 61, 67). A comparison of the sequence of 80 PLA2s shows that none of these residues is absolutely conserved (see also Figure 2), but the hydrophobic and basic character of these residues is maintained, with the exception of a few acidic residues.

In porcine pancreatic PLA2, several of the residues mentioned above have been modified by chemical methods, semi-synthesis and site-directed mutagenesis. Furthermore, PLA2s from other sources have been modified at corresponding sites. In most cases the result of such modifications has been a reduction of the

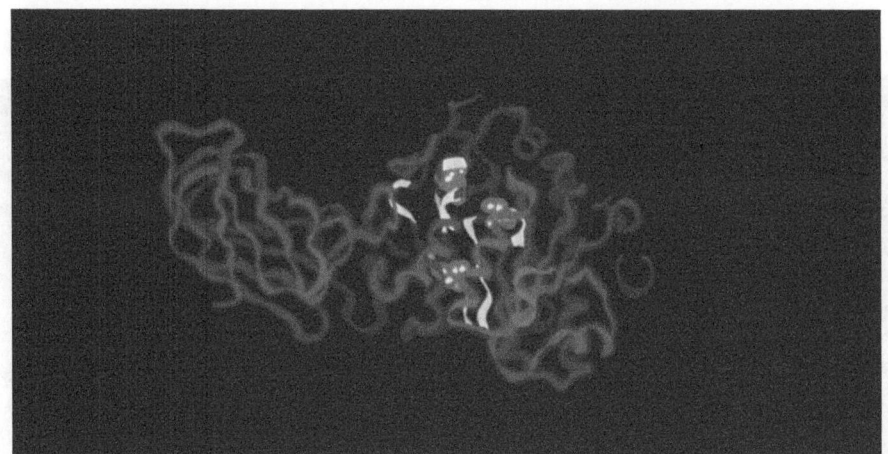

Plate 1. The conserved motifs identified by the program MULTIM in the lipoprotein, hepatic and pancreatic lipases (see Chapter 2, Figure 2) are highlighted, with the three-dimensional structure of human pancreatic lipase used as reference. Only the backbone of the protein is shown. The active-site residues are highlighted by a spherical (CPK) representation. Conserved residues in motifs identified by MULTIM are shown in yellow.

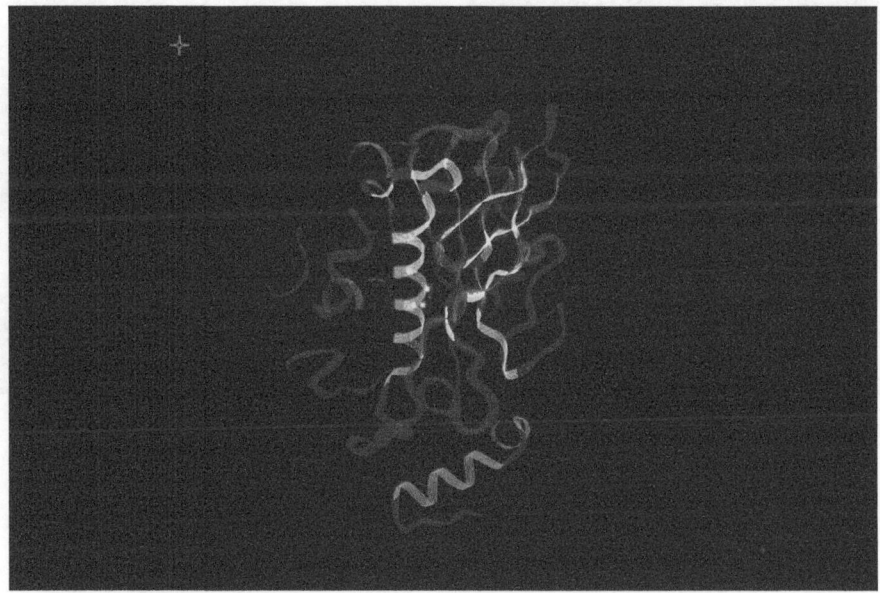

Plate 2. The conserved motifs identified by MULTIM in the esterase group of enzymes (see Chapter 2, Figure 4) are highlighted using the three-dimensional structure of acetylcholinesterase as reference. Only the backbone of the protein is shown. The active-site residues are highlighted by a spherical (CPK) representation. Conserved residues in motifs identified by MULTIM are shown in yellow.

This plate section is available for download in colour from www.cambridge.org/9780521207997

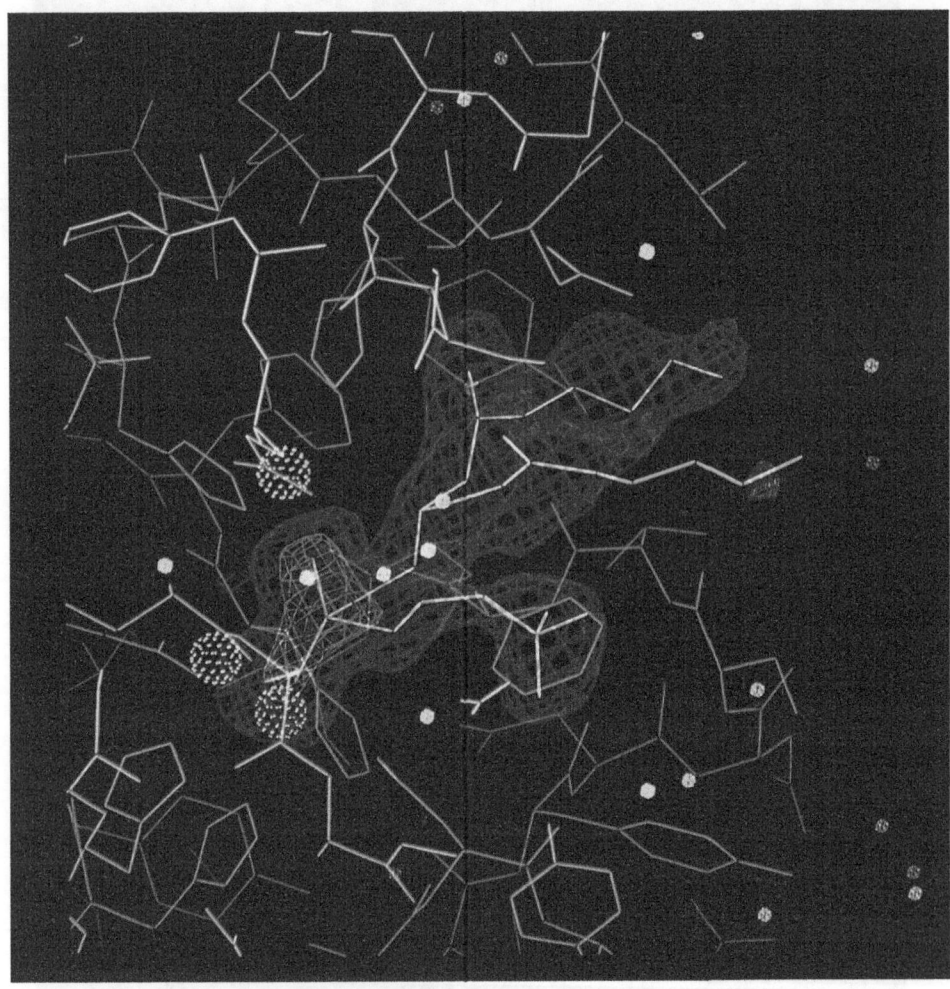

Plate 3. Difference electron density at the active site of phospholipase C (Chapter 5) after soaking it with the phosphonate inhibitor 4-dihexanoyl-oxybutyl-1-phosphonylcholine. The enzyme structure is shown in pale blue, zinc ions as large dotted spheres and water molecules as small spheres. The high-level (10σ) phosphonate electron density is shown as a red net, the lower-level (4σ) as a blue net and the phosphonate skeleton in orange.

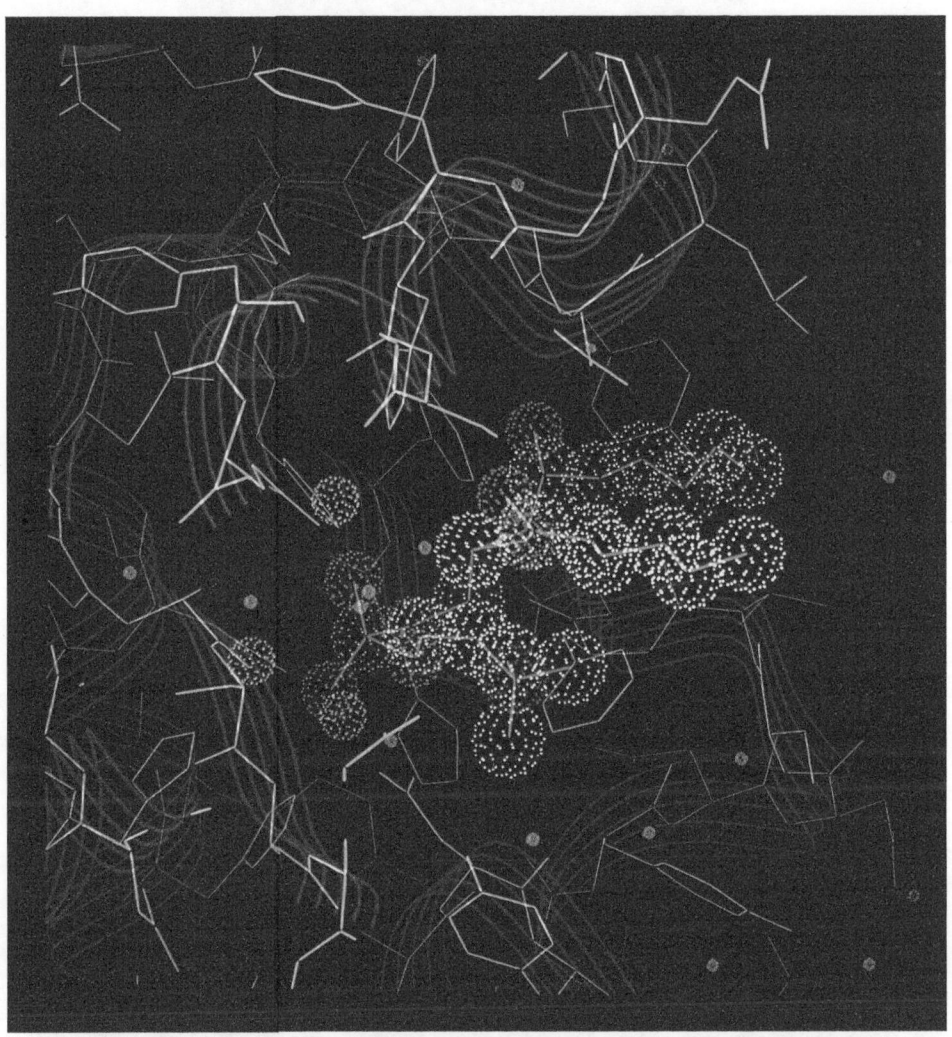

Plate 4. Space-filling representation of the phosphonate in the active-site cleft of phospholipase C (Chapter 5), viewed along the aliphatic chains of the phosphonate molecule. The protein structure is shown in yellow with the backbone emphasized as a blue ribbon. The zinc ions are green spheres and the water molecules small red spheres. In the phosphonate molecule, atoms are: oxygen, red; phosphorus, purple; nitrogen, blue; carbon, white. (Phosphonate atoms are drawn at 40% of their Van der Waals radius and zinc ions at 30%.)

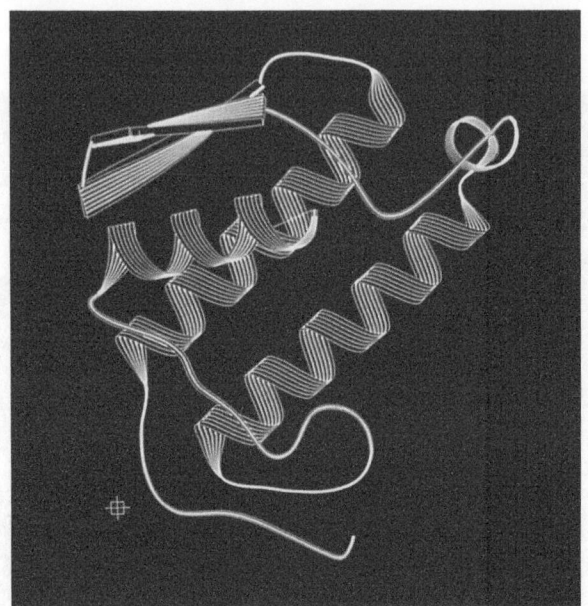

Plate 5. Schematic representation of the three-dimensional structure of bovine phospholipase A_2. See Chapter 6, in which the structure is sketched in the same orientation (Figure 4). The C terminus is at the bottom.

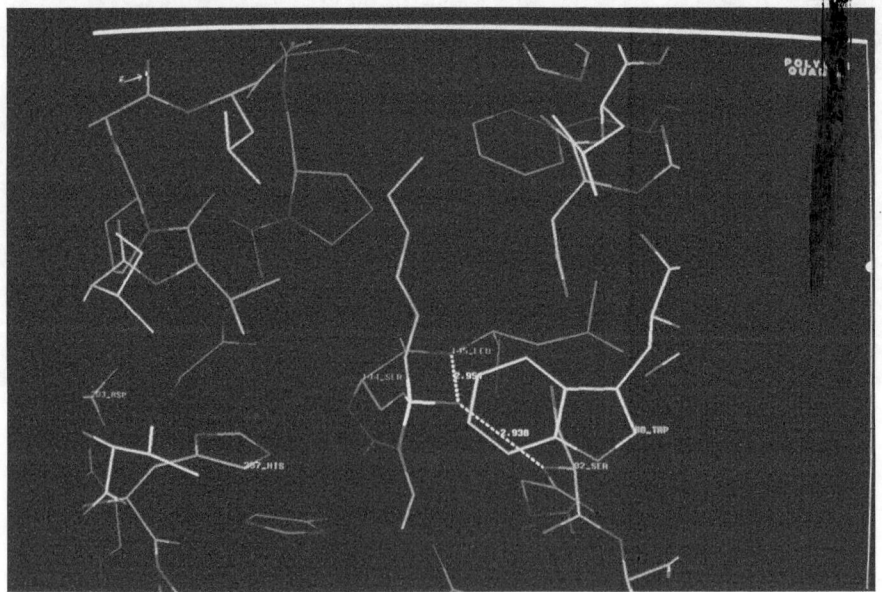

Plate 6. Skeletal representation of the X-ray structure of the inhibitor $C_6H_{13}PO(OC_2H_5)Cl$ bound covalently to the active serine of triglyceride lipase from *Rhizomucor miehei*. Compare Chapter 10, Figure 3.

enzymic activity or the interfacial binding, but in a limited number of cases the modification improved the enzyme.

Mononitration of Tyr69 in porcine, bovine and equine PLA2 (68) generated enzymes with nearly unchanged properties at low pH values (*ca.* 6) where the nitroTyr phenolic OH group is protonated. At higher pH values the ionisation of nitroTyr69 hampered the binding of monomeric as well as aggregated substrates and substrate analogues. Reduction of the nitroTyr to aminoTyr and the specific introduction of a dansyl group onto this aromatic amino group yielded highly active enzymes, which bound 10–50 times more tightly to aggregated lipids. Similarly, monodansylation of Tyr19 in equine and of Tyr123 in porcine PLA2 improved interfacial binding significantly. In an elegant series of studies by Slotboom and co-workers, a combination of selective chemical modification and semi-synthesis was employed to produce PLA2s with only one free lysine, and these were used to introduce fatty acyl chains of variable chain lengths at Lys6 and Lys116 in the porcine PLA2 and at Lys10 in bovine PLA2. In all cases, the activity of the acylated enzymes was higher than that of the respective non-acylated form, the effect being greater with increasing chain length. The affinity for aggregated substrate analogue micelles increased 10–25-fold and the capacity of the acylated enzymes to penetrate into densely packed mono- and bilayers increased significantly, although no increase in activity on monolayers was seen (69, 70). In the previous studies, acylation was brought about by specific chemical reactions, but it has been reported (40, 71) that acylation of a number of native PLA2s is connected with the catalysis of pseudosubstrates such as fatty acyl esters of hydroxy-nitrobenzoic acid and of the natural substrate egg yolk lecithin. The site of modification varies with the PLA2 source. In *Agkistrodon piscivorus* PLA2, the incorporation is one or two fatty acids per PLA2 molecule, and these fatty acyl chains are attached to Lys7 and Lys10 in the N-terminal helix. In porcine PLA2, only Lys56 was reported to have reacted. Activation was observed on monomeric substrates and on monolayers, and this activation was explained on the basis of an increased tendency of the acylated enzymes to form dimers in solution; it was suggested that this dimer is the activated species. This conclusion opposes the conclusion by Jain and co-workers (72) that PLA2s from many different sources are active as monomers in vesicular interfaces. In bovine PLA2, Lys56 has been changed into Met by site-directed mutagenesis (73) and this modification significantly improved the activity on and the affinity for micellar dioctanoyllecithin, resulting in an increase of k_{cat}/K_m by a factor of 20–25.

Although porcine and bovine PLA2s have similar K_m values for the substrate dioctanoyllecithin, direct binding techniques show that the porcine enzyme has a much greater (about 20-fold) affinity for lipid–water interfaces. Moreover, the porcine enzyme can degrade more densely packed monolayers of lecithin than the bovine enzyme can (74). Between the bovine and porcine enzyme several substitutions have occurred in the binding site for aggregated substrates: Asn6Arg, Glu17His, Leu20Met and Lys120Thr. In addition, the different conformations of the loop 59–72 (*vide infra*) in these pancreatic enzymes could affect the availability of the residues in this loop for interfacial interactions. Also the C-terminal loop has a different conformation in both enzymes, caused by a deletion in the bovine PLA2. The conformational differences between the bovine and porcine PLA2 might explain their differing affinity for aggregated substrates. On the other hand, the substitutions Asn6Arg and Glu17His make the binding site more basic in the porcine enzyme, and this suggests a simple electrostatic explanation for its tighter binding to the polar surface layer of the substrate. This is supported by the fact that a His17Asp mutation in the porcine enzyme lowers this enzyme's activity by a factor of two and reduces its affinity for micelles by a factor of three to five (H.M.V., unpublished results). Furthermore, a semi-synthetic bovine PLA2 was obtained which has an Arg at position 6 instead of an Asn (74). This modified enzyme has an affinity for aggregated substrates comparable with that of the porcine enzyme, and its activity on dioctanoyllecithin doubles as a result of this substitution. The fact that the substitution of Asn for Arg6 in porcine PLA2 does not change the properties of this enzyme underlines the fact that our understanding of the underlying principles relating structure and activity is still scanty.

To investigate the effect of a deletion of residues 62–66, which form the surface loop that is supposed to take part in interfacial recognition (see IRS model above), a mutant porcine PLA2 was made that lacked these residues (59). The rationale behind this modification is that sequences of snake-venom PLA2 show a deletion of five (Elapidae) to eight (Crotalidae) residues in this loop. Also, the snake venom PLA2s have in general higher turnover numbers and a greater affinity for phospholipid molecules aggregated in micelles than the pancreatic phospholipases do. Three substitutions were introduced to make the mutant more similar to snake-venom PLA2s in this region. The mutant PLA2 shows a twofold increase in k_{cat}/K_m for monomers whereas the increase is 28-, 19- and 4-fold for micellar dihexanoyl-, diheptanoyl- and dioctanoyllecithin, respectively, in comparison with wild-type porcine PLA2. A comparison of the three-dimensional structures of this mutant and the wild-type PLA2 by X-ray crystallography shows that the largest differences are found near the area of the deletion, residues 59–70. The most conspicuous change is around residue 63. In wild-type porcine PLA2 this residue is a phenylalanine, and it occupies a hydrophobic pocket. In the mutant, Phe63 is deleted, but the hydrophobic pocket is now filled with the disulphide bridge between Cys61 and Cys91. The C^α atom of Cys61 is thereby displaced by 4.4Å, and the side-chain of Cys91 has to rotate 180° around the C^α–C^β bond in order to make a proper disulphide bridge. The rest of Cys91 does not change, and also the hydrophobic pocket has hardly changed in conformation. The biggest change seen here is that Pro68 has moved closer to the disulphide bridge, pulling the C^α atom of Tyr69 more towards the outside of the protein; at the same time, the Tyr69 side-chain is now oriented more towards the solvent. The active-site residues are hardly affected by the mutation. The small increase in activity towards monomeric substrates could be caused by small movements of the active-site residues that are not significant in the electron-density map. Alternatively, the altered position of Tyr69, which is important for orienting the substrates' phosphate group, might explain the slight increase in activity. One possible explanation for the increased activity on aggregated substrates could be that, while in the wild-type enzyme residues 64–66 bulge out of the plane that interacts with the aggregated substrates, in the mutant this part of the molecule now forms a smooth surface. This could result in a more favourable orientation of the enzymes' active site towards the lipid aggregates. Alternatively, the more exposed orientation of the Tyr69 side-chain might facilitate interaction with the phospholipid molecules in the aggregate. Thirdly, in the mutant two Asp residues have been removed (Asp59 was changed to Ser, and Asp66 has been deleted), making the enzyme more basic and perhaps stimulating interaction with the negatively charged lipid surface. It is clear that more research is needed before a definite explanation can be given for the increase in enzymic activity in the presence of aggregated substrates.

5. CONCLUDING REMARKS

It is clear that PLA2 exerts its high activity only in the presence of aggregated substrates, and it is also clear that the enzyme is inhibited efficiently only when the inhibitor molecules are incorporated into the aggregate. Among the hypotheses that have been put forward to explain interfacial catalysis is the one that states that the enzymes changes conformation, either as a monomer or as a dimer, when it is in contact with aggregated lipids. It has been argued (5) that PLA2 is too rigid to undergo substantial conformational changes. On the other hand, the same authors (19) observed a change in the acyl-binding pocket upon binding of human synovial-fluid PLA2. It could well be that the side-chains of residues 6, 31 and 69, which are at the entrance to the active site (*vide supra*) indeed change position in the presence of aggregated substrates so as to increase the accessibility of the active site. Compared with triglyceride lipases, where a large flap has to move in order to open the way to the active-site residues (75), the movements in PLA2 are probably small. It remains difficult to ascertain whether or not this movement is accompanied by rearrangements of less than 0.5 Å of the true active site residues. In addition to the effects imposed by the enzyme's conformation, the contributions that are stressed by the various substrate models may very well reinforce the catalytic efficiency. The exchange rates of substrates and products between the interface and the active site will undoubtedly be dependent on the relative orientation of enzyme and interface, and thus on the shape and character (hydrophobic:polar ratio) of the lipid-binding domain.

In recent years, a wealth of information on the structure of PLA2, both alone and in the presence of a single strong competitive inhibitor bound to the active site, has become available. To gain information on the structure of the enzyme in the presence of aggregated lipids, the technique of choice is NMR. The current advances in NMR, and the possibilities of introducing ^{15}N and/or ^{13}C labels into PLA2 by recombinant-DNA techniques, will be key elements in future research to solve the mystery of lipolysis.

REFERENCES

1. Waite, M. (1987). *Handbook of Lipid Research*, vol. 5, ed. D.J. Hanahan. New York: Plenum Press.
2. van Deenen, L.L.M. & de Haas, G.H. (1963). The substrate specificity of phospholipase A. *Biochim. Biophys. Acta*, 70, 538–553.
3. Verheij, H.M., Slotboom, A.J. & de Haas, G.H. (1981). Structure and function of phospholipase A2. *Rev. Physiol. Biochem. Pharmacol.*, 91, 91–203.
4. Volwerk, J.J. & de Haas, G.H. (1982). Pancreatic phospholipase A2: a model for membrane-bound enzymes? In *Molecular biology of lipid–protein interactions*, ed. Griffith, O.H. & Jost, P.C., pp. 69–149. New York: Wiley.
5. Scott, D.L., White, S.P., Otwinowski, Z., Yuan, W., Gelb, M.H. & Sigler, P.B. (1990). Interfacial catalysis: the mechanism of phospholipase A2. *Science*, 250, 1541–1546.
6. Jain, M.K. & Berg, O. (1989). Kinetics of interfacial catalysis by phospholipase A2 and regulation of interfacial activation and inhibition. *Biochim. Biophys. Acta*, 1002, 127–156.
7. van den Bosch, H. (1980). Intracellular phospholipases. *Biochim. Biophys. Acta*, 604, 191–246.
8. Vadas, P. & Pruzanski, W. (1986). Biology of disease. Role of secretory phospholipases A2 in the pathobiology of disease. *Lab. Invest.*, 4, 391–404.
9. Flower, R.J. (1981). Glucocorticoids, phospholipase A2 and inflammation. *Advan. Tr. Pharmacol. Sci.*, 2, 186–188.
10. Elsbach, P. & Weiss, J. (1988). Phagocytosis of bacteria and phospholipid degradation. *Biochim. Biophys. Acta*, 947, 29–52.
11. Aarsman, A.J., de Jong, J.G.N., Arnoldussen, E., Neys, F.W., van Wassenaar, P.D. & van den Bosch, H. (1989). Immunoaffinity purification, partial sequence, and subcellular localization of rat liver phospholipase A2. *J. Biol. Chem.*, 264, 10008–10014.
12. Hayakawa, M., Kudo, I., Tomita, M., Nojima, S. & Inoue, K. (1988). The primary structure of rat platelet phospholipase A2. *J. Biochem.*, 104, 767–772.
13. Kanda, A., Ono, T., Yoshida, N., Tojo, H. & Okamoto, M. (1989). The primary structure of a membrane-associated phospholipase A2 from human spleen. *Biochim. Biophys. Res. Commun.*, 163, 42–48.
14. Seilhamer, J.J., Vadas, P., Plant, S., Millar, J.A., Kloss, J., Pruzanski, W. & Johnson, L.K. (1989). Cloning and recombinant expression of phospholipase A2 present in rheumatoid arthritic synovial fluid. *J. Biol. Chem.* 264, 5335–5338.
15. Kramer, R.M., Hession, C., Johansen, B., Hayes, G., McGray, P., Chow, E.P., Tizzard, R. & Pepinsky, R.B. (1989). Structure and properties of a human non-pancreatic phospholipase A2. *J. Biol. Chem.* 264, 5768–5775.
16. Komada, M., Kudo, I., Mizushima, H., Kitamura, N. & Inoue, K. (1989). Structure of

cDNA coding for rat platelet phospholipase A_2. *J. Biochem.*, 106, 545–547.

17. Ishizaki, J., Ohara, O., Nakamura, E., Tamaki, M., Ono, T., Kanda, A., Yoshida, N., Teraoka, H., Tojo, H. & Okamoto, M. (1989). cDNA cloning and sequence determination of rat membrane-associated phospholipase A_2. *Biochim. Biophys. Res. Commun.*, 162, 1030–1036.

18. Wery, J.-P., Schevitz, R.W., Clawson, D.K., Bobbitt, J.L., Dow, R.L., Gamboa, G., Goodson, T. Jr, Hermann, R.B., Kramer, R.M., McClure, D.B., Mihelich, E.D., Putnam, J.E., Sharp, J.D., Stark, D.H., Teater, C., Warrick, M.W. & Jones, N.D. (1991). Structure of recombinant human rheumatoid arthritic synovial fluid phospholipase A_2 at 2.2 Å resolution. *Nature*, 352, 79–82.

19. Scott, D.L., White, S.P., Browning, J.L., Rosa, J.J., Gelb, M.H. & Sigler, P.B. (1991). Structures of free and inhibited human secretory phospholipase A_2 from inflammatory exudate. *Science*, 254, 1007–1010.

20. Howard, B.D. & Gundersen, C.B. Jr (1980). Effects and mechanisms of polypeptide neurotoxins that act presynaptically. *Annu. Rev. Pharmacol. Toxicol.*, 20, 307–326.

21. Kuchler, K., Gmachl, M., Sippl, M. & Kreil, G. (1989). Analysis of the cDNA for phospholipase A_2 from honeybee venom glands. *Eur. J. Biochem.*, 184, 249–254.

22. van den Bergh, C.J., Slotboom, A.J., Verheij, H.M. & de Haas, G.H. (1989). The role of Asp-49 and other conserved amino acids in phospholipases A_2 and their importance for enzymatic activity. *J. Cellular Biochem.*, 39, 379–390.

23. Dufton, M.J. & Hider, R.C. (1983). Classification of phospholipases A_2 according to sequence. Evolutionary and pharmacological implications. *Eur. J. Biochem.*, 137, 544–551.

24. Franken, P.A., van den Berg, L., Huang, J., Gunyuzlu, P., Lugtigheid, R.B., Verheij, H.M. & de Haas, G.H. (1992). Purification and characterization of a mutant human platelet phospholipase A_2 expressed in *Escherichia coli*. cleavage of a fusion protein with cyanogen bromide. *Eur. J. Biochem.*, 203, 89–98.

25. Pieterson, W.A., Vidal, J.C., Volwerk, J.J. & de Haas, G.H. (1974). Zymogen-catalysed hydrolysis of monomeric substrates and the presence of a recognition site for lipid–water interfaces in phospholipase A_2. *Biochemistry*, 13, 1455–1459.

26. van Eijk, J.H., Verheij, H.M., Dijkman, R. & de Haas, G.H. (1983). Interaction of phospholipase A_2 from *Naja melanoleuca* snake venom with monomeric substrate analogs. Activation of the enzyme by protein–protein or lipid–protein interactions? *Eur. J. Biochem.*, 132, 183–188.

27. Roberts, M.F., Deems, R.A. & Dennis, E.A. (1977). Dual role of interfacial phospholipid in phospholipase A_2 catalysis. *Proc. Natl. Acad. Sci. USA*, 74, 1950–1954.

28. van Oort, M.G., Dijkman, R., Hille, J.D.R. & de Haas, G.H. (1983). Kinetic behaviour of porcine pancreatic phospholipase A_2 on zwitterionic and negatively charged double-chain substrates. *Biochemistry*, 22, 5353–5358.

29. van Oort, M., Dijkman, R., Deveer, A.M.Th.J., Leuveling Tjeenk, M., Verheij, H.M., de Haas, G.H., Wenzing, E. & Götz, F. (1989). Purification and substrate specificity of *Staphylococcus hyicus* lipase. *Biochemistry*, 28, 9278–9285.

30. Brockman, H.L., Law, J.H. & Kézdy, F.J. (1973). Hydrolysis of tripropionin by pancreatic lipase adsorbed to siliconized glass beads. *J. Biol. Chem.*, 248, 4965–4970.

31. Wells, M.A. (1972). A kinetic study of the phospholipase A_2 (*Crotalus adamanteus*) catalysed hydrolysis of 1, 2 dibutyryl-*sn*-glycero-3-phosphorylcholine. *Biochemistry*, 11, 1030–1041.

32. Wells, M.A. (1974). The mechanism of interfacial activation of phospholipase A_2. *Biochemistry*, 13, 2248–2257.

33. Brockerhoff, H. (1968). Substrate specificity of pancreatic lipase. *Biochim. Biophys. Acta*, 159, 296–303.

34. Brockerhoff, H. & Jensen, R.G. (1974). *Lipolytic Enzymes*. New York: Academic Press.

35. Allgyer, T.T. & Wells, M.A. (1979). Thermodynamic model for micelle formation by phosphatidylcholines containing short-chain fatty acids. Correlation with physical chemical data and the effects of concentration on the activity of phospholipase A_2. Biochemistry, 18, 4354–4361.
36. Jain, M.K., Rogers, J. & de Haas, G.H. (1988). Kinetics of binding of phospholipase A_2 to lipid/water interfaces and its relationship to interfacial activation. Biochim. Biophys. Acta, 940, 51–62.
37. Tinker, D.O., Law, R. & Lucassen, M. (1980). Heterogeneous catalysis by phospholipase A_2: mechanism of gel phase phosphatidylcholine. Can. J. Biochem., 58, 898–912.
38. Dennis, E.A. (1973). Phospholipase A_2 activity towards phosphatidylcholine in mixed micelles: surface dilution kinetics and the effect of thermotropic phase transition. Arch. Biochim. Biophys., 158, 485–493.
39. Bell, J.D. & Biltonen, R.L. (1989). The temporal sequence of events in the activation of phospholipase A_2 by lipid vesicles. J. Biol. Chem., 264, 12194–12200.
40. Cho, W., Tomasselli, A.G., Heinrikson, R.L. & Kézdy, F.J. (1988). The chemical basis for interfacial activation of monomeric phospholipases A_2. J. Biol. Chem., 263, 11237–11241.
41. Verger, R., Mieras, M.C.E. & de Haas, G.H. (1973). Action of phospholipase A_2 at interfaces. J. Biol. Chem., 248, 4023–4034.
42. Verger, R. & de Haas, G.H. (1976). Interfacial enzyme kinetics of lipolysis. Annu. Rev. Biophys. Bioengin., 5, 77–119.
43. Dijkstra, B.W., Kalk, K.H., Hol, W.G.J. & Drenth, J. (1981). Structure of bovine pancreatic phospholipase A_2 at 1.7 Å resolution. J. Mol. Biol., 147, 97–123.
44. Dijkstra, B.W., van Nes, G.J.H., Kalk, K.H., Brandenburg, N.P., Hol, W.G.J. & Drenth, J. (1982). The structure of bovine pancreatic prophospholipase A_2 at 3.0 Å resolution. Acta Crystallogr., B38, 793–799.
45. Finzel, B.C., Weber, P.C., Ohlendorf, D.H. & Salemme, F.R. (1991). Crystallographic refinement of bovine pro-phospholipase A_2 at 1.6 Å resolution. Acta Crystallogr., B47, 814–816.
46. Dijkstra, B.W., Kalk, K.H., Drenth, J., de Haas, G.H., Egmond, M.R. & Slotboom, A.J. (1984). Role of the N-terminus in the interaction of pancreatic phospholipase A_2 with aggregated substrates. Properties and crystal structure of transaminated phospholipase A_2. Biochemistry, 23, 2759–2766.
47. Renetseder, R., Dijkstra, B.W., Huizinga, K., Kalk, K.H. & Drenth, J. (1988). Crystal structure of bovine pancreatic phospholipase A_2 covalently inhibited by p-bromophenacylbromide. J. Mol. Biol., 200, 181–188.
48. Dijkstra, B.W., Renetseder, R., Kalk, K.H., Hol, W.G.J. & Drenth, J. (1983). Structure of porcine pancreatic phospholipase A_2 at 2.6 Å resolution and comparison with bovine phospholipase A_2. J. Mol. Biol., 168, 163–179.
49. Finzel, B.C., Ohlendorf, D.H., Weber, P.C. & Salemme, F. (1991). An independent crystallographic refinement of porcine phospholipase A_2 at 2.4 Å resolution. Acta Crystallogr., B47, 558–559.
50. Thunnissen, M.M.G.M., Kalk, K.H., Drenth, J. & Dijkstra, B.W. (1990). Structure of an engineered porcine phospholipase A_2 with enhanced activity at 2.1 Å resolution. Comparison with the wild-type porcine and Crotalus atrox phospholipase A_2. J. Mol. Biol., 216, 425–439.
51. Thunnissen, M.M.G.M., AB, E., Kalk, K.H., Drenth, J., Dijkstra, B.W., Kuipers, O.P., Dijkman, R., de Haas, G.H. & Verheij, H.M. (1990). X-ray structure of phospholipase A_2 complexed with a substrate-derived inhibitor. Nature, 347, 689–691.
52. Keith, C., Feldman, D.S., Deganello, S., Glick, J., Ward, K.B., Jones, O. & Sigler, P.B. (1981). The 2.5 Å crystal structure of a dimeric phospholipase A_2 from the venom of Crotalus atrox. J. Biol. Chem., 256, 8602–8607.

53. Brunie, S., Bolin, J., Gewirth, D. & Sigler P.B. (1985). The refined crystal structure of dimeric phospholipase A_2 at 2.5Å. Access to a shielded catalytic center. *J. Biol. Chem.*, 260, 9742–9749.
54. White, S.P., Scott, D.L., Otwinowski, Z., Gelb, M.H. & Sigler, P.B. (1990). Crystal structure of cobra-venom phospholipase A_2 in a complex with a transition-state analogue. *Science*, 250, 1560–1563.
55. Holland, D.R., Clancy, L.L., Muchmore, S.W., Ryde, T.J., Einspahr, H.J., Finzel, B.C., Heinrikson, R.L. & Watenpaugh, K.D. (1990). The crystal structure of a lysine 49 phospholipase A_2 from the venom of the cottonmouth snake at 2.0Å resolution. *J. Biol. Chem.*, 265, 17649–17656.
56. Scott, D.L., Otwinowski, Z., Gelb, M.H. & Sigler, P.B. (1990). Crystal structure of bee-venom phospholipase A_2 in a complex with a transition-state analogue. *Science*, 250, 1563–1566.
57. Renetseder, R., Brunie, S., Dijkstra, B.W., Drenth, J. & Sigler, P.B. (1985). A comparison of the crystal structures of phospholipase A_2 from bovine pancreas and *Crotalus atrox* venom. *J. Biol. Chem.*, 260, 11627–11634.
58. Verheij, H.M., Volwerk, J.J., Jansen, E.H.J.M., Puijk, W.C., Dijkstra, B.W., Drenth, J. & de Haas, G. H. (1980). Methylation of histidine-48 in pancreatic phospholipase A_2. Role of histidine and calcium ion in the catalytic mechanism. *Biochemistry*, 19, 743–750.
59. Kuipers, O.P., Thunnissen, M.M.G.M., de Geus, P., Dijkstra, B.W., Drenth, J., Verheij, H.M. & de Haas, G.H. (1989). Enhanced activity and altered specificity of phospholipase A_2 by deletion of a surface loop. *Science*, 244, 82–85.
60. Kuipers, O.P., Dijkman, R., Pals, C.E.G.M., Verheij, H.M. & de Haas, G.H. (1989). Evidence for the involvement of Tyr-69 in the control of stereospecificity of porcine pancreatic phospholipase A_2. *Prot. Engin.*, 2, 467–471.
61. Kuipers, O.P., Dekker, N., Verheij, H.M. & de Haas, G.H. (1990). The activity of native and Tyr-69 porcine pancreatic phospholipase A_2 mutants on phospholipid analogs. A re-evaluation of the minimal substrate requirements. *Biochemistry*, 29, 6094–6102.
62. Bruzik, K., Jiang, R.-T. & Tsai, M.-D. (1983). Phospholipids chiral at phosphorous. Preparation and spectral properties of chiral thiophospholipids. *Biochemistry*, 22, 2478–2486.
63. Hitchcock, P.B., Mason, R., Thomas, K.M. & Shipley, G.G. (1974). Structural chemistry of 1,2 dilauroyl-DL-phosphatidylethanolamine: molecular conformation and intermolecular packing of phospholipids. *Proc. Natl. Acad. Sci. USA*, 71, 3036–3040.
64. Seelig, J. & Browning, J.L. (1978). General features of phospholipid conformation in membranes. *FEBS Letters*, 92, 41–44.
65. De Haas, G.H., Dijkman, R., Ransac, S. & Verger, R. (1990). Competitive inhibition of lipolytic enzymes. IV. Structural details of acylamino phospholipid analogues important for the potent inhibitory effects on pancreatic phospholipase A_2. *Biochim. Biophys. Acta*, 1046, 249–257.
66. Yuan, W., Quinn, D.M., Sigler, P.B. & Gelb, M.H. (1990). Kinetic and inhibition studies of phospholipase A_2 with short-chain substrates and inhibitors. *Biochemistry*, 29, 6082–6094.
67. Kuipers, O.P., Kerver, J., van Meersbergen, J., Vis, R., Dijkman, R., Verheij, H.M. & de Haas, G.H. (1990). Influence of size and polarity of residue 31 in porcine pancreatic phospholipase A_2 on catalytic activity. *Prot. Engin.*, 3, 599–603.
68. Meyer, H., Puijk, W.C., Dijkman, R., Foda-van der Hoorn, M.M.E.L., Pattus, F., Slotboom, A.J. & de Haas, G.H. (1979). Comparative studies of tyrosine modification in pancreatic phospholipases. 2. Properties of nitrotyrosyl, aminotyrosyl, and dansylaminotyrosyltyrosyl derivatives of pig, horse and ox phospholipases A_2 and their zymogens. *Biochemistry*, 16, 3589–3597.
69. van der Wiele, F.Chr., Atsma, W., Roelofsen, B., van Linde, M., van Binsbergen, J.,

Radvany, F., Raykova, D., Slotboom, A.J. & de Haas, G.H. (1988). Site-specific ε-NH$_2$ mono-acylation of pancreatic phospholipase A$_2$. Transformation of soluble phospholipase A$_2$ into a highly penetrating 'membrane-bound' form. *Biochemistry*, 27, 1688–1694.

70. van Binsbergen, J. (1990). Protein engineering of pancreatic phospholipase A$_2$ by enzymatic semisynthesis. Ph.D. Thesis, University of Utrecht, The Netherlands.

71. Tomasselli, A.G., Hui, J., Fisher, J., Zürcher-Neely, H., Reardon, I.M., Oriaku, E., Kézdy, F.J. & Heinrikson, R.L. (1989). Dimerization and activation of porcine pancreatic phospholipase A$_2$ *via* substrate level acylation of lysine 56. *J. Biol. Chem.*, 264, 10041–10047.

72. Jain, M.K., Ranadive, G., Yu, B.-Z. & Verheij, H.M. (1991). Interfacial catalysis by phospholipase A$_2$: monomeric enzyme is fully catalytically active at the bilayer interface. *Biochemistry*, 30, 7330–7340.

73. Noel, J.P., Deng, T., Hamilton, K.J. & Tsai, M.-D. (1990). Phospholipase A$_2$ engineering. 3. Replacement of Lysine-56 by neutral residues improves catalytic potency significantly, alters substrate specificity, and clarifies the mechanism of interfacial recognition. *J. Am. Chem. Soc.*, 112, 3704–3706.

74. van Scharrenburg, G.J.M., Puijk, W.C., de Haas, G.H. & Slotboom, A.J. (1983). Semisynthesis of phospholipase A$_2$. The effect of substitution of amino-acid residues at positions 6 and 7 in bovine and porcine pancreatic phospholipases A$_2$ on catalytic and substrate binding properties. *Eur. J. Biochem.*, 133, 83–89.

75. Brzozowski, A.M., Derewenda, U., Derewenda, Z.S., Dodson, G.G., Lawson, D.M., Turkenburg, J.P., Björkling, F., Huge-Jensen, B., Patkar, S.A. & Thim, L. (1991). A model for interfacial activation in lipases from the structure of a fungal lipase-inhibitor complex. *Nature*, 351, 491–494.

7.

Structure and mechanism of human pancreatic lipase

FRITZ K. WINKLER and KLAUS GUBERNATOR

Pancreatic lipase (PL) hydrolyses the water-insoluble triacylglycerols in the intestinal lumen and thereby plays an important rôle in dietary fat absorption. Many of the pioneering biochemical studies in the field of interfacial enzymology have been carried out with porcine pancreatic lipase, and its chemical and functional characteristics have been the focus of several reviews on lipases (1–3). Pancreatic lipase has a molecular weight of approximately 50 kDa and has been isolated from a number of mammalian species. The amino-acid sequences of the human, pig, rat, dog and mouse enzymes show clear homology. Triglyceride hydrolysis by PL is inhibited by physiological concentrations of bile salts. This inhibition can be overcome by the addition of colipase, a small pancreatic protein that binds to the lipase and to lipid micelles. It is thought that colipase is needed to anchor pancreatic lipase to micelles containing bile salt in the intestinal milieu (for a general review on pancreatic colipase, see reference 4). Inhibition of the human enzyme is of medical interest, as a reduction of fatty-acid adsorption is potentially useful for the treatment of obesity. Tetrahydrolipstatin (THL; see below, Figure 6), a hydrogenated derivative of lipstatin originally isolated from *Streptomyces toxytricini* (5), is a potent inhibitor of pancreatic lipase and is currently being evaluated in clinical trials.

Despite extensive biochemical studies (2, 6), the function of pancreatic lipase remained poorly understood at the molecular level until quite recently. A major step forward was made when the crystal structure of human pancreatic lipase, hPL, was determined (7). It has led to the identification of the catalytic residues and to an understanding of the essential features and stereochemistry of the hydrolytic mechanism, and to the postulate that a major conformational change occurs upon activation of the enzyme at the interface. In addition, it is found that the three lipases whose three-dimensional structures are known, hPL and the fungal lipases from *Rhizomucor miehei*, RmL (8) and *Geotrichum candidum*, GcL (9), all share structural features that could not at present have been predicted from a comparison of their primary structures. New functional insights gained for one lipase may therefore be of relevance for other lipases as well. Knowledge of their structures is, however, a prerequisite for the establishment of such general structure–function relationships.

Many important questions concerning the activation and action of hPL at the substrate interface remain to be answered. However, the structure provides a framework to define some of these questions more precisely, and leads naturally to the design of experiments to resolve them. In this chapter, we review our current understanding of the hydrolytic mechanism of pancreatic lipase and address some of the open questions with regard to interfacial activation and substrate recognition.

1. OVERALL STRUCTURE

The crystal structure of human pancreatic lipase (hPL) has been refined to a resolution of 2.3 Å (7, 10). Its amino-acid sequence, comprising 449 residues for the mature enzyme, was deduced from the cloned cDNA sequence (7) and has 85% identity with that of porcine pancreatic lipase (pPL). Two independent copies of the lipase are present in the asymmetric unit of the monoclinic crystals. They are related by a non-crystallographic twofold axis and have essentially identical structures. Figure 1 gives a schematic representation of the overall protein structure and indicates the approximate location of the major functional sites. The protein is folded into two domains, a larger N-terminal domain (N domain) comprising residues 1–335 and a smaller C-terminal domain (C domain). The core of the N domain is formed by a nine-stranded β-pleated sheet in which most of the strands run parallel to one another. Seven α-helical segments of varying length occur in the strand connections, and six of them pack against the two faces of the core sheet. The C domain is formed by two layers of antiparallel sheet, the strands of which are connected by loops of varying length.

The N domain contains the active site, a glycosylation site, a Ca^{2+}-binding site and possibly a heparin-binding site. In the structure of crystalline hPL, the active site is buried beneath a surface loop, termed the flap. In this form, the enzyme cannot be enzymically active, and we shall refer to it as the closed or inactive state of the enzyme. It is probable that the structure observed in the crystalline state represents the structure of the protein in aqueous solution, that is, when it is not bound to a substrate interface. The porcine enzyme has been shown to be glycosylated at Asn166, and the glycan's primary structure has been determined (11). The human enzyme is also glycosylated (12), but the attachment site has not been determined. In the structure of hPL, Asn166 lies exposed at the surface, and significant electron density is seen extending from it. As is often observed with glycoproteins, the carbohydrate is mostly disordered, and the observed density accounts for only the first two or three sugar residues. The glycosylation site is remote from the active site, and the carbohydrate is unlikely to have significant

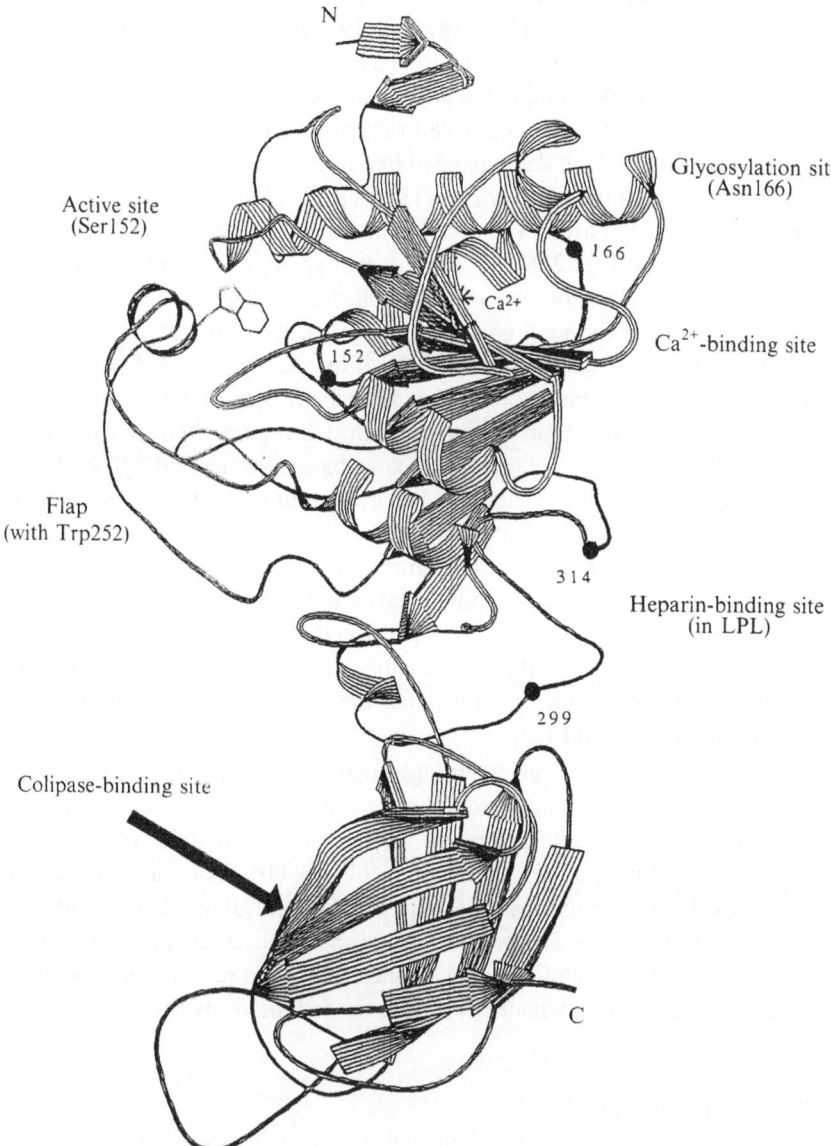

Figure 1. Schematic ribbon diagram drawn using the program RIBBON (47) of the structure of human pancreatic lipase, with indications of the approximate location of functional sites. The substrate interface needs to be on the left of the molecule in order to contact both the active-site region and the bound colipase. The Ca^{2+}-binding site is on the rear side of the molecule and is marked by an asterisk (partly hidden). Residues 299 and 314 mark the positions of two segments that are thought to be involved in heparin-binding in the homologous lipoprotein lipases.

influence on the catalytic activity. The Ca^{2+}-binding site is formed by a short segment of chain between residues 187 and 195. It provides two main-chain oxygen atoms and the two carboxylate groups of Asp192 and Asp195 as ligands for the divalent cation. Again, this site is rather remote from the active site, and it appears to fulfil a purely structural rôle in stabilising the conformation of this surface segment. It has been demonstrated that intestinal heparin binds hPL in a receptor-like manner at the brush border (13). Heparin-binding is well established for the homologous lipoprotein lipase (LPL), and conserved basic amino-acid residues between residues 278 and 306 are thought to participate in heparin-binding (14). This region corresponds to residues 299–330 in hPL. It represents the C-terminal part of the long connection between the two last strands of the central β sheet, and it contains two of the three disulphide bridges that are conserved among pancreatic, lipoprotein and hepatic lipases in the N domain. If it is supposed that LPL has a similar structure in this region, then two segments with exposed basic residues (279–282 with three basic residues and 292–297 with four) appear well suited to contribute to heparin-binding by LPL. Seen from the active site, these segments are located on the opposite face of the N domain. In this way, the enzyme can bind at the same time to two surface structures, to heparan sulphate on the surface of endothelial cells and to a substrate interface. In hPL, however, only three basic residues are found in the corresponding region, and only two of these are in equivalent positions in the second basic segment of LPL. It therefore appears uncertain whether the location of this binding site is conserved between PL and LPL.

Biochemical evidence suggests strongly that the C-terminal, chymotryptic fragment of pPL, corresponding to domain C, contains the colipase-binding site (15). This has now been confirmed by the very recent determination of the crystal structure of the complex of hPL and human colipase (16). The same proteolytic fragment was also shown to have a site capable of hydrolysing the activated ester *p*-nitrophenyl acetate (17). The presence of this non-conserved site in pPL, now known to be functionally non-essential, is one of the main reasons why there has long been some doubt as to whether Ser152 is needed for catalysis.

2. THE ACTIVE SITE

The presence of a trypsin-like Asp...His...Ser constellation in hPL, including Ser152, which is known from chemical modification studies to be essential for triglyceride hydrolysis (18), was strong evidence that this represented the hydrolytic site of the enzyme. This is now established beyond doubt through experimental results obtained with hPL itself (10, 19) as well as from structural and functional studies with other lipases that share the same hydrolytic site with

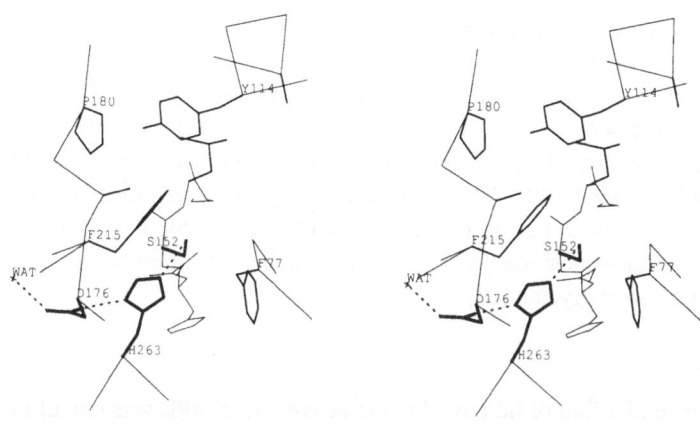

Figure 2. Stereo diagram showing the residues (very thick lines) and the hydrogen bonds (dotted lines) of the catalytic triad in the crystal structure of human pancreatic lipase together with non-polar residues in its immediate vicinity (medium thick lines). Flap residues have been omitted. Except for the chain segment around the essential Ser152, main-chain segments are represented by C^α positions only (thin lines).

the essential serine at the centre of the structurally homologous consensus peptide GX_1SX_2G. In Figure 2, the hydrogen-bonding network of the catalytic triad is shown together with residues in its immediate vicinity, which are almost exclusively non-polar (Phe77, Tyr114, Ala117, Ile153, Ala178, Pro180, Phe215). As discussed later, comparison of the geometry of the hPL triad (Asp176...His263...Ser152) with that of trypsin-like proteases reveals striking similarities, but also important differences. The second carboxylate oxygen atom of Asp176 is within hydrogen-bonding distance of two main-chain nitrogen atoms and one internal water molecule. It is not connected to bulk water through a chain of hydrogen bonds.

As first noticed by Henderson (20) for chymotrypsin, another essential feature of the active site of serine proteases is the presence of the so-called oxyanion hole. Two hydrogen-bond donors are oriented in such a way as to stabilise the oxyanion of the tetrahedral intermediate. Identification of two such hydrogen-bond donors in hPL, expected by analogy with the serine proteases, was not straightforward. However, modelling studies with substrate and inhibitor molecules (section 4), and the structural information now available for other lipases make it almost certain that they are provided by the main-chain nitrogen atoms of residues 77 and 153. The NH of residue 153 is at the N terminus of an α helix, which makes it an ideal donor for a strong hydrogen bond to an anion.

The lipase consensus peptide forms a tight turn between the C terminus of the central strand of the core sheet and the N terminus of an α helix. The turn (residues 151–154) can be classified as II′ (21), and residue 152 is in the rare ε conformation ($\phi = 60°$, $\psi = -120°$). As this structural motif has turned out to be a characteristic feature of lipases and other hydrolytic enzymes, it has been termed the β-εSer-α motif (22). The two invariant glycine residues flanking the pentapeptide appear to be conserved because of unavoidable steric conflicts with side-chains in these positions, and not because their ϕ, ψ angles are disallowed for residues other than glycine (22).

3. THE FLAP

The presence of a flap or lid covering the active site of hPL was one of the most interesting initial observations (7). It suggested immediately that a large conformational change would have to take place before the substrate could diffuse into the active site, and that this might be related to the phenomenon of interfacial activation. At the same time, a similarly positioned lid was reported for the structure of RmL and suggested the same general conclusion (8). In hPL, the flap is formed by the chain segment between the two disulphide-linked residues 237 and 261 (Figure 1). The residues of this segment interact with the rest of the N domain mainly through hydrophobic contacts, and only three hydrogen-bond interactions are observed. Trp252, located in the short α helix at the very tip of the loop, is embedded directly over the essential serine, in a hydrophobic pocket formed by the side-chains of Phe77, Ala117, Leu153 and Phe215. Its N^ε atom makes a hydrogen bond to the main-chain oxygen atom of residue 112 and possibly to the O^γ atom of Ser110. Another hydrogen bond linking the flap with the chain segment 110–114, located just in front of the long helix α3, occurs between Asp249 and the main-chain nitrogen atom of residue 114. Further away from the active site, extensive hydrophobic contacts are made by the flap residues Ile248, Ile251 and Phe258. Their side-chains pack against each other and against hydrophobic side-chains of the segment 209–215 (Ile209, Val210, Leu213 and Phe215).

Although it is certain that at least the tip of the loop, the short helix with Trp252, must reorient upon activation of the enzyme, there is at present no direct information about how this might occur. Attempts in the authors' laboratory to crystallise the enzyme complexed with inhibitors that would force Trp252 out of the way have not yet been successful. After reorientation of the flap, additional conformational adjustments of other residues in the active-site region appear necessary to generate enough room for a triglyceride molecule.

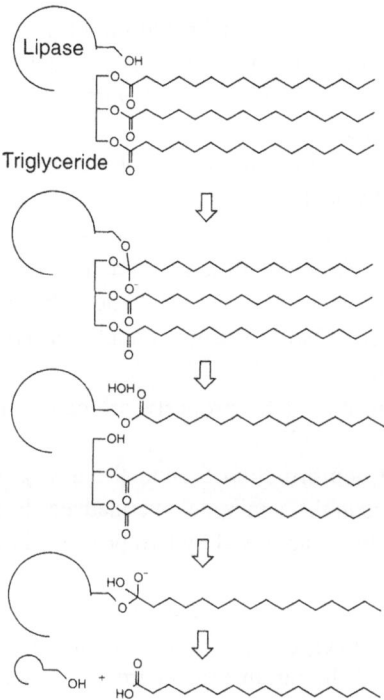

Figure 3. Elementary steps of the cleavage of a trilaurin substrate by lipase, as described in the text.

4. CATALYTIC MECHANISM

We have undertaken molecular modelling studies with the goal of producing a chemically sound and stereochemically acceptable reaction pathway for ester hydrolysis by pancreatic lipase. In particular, we wanted to predict the approximate exit directions of the substituents of the attacked ester from the active site and to investigate how a triglyceride substrate, attacked at the preferred 1- or 3-position (23), could be fitted into the active-site cavity.

The presence of the Asp...His...Ser triad suggested strongly that the salient stereochemical features of the hydrolytic mechanism might be analogous to those established for serine proteases of the trypsin family, whose catalytic mechanism has been extensively studied (24). Following this analogy, the catalytic reaction was subdivided into the following elementary reaction steps, as illustrated in Figure 3.

1. Formation of the non-covalent Michaelis complex.
2. Attack of the nucleophilic serine O^γ to form the tetrahedral, hemiacetal intermediate.

3. Cleavage of the substrate ester bond, breakdown of the tetrahedral intermediate to the acyl-enzyme, protonation and dissociation of the diglyceride.
4. Attack of the serine ester by an activated water molecule, formation of the deacylation intermediate.
5. Cleavage of the acyl-enzyme, protonation of the leaving group (Ser152) and dissociation of the fatty acid.

Models for the Michaelis complex and for the tetrahedral intermediate of serine proteases are available from structural studies of complexes with protein inhibitors (24) and transition-state analogues (24–27). Guided by these, we have tried to model a tetrahedral intermediate for triglyceride hydrolysis with the following restrictions in mind (10):

I.   The leaving group should be properly oriented for protonation by the histidine.
II. The geometry and conformation around the new covalent bond between enzyme and substrate in the hemiacetal intermediate should be acceptable.
III. The oxyanion should be stabilised by two properly oriented hydrogen-bond donors.
IV. The incoming ester group should be *trans* prior to the attack; the resulting acyl enzyme should also emerge in a *trans* conformation with respect to the ester bond. Breakdown of the tetrahedral intermediate should be possible with minimal conformational adjustments.
V.   Sterically acceptable models of the intermediate steps should be produced without major structural rearrangements in either protein or substrate.

Initial attempts to transfer the known stereochemistry of the tetrahedral intermediate in trypsin into the active site of the lipase by straightforward superposition methods led to severe contradictions to some of the restrictions stated above. It was then realised that inversion of the chirality of the tetrahedral intermediate avoided these problems while remaining chemically analogous to the trypsin case. With this new starting-point, the construction of a working model of the catalytic domain (N domain) of the lipase and of the reaction steps of substrate cleavage proceeded as follows:

A. The 238–260 loop (flap) was omitted from the model, because it was considered certain that in the catalytically active state the tip of this loop could not remain in the immediate vicinity of the active site. In addition, the C domain was omitted, because of its remoteness from the catalytic centre.
B. Two aromatic side-chains (Tyr114 and Phe215), situated close to the active site and making contacts with flap residues in the observed structure, were reoriented conservatively into alternative preferred conformations. This was the easiest way to make sufficient room for the substrate in this part of the active-site cavity.
C. The chain segment around Phe77, obstructing the active site severely, was

Lipase   Trypsin

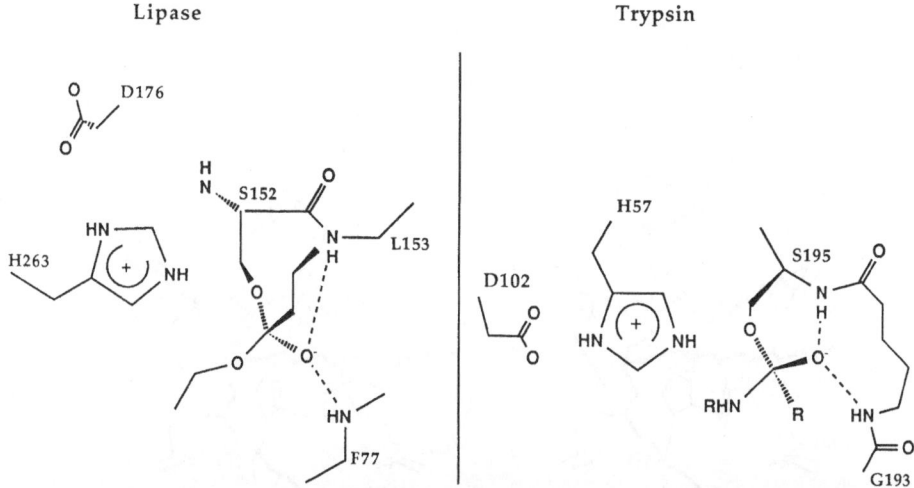

Figure 4. Schematic representation of the modelled tetrahedral intermediate of lipase with an ethyl butyrate model substrate (left) and of trypsin with a generic amide substrate (right). The tetrahedral centres generated by the attack of the serine Oγ (stippled bonds lying in the plane of the diagram) have the opposite handedness.

readjusted manually by moving it about 2 Å, so as to enlarge the cavity yet further. Since this segment is on the surface of the protein and interacts only weakly with the rest of the protein once the flap is removed, such a rearrangement appears quite reasonable.

D. A truncated substrate model (ethyl butyrate) was attached covalently to Oγ of Ser152 in the hemiacetal form, and the two hydrogen bonds postulated to stabilise the oxyanion were initially enforced (Figure 4). The complete model (N domain with model substrate but without flap) was energy-minimised using the Moloc force field (P. Gerber, unpublished results). The resulting model is called the 'activated' lipase model. Similar results could be obtained with other appropriately parametrised force fields (such as AMBER, CHARMM *etc.*).

Starting from the 'activated' lipase model and the X-ray structure of trilaurin (28), the intermediate states in the cleavage of this triglyceride substrate were constructed successfully (Michaelis complex, hemiacetal- and acyl-enzyme, water attack, product). The resulting model of the hemiacetal structure is shown in Figure 5. The structural modifications described (removal of the flap, adjustment of the segment around Phe77, side-chain reorientation for Tyr114 and Phe215) were sufficient to produce stereochemically satisfactory models of all these intermediate stages. In particular, the assumption that the backbone NH groups of Leu153 and Phe77 stabilise the oxyanion through hydrogen bonding led to very

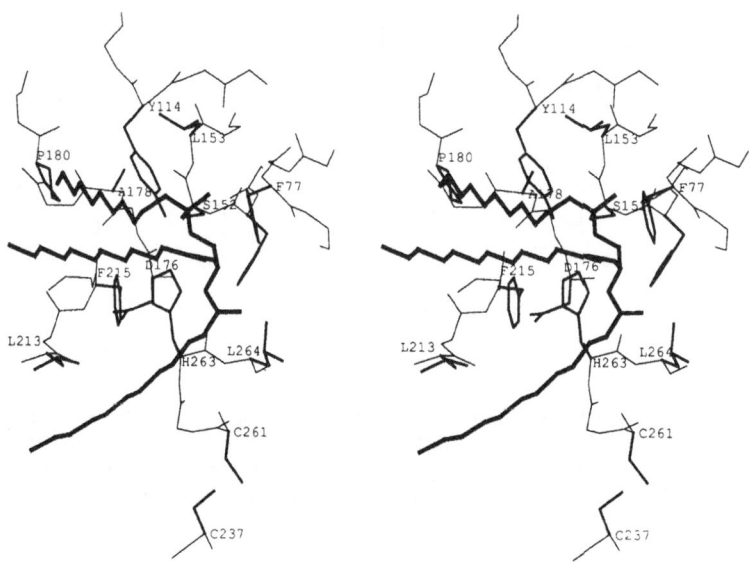

Figure 5. Stereo diagram of a model of the hemiacetal intermediate in the cleavage of trilaurin (very thick lines) by 'activated' lipase. The latter is derived from the X-ray structure of the closed form (see Figure 2 for comparison) as described in the text. Apart from the residues of the catalytic triad, only the side-chains in direct contact with the substrate are displayed. The fatty-acid chains were positioned arbitrarily in an extended conformation and may actually bend more towards the viewer, that is, they may still be anchored in the substrate phase by their tails. As the position of the flap residues is unknown in the 'activated' state, no attempt has been made to model the interactions with the aliphatic tails of trilaurin beyond the immediate vicinity of the active site.

satisfactory results. The mechanistic model also suggests that THL (19) and other β-lactones form stable acyl-enzyme complexes because the alcoholic leaving group stays in place near the catalytic serine and prevents water from attacking the acyl-enzyme complex (Figure 6). However, the model is not suited to the prediction of details of the interactions with the hydrophobic alkyl chains further away from the active site. For this, at least the position of the flap in the activated state would have to be known.

In summary, the proposed mechanism is chemically analogous to that of the trypsin-like serine proteases, but it differs in the handedness of the hemiacetal intermediate (Figure 4). This is reflected in a different three-dimensional

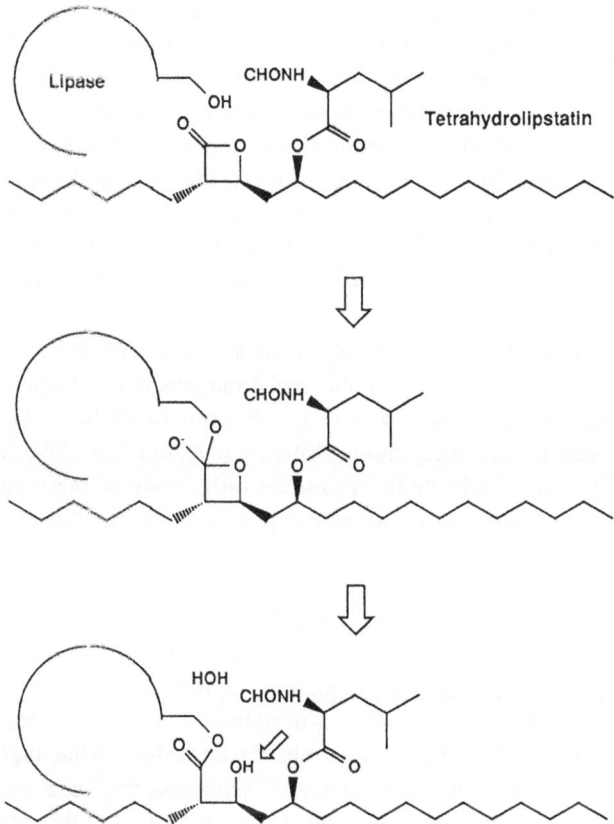

Figure 6. Postulated mechanism of inhibition of the lipase by THL. A stable acyl enzyme is formed by blocking the access of water through the presence of the alcoholic leaving group (see arrow).

arrangement of some of the functional groups involved in catalysis. The proposed inverse handedness has actually been observed in the complex of Rm lipase with a covalently attached phosphonate inhibitor (29). Interestingly, it had been observed much earlier that the tetrahedral intermediate of the cysteine protease papain is also of opposite hand to that of trypsin (30).

5. RELATIONSHIP WITH OTHER LIPASES AND HYDROLASES

There is increasing evidence that human pancreatic lipase is a member of a superfamily of enzymes that share what has been called the 'hydrolase fold' (31). This superfamily contains lipases that are both homologous to and very similar in structure to hPL, and enzymes that have no detectable sequence homology, but still share the overall fold and the mechanism. Closely related lipases comprise pancreatic lipases from other species, lipoprotein lipases and hepatic lipases. The sequences of these enzymes show significant similarity (32, 33), the hydrophobic core and the active site's environment being most clearly conserved (34). This provides strong support for the hypothesis that these enzymes have very similar structures and follow the mechanism of substrate cleavage outlined in section 4. Detailed analysis of naturally occurring and artificial mutants of lipoprotein lipase (34, 35) revealed that the effects of the mutations can be rationalised by disturbance of the catalytic mechanism in some cases and by interference with proper folding in others.

In addition to the fungal lipases RmL (8) and GcL (9), the structures of a number of other enzymes containing the triad arrangement (nucleophile-histidine-acid) show structural homology with the N domain of hPL. They include carboxypeptidase II (36), acetylcholine esterase (37), cutinase (38), dienelactone hydrolase (39) and haloalkane dehalogenase (40). None of these enzymes has detectable sequence homology with human pancreatic lipase. Nevertheless, they share, with modifications in the strand connections, the same overall topology of the core β sheet with the triad assembled on the C-terminal side of its central strands. The topological position of the acidic triad residue in hPL (Asp) differs from that in RmL (Asp), GcL (Glu) and acetylcholinesterase (Glu). Interestingly, hPL has another Asp at the position equivalent to the acidic function in the other three enzymes, but in a different conformation. It has been speculated that pancreatic lipase may represent an evolutionary intermediate in the migration of the acidic function from what is found in the pure 'hydrolase fold' enzymes to what is found in pancreatic, lipoprotein and hepatic lipases (41). In the latter two, the second acidic residue is no longer conserved.

A more detailed comparison of the environment of the catalytic serine in hPL, RmL and acetylcholinesterase (AChE) reveals close similarities for the backbone NH positions that have been postulated to form the oxyanion hole. The NH groups of the residues after the catalytic serine (Leu153 in hPL, Leu145 in RmL and Ala201 in AChE) occupy equivalent positions: the NH of Phe77 in hPL in our active hPL model corresponds to NH of Ser82 in RmL and to Gly118 in AChE, while in the original hPL structure this NH is about 1.5 Å closer to Ser152. Interestingly, the NH of Gly119 in AChE may contribute a third partner for the oxyanion hole which is not observed in any of the other enzymes; in RmL, this rôle could be adopted by the O^γ of Ser82.

In view of the observed similarity in the active site, it is tempting to speculate that these distantly related enzymes not only share the triad and the general fold, but also the catalytic mechanism in its stereochemical details. The non-conserved loops around the active site would then be responsible for the different substrate specificities.

6. CONFORMATION AT THE INTERFACE

A number of hypotheses have been put forward to explain interfacial activation, which is characterised by an often dramatic increase in the activity of lipolytic enzymes that is seen when the critical micellar concentration of partly water-soluble substrate molecules is exceeded. These hypotheses include: increased effective substrate concentration at the interface, favourable pre-orientation of the substrate's reactive group, enhanced diffusion of the water-insoluble products from the enzyme and conformational changes of the enzyme on binding to the interface (for sources, see reference 42). All these factors are likely to contribute, but the present discussion will focus on the conformational change indicated in the case of pancreatic lipase. It has already been stressed that the flap residues, and in particular Trp252, must move out of the active-site region, and that additional, smaller conformational adjustments in other parts of the molecule are required to generate enough room for a triglyceride substrate. The inevitable change in the location of Trp252 has stimulated studies by spectroscopic methods sensitive to a change in the microenvironment of tryptophans (43). As a substitute for activation at a substrate interface, hPL was acylated with the inhibitor tetrahydrolipstatin in the presence of bile salts at concentrations above their critical micellar concentration. Indeed, upon inactivation, large changes were measured in intrinsic tryptophan fluorescence and in its quenching by acrylamide, as well as in its near-ultraviolet circular dichroism (CD) spectrum. Removal of the bile salts leads to recovery of the activity and reversal of the spectroscopic changes. No significant alterations in secondary structure occur, as indicated by the absence of changes in the far-ultraviolet CD spectrum. This provides strong experimental evidence that a local, reversible conformational change takes place. However, the spectroscopic effects do not tell us which tryptophans are affected and what the structural change is.

When ones tries to imagine how the enzyme works at the interface, a number of structural questions immediately arise:
- What is the orientation of the enzyme at the interface, and which residues are involved in interface interactions?
- Does binding to the interface induce structural changes?
- Does any part of the adsorbed, activated enzyme penetrate physically into the substrate phase?
- What are the paths of substrate entry and product release, and are they the same?
- How does the water needed for catalysis gain access to the active site?
- Does the flap assume a defined alternative conformation in the active form of the enzyme, and do flap residues participate in substrate recognition?
- Are there additional conformational changes in other segments?

The last point has been addressed in the preceding section, but in general no detailed answers can at present be given to these questions. However, the available structural information on pancreatic and other lipases does allow some interesting comparisons and speculations. The phospholipases A_2 from mammalian and venom species are the best-understood lipases as far as the orientation and conformation of the enzyme at the substrate interface are concerned (44, 45). Except for some adjustments to the side-chains, no large changes in the backbone structure are observed between the structures of the apoenzymes and their complexes with inhibitors that closely mimic substrate molecules. In the apoenzymes, the cavity or hydrophobic channel that binds the aliphatic chains of the inhibitors is pre-formed and filled with solvent molecules. The cavity has two large openings at the surface of the protein. One is thought to serve as an entry and exit for substrate and product, respectively, and will be sealed when the enzyme becomes adsorbed at an interface. The other remains solvent-accessible and leads to the binding site of the polar phosphoglyceride head group. This means that solvent molecules could easily diffuse in and out of the cavity, in concert with substrate entry and product release.

In contrast, the three triglyceride lipase apoenzyme structures known to date have deeply buried, solvent-inaccessible active sites and no open, solvent-filled cavities. Instead, the space around the active site is occupied by hydrophobic side-chains of the protein in such a way that major rearrangements of the backbone are needed to generate the substrate-binding site. The structure of the fungal lipase from *R. miehei* complexed with covalently attached inhibitors has provided first information on the nature of the backbone rearrangement of this lipase upon activation (29, 46). The short α helix forming the lid rolls across the protein surface and moves about 8 Å, as indicated in Figure 7(*a*). The concomitant rotation around the helical axis exposes the previously buried, hydrophobic side-chains of the lid and generates a large non-polar surface around the active site. It appears plausible that this surface could face the substrate interface. In the view chosen in Figure 7(*a*), we look from the interface onto the interfacial contact area of the protein. Pancreatic lipase has a short helix in about the same position as RmL does (Figure 7(*b*)), which seems to indicate a similar mechanism. Closer examination, however, shows that an analogous rolling motion is not possible in hPL. Figure 7(*c*) shows a superposition of the surface segments around the active sites in the two apoenzyme structures. The superposition is based on the conserved β-εSer-α motif (not shown in Figure 7). As can be seen, the two helical segments are joined in very different ways to the rest of the catalytic domain. In the fungal lipase, the segments on the N- and C-terminal sides of the lid helix are oriented in an ideal way for them to act as hinges, to produce the rolling motion (46). In hPL, an analogous motion is not only more difficult to achieve and sterically forbidden by the presence of chain segment 111–115; it would even fail to make the active site more accessible, as other parts of the flap would come to lie over the active site. A much easier and more plausible displacement of the flap helix can, however, occur

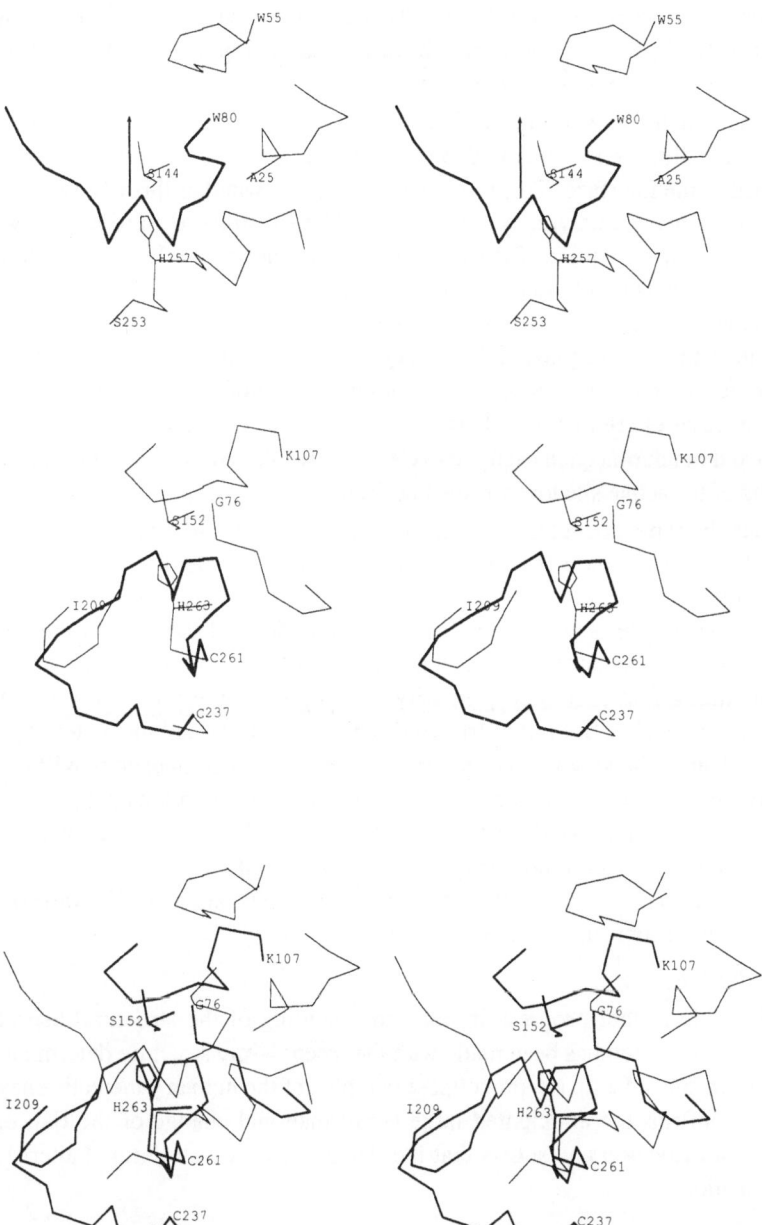

Figure 7: Stereo diagrams of the flap positions in RmL (*a*) and hPL (*b*). The α-carbon flap segments (thick lines) are shown together with adjacent C$^\alpha$ segments and the catalytic serine and histidine residues. (*c*) A superposition of RmL (thin lines) with hPL (thick lines, selected residues labelled) is shown.

towards the left and downwards. If hPL is positioned at the interface in a way similar to RmL, such a motion will have its main component in the interfacial plane and a minor component away from it.

In contrast to phospholipase A_2, the hydrophobic substrate cavity of the triglyceride lipases is generated only when the enzyme is conformationally activated at the interface. This raises the question of whether this cavity remains open during many catalytic cycles, that is, while substrate and product move in and out. Water is required for hydrolysis and cannot be excluded completely. However, there is no obvious pore or channel connecting the substrate cavity to bulk water such that solvent could transiently fill the active-site cavity, as appears possible in the case of phospholipase A_2. As an alternative, substrate entry and product release could take place in a co-ordinated fashion in such a way that large transient cavities are avoided. This would seem more efficient than moving the flap in and out during each catalytic cycle, but it would require a sufficiently large opening of the active site towards the interface.

It must be remembered that not all lipases need necessarily work in the same general way. The lipase from *G. candida* (9), for example, has more than one surface loop covering the active site. Cutinase, another lipolytic enzyme capable of hydrolysing soluble as well as emulsified triglycerides and showing no interfacial activation, has no analogous flap and its catalytic site is solvent-accessible (38). It may be that a flap is needed primarily not to protect the potentially unstable hydrophobic cavity, but rather to avoid the hydrolysis of soluble esters by the lipase. Clearly, the way to answer some of the many open questions will be the construction, expression and biochemical analysis of recombinant lipases with suitably chosen amino-acid substitutions. Obviously, the structural analysis of an active-site-directed inhibitor–enzyme complex will also be needed in order to establish the conformation of the flap in the active state and to verify some of the other proposed conformational adjustments.

Note added in proof

Considerable further progress in our understanding of the structural aspect of interfacial activation has been made with the recent X-ray structure determination of the open state of a lipase–procolipase complex of the human pancreatic enzyme (48). It confirms the anticipated large conformational change of the flap upon interfacial activation and shows that the flap assumes a well-defined alternative conformation.

REFERENCES

1. Desnuelle, P. (1972). The lipases. In *The Enzymes*, 3rd edn., vol. 7, ed. Boyer, P.D., pp. 575–616. New York: Academic Press.

2. Verger, R. (1984). Pancreatic lipases. In *Lipases*, ed. Borgström, B. & Brockman, H.L., pp. 84–150, Amsterdam: Elsevier.

3. Desnuelle, P. (1986). Pancreatic lipase and phospholipase. In *Molecular and Cellular Basis of Digestion*, ed. Desnuelle, P., Sjöström, H. & Noren, O., pp. 275–296. Amsterdam: Elsevier.

4. Borgström, B. & Erlanson-Albertsson, C. (1984). Pancreatic colipase. In *Lipases*, ed. Borgström, B. & Brockman, H.L., pp. 152–183, Amsterdam: Elsevier.

5. Weibel, E.K., Hadváry, P., Hochuli, E., Kupfer, E. & Lengsfeld, H. (1987). Lipstatin, an inhibitor of pancreatic lipase produced by *Streptomyces toxytricini*. *J. Antibiot. (Tokyo)*, 40, 1081–1085.

6. Chapus, C., Rovery, M., Sarda, L. & Verger, R. (1988). Minireview on pancreatic lipase and colipase. *Biochimie*, 70, 1223–1234.

7. Winkler, F.K., D'Arcy, A. & Hunziker, W. (1990). Structure of human pancreatic lipase. *Nature*, 343, 771–774.

8. Brady, L., Brzozowski, A.M., Derewenda, Z.S., Dodson, E., Dodson, G., Tolley, S., Turkenburg, J., Christiansen, L., Huge-Jensen, B., Norskou, L., Thim, L. & Menge, U. (1990). A serine protease triad forms the catalytic centre of a triacylglycerol lipase. *Nature*, 343, 767–770.

9. Schrag, J.D., Li, Y., Wu, S. & Cygler, M. (1991). Ser-His-Glu triad forms the catalytic site of the lipase from *Geotrichum candidum*. *Nature*, 351, 761–764.

10. Gubernator, K., Müller, K. & Winkler, F.K. (1991). The structure of human pancreatic lipase suggests a locally inverted, trypsin-like mechanism. In *Lipases: Structure, Mechanism and Genetic Engineering*, GBF monographs, vol. 16, ed. Alberghina, L., Schmid, R.D. & Verger, R., pp. 9–16. Weinheim: VCH.

11. Fournet, B., Leroy, Y., Montreuil, J., DeCaro, J., Rovery, M., van Kuik, J.A. & Vliegenthart, J.F.G. (1987). Primary structure of the glycans of porcine pancreatic lipase. *Eur. J. Biochem.*, 170, 369–371.

12. DeCaro, A., Figarella, C., Amic, J., Michel, R. & Guy, O. (1977). Human pancreatic lipase: a glycoprotein. *Biochim. Biophys. Acta*, 490, 411–419.

13. Bosner, M.S., Gulick, T., Riley, D.J.S., Spilburg, C.A. & Lange, L.G. (1989). Heparin-modulated binding of pancreatic lipase and uptake of hydrolysed triglycerides in the intestine. *J. Biol. Chem.*, 264, 20261–20264.

14. Yang, C.-Y., Gu, Z.-W., Yang, H.-X., Rohde, M.F., Gotto, A.M. & Pownall, H.J. (1989). Structure of bovine milk lipoprotein lipase. *J. Biol. Chem.*, 264, 16822–16827.

15. Mahe-Gouhier, N. & Leger, C.L. (1988). Immobilized colipase affinities for lipases B, A, C and their terminal peptide (336–449): the lipase recognition site lysine residues are located in the C-terminal region. *Biochim. Biophys. Acta*, 962, 91–97.

16. van Tilbeurgh, H., Sada, L., Verger, R. & Cambillau, C. (1992). Structure of the lipase-prolipase complex. *Nature*, 359, 159–162.

17. DeCaro, J.D., Rouimi, P. & Rovery, M. (1986). Hydrolysis of *p*-nitrophenyl acetate by the peptide chain fragment (336–449) of porcine pancreatic lipase. *Eur. J. Biochem.*, 158, 601–607.

18. Guidoni, A., Benkouka, F., DeCaro, J.D. & Rovery, M. (1981). Characterization of the serine reacting with diethyl *p*-nitrophenyl phosphate in porcine pancreatic lipase. *Biochim. Biophys. Acta*, 660, 148–150.

19. Lüthi-Peng, Q., Märki, H.P. & Hadváry, P. (1992) Identification of the active-site serine in human pancreatic lipase by chemical modification with tetrahydrolipstatin. *FEBS Letters*,

299, 111–115.
20. Henderson, R. (1970). Structure of crystalline α-chymotrypsin. *J. Mol. Biol.*, 54, 341–354.
21. Sibanda, B.L., Blundell, T.L. & Thornton, J.M. (1989). β-Hairpins in protein structures. *J. Mol. Biol.*, 206, 759–777.
22. Derewenda, Z.S. & Derewenda, U. (1991). Relationship among serine hydrolases: evidence for a common structural motif in triacylglyceride lipases and esterases. *Biochem. Cell Biol.*, 69, 842–851.
23. Rogalska, E., Ransac, S. & Verger, R. (1990). Stereoselectivity of lipases. *J. Biol. Chem.*, 265, 20271–20276.
24. Bode, W. & Huber, R. (1986). Crystal structures of pancreatic serine endopeptidases. In *Molecular and Cellular Basis of Digestion*, ed. Desnuelle, P., Sjöström, H. & Noren, O., pp. 213–234. Amsterdam: Elsevier.
25. Marquart, M., Walter, J., Deisenhofer, J., Bode, W. & Huber, R. (1983). The geometry of the reactive site and of the peptide groups in trypsin, trypsinogen and its complexes with inhibitors. *Acta Crystallogr.*, B39, 480–490.
26. Bone, R., Shenvi, A.B., Kettner, C.A. & Agard, D.A. (1987). Serine protease mechanism: structure of an inhibitory complex of α-lytic protease and a tightly bound peptide boronic acid. *Biochemistry*, 26, 7609–7614.
27. Bone, R., Frank, D., Kettner, A. & Agard, D.A. (1989). Structural analysis of specificity: α-lytic protease complexes with analogues of reaction intermediates. *Biochemistry*, 28, 7600–7609.
28. Gibon, V., Blanpain, P., Norberg, B. & Durant, F. (1984). New data about molecular structure of β-trilaurin. *Bull. Soc. Chim. Belg.*, 93, 27–34.
29. Brzozowski, A.M., Derewenda, U., Derewenda, Z.S., Dodson, G.G., Lawson, D.M., Turkenburg, J.P., Björkling, F., Huge-Jensen, B., Patkar, S.A. & Thim, L. (1991). A model for interfacial activation in lipases from the structure of a fungal lipase–inhibitor complex. *Nature*, 351, 491–494.
30. Garavito, R.M., Rossmann, M.G., Argos, P. & Eventoff, W. (1977). Convergence of active center geometries. *Biochemistry*, 16, 5065–5071.
31. Ollis, D.L., Cheah, E., Cygler, M., Dijkstra, B., Frolow, F., Franken, S.M., Harel, M., Remington, S.J., Silman, I., Schrag, J., Sussman, J.L., Verschueren, K.H.G. & Goldman, A. (1992). The α/β hydrolase fold. *Prot. Engin.*, 5, 197–211.
32. Datta, S., Luo C.-C., Li, W.-H., van Tuinen, P., Ledbetter, D.H., Brown, M.A., Chen, S.-H., Liu, S.-W. & Chan, L. (1988). Human hepatic lipase: cloned cDNA sequence, restriction fragment length polymorphisms, chromosomal localization, and evolutionary relationships with lipoprotein lipase and pancreatic lipase. *J. Biol. Chem.*, 263, 1107–1110.
33. Kirchgessner, T.G., Chuat, J.-C., Heinzmann, C., Etienne, J., Guilhot, S., Svenson, K., Ameis, D., Pilon, C., D'Auriol, L. Andalabi, A., Schotz, M.C., Galibert, F. & Lusis, A.J. (1989). Organization of the human lipoprotein lipase gene and evolution of the lipase gene family. *Proc. Nat. Acad. Sci. USA*, 86, 9647–9651.
34. Derewenda, Z.S. & Cambillau, C. (1991). Effects of gene mutations in lipoprotein and hepatic lipases as interpreted by a molecular model of the pancreatic triglyceride lipase. *J. Biol. Chem.*, 266, 23112–23119.
35. Henderson, H.E., Hassan, M.F., Monsalve, M.V., Marais, A.D., Winkler, F., Gubernator, K., Peterson, J., Brunzell, J.D. & Hayden, M.R. (1991). Amino acid substitution (Ile194-Thr) in exon 5 of the lipoprotein lipase gene causes lipoprotein lipase deficiency in three unrelated probands. Support for a multicentric origin. *J. Clin. Invest.*, 87, 2005–2011.
36. Liao, D.-I. & Remington, S.J. (1990). Structure of wheat serine carboxypeptidase II at 3.5 Å resolution. *J. Biol. Chem.*, 265, 6528–6531.

37. Sussman, J.L., Harel, M., Frolow, F., Oefner, C., Goldman, A., Toker, L. & Silman, I. (1991). Atomic structure of acetylcholinesterase from *Torpedo californica*: a prototypic acetylcholine-binding protein. *Science*, 253, 872–879.

38. Martinez, C., De Geus, P., Lauwereys, M., Matthyssens, G. & Cambillau, C. (1992). *Fusarum solani* cutinase is a lipolytic enzyme with a catalytic serine accessible to solvent. *Nature*, 356, 615–618.

39. Pathak, D. & Ollis, D. (1990). The refined structure of dienelactone hydrolase at 1.8 Å. *J. Mol. Biol.*, 214, 497–525.

40. Franken, S.M, Rozeboom, H.J., Kalk, K.H. & Dijkstra, B.W. (1991). Crystal structure of haloalkane dehalogenase: an enzyme to detoxify halogenated alkanes. *EMBO J.*, 10, 1297–1302.

41. Schrag, J.D., Winkler, F.K. & Cygler, M. (1992). Pancreatic lipases: evolutionary intermediates in a positional change of catalytic carboxylates. *J. Biol. Chem.*, 267, 4300–4303.

42. Derewenda, Z.S. & Sharp, A.M. (1993). News from the interface: the molecular structures of triglyceride lipases. *Trends Biochem. Sci.*, 18, 20–25.

43. Lüthi-Peng, Q. & Winkler, F.K. (1992). Large spectral changes accompany the conformational transition of human pancreatic lipase induced by acylation with the inhibitor tetrahydrolipstatin. *Eur. J. Biochem.*, 205, 383–390.

44. Thunnissen, M.M.G.M., Ab, E., Kalk, K.H., Drenth, J., Dijkstra, B.W., Kuipers, O.P., Dijkman, R., de Haas, G.H. & Verheij, H.M. (1990). X-ray structure of phospholipase A_2 complexed with a substrate-derived inhibitor. *Nature*, 347, 689–691.

45. Scott, D.L., White, S.P., Otwinowski, Z., Yuan, W., Gelb, M.H. & Sigler, P.B. (1990). Interfacial catalysis; the mechanism of phospholipase A_2. *Science*, 250, 1541–1546.

46. Derewenda, U., Brzozowski, A.M., Lawson, D.M. & Derewenda, Z.S. (1992). Catalysis at the interface: the anatomy of a conformational change in a triglyceride lipase. *Biochemistry*, 31, 1532–1541.

47. Priestle, J.P. (1988). RIBBON: a stereo cartoon drawing program for proteins. *J. Appl. Crystallogr.*, 21, 572–576.

48. van Tilbeurgh, H., Egloff, M.-P., Martinez, C., Rugani, N., Verger, R. & Cambillau, C. (1993). Interfacial activation of the lipase–procolipase complex by mixed micelles, revealed by X-ray crystallography. *Nature*, 362, 814–820.

8.

Kinetics of triglyceride lipases

MATS MARTINELLE and KARL HULT

Triglyceride lipases are enzymes that hydrolyse triacylglycerides to fatty acids and glycerol. One characteristic of lipases is that they are more active with insoluble than with soluble ester substrates. The factors that trigger this activation at the lipid–water interface have been discussed for a long time. Hypotheses put forward involve (i) increased substrate availability by displacement of the water shell around the ester molecules (1), (ii) increased substrate concentration at the interface (2), (iii) better orientation of the ester bond to be cleaved (3) and, as suggested by Sarda and Desnuelle (4, 5), (iv) a conformational change undergone by the enzyme and leading to increased activity.

The determination of the crystal structure of the lipase from *Rhizomucor miehei* complexed with the active-site inhibitor *n*-hexylphosphonate ethyl ester (6) has, at least in one case, supported the interfacial activation hypothesis based on conformational changes. The structure reveals that the helical lid that covers the active site in the uncomplexed enzyme (7) is displaced, exposing the active site to the substrate. Concomitantly, a hydrophobic surface around the active site becomes exposed to the surroundings, forming a strong binding site for a hydrophobic interface.

Lids covering the active sites of lipases seem to be common. The two other lipases with known structures, human pancreatic lipase (8) and the lipase from the fungus *Geotrichum candidum* (9), both have their active sites deep in the molecule's structure, inaccessible to substrates without conformational changes of the protein. In contrast to the triacylglyceride hydrolases, discussed here, interfacial activation of phospholipase A_2 is not due to a substantial conformational change of the protein. Instead, the first hypothesis, which postulates increased substrate availability by displacement of the water shell around the substrate molecules, seems to be the most relevant (10).

Lipases are today used as versatile chiral catalysts in organic synthesis. They are employed not only in traditional water-based systems with insoluble substrates, but also in non-aqueous systems with dissolved substrates and immobilised enzyme. For an understanding of lipases, and for optimisation of industrial processes, a thorough knowledge of lipase kinetics in different systems will be needed. The known lipase structures mentioned above will be very useful for the understanding of the activation and catalytic processes of lipase action. Furthermore, these structures can be used as models for other, closely related lipases. The widely used porcine pancreatic lipase is very similar to human pancreatic lipase (8). The lipase from *Candida rugosa* (*cylindracea*) was recently shown to have a structure close to that of the homologous enzyme of *Geotrichum candidum* (9). In contrast to the *G. candidum* lipase, the structure of the *C.rugosa* lipase was determined in an open conformation, and it was pointed out that the two structures could suggest a mechanism for their interfacial activation (11).

Below, we shall discuss the kinetics of triglyceride lipases in different systems. We shall see how complicated it is to distinguish between activation and catalysis in water with insoluble substrates, and shall point out some of the advantages of studying lipases in less conventional media.

1. LIPASE INTERFACIAL KINETICS

A characteristic feature of lipases is their activation by interfaces. Desnuelle and co-workers have shown that the activity of lipases increases greatly when the concentration of the substrate reaches its limit of solubility and forms a second phase (4). This indicates a fundamental difference between ordinary esterases and lipases. The esterases follow normal Michaelis-Menten kinetics with soluble substrates and have no activity toward insoluble esters. The lipases, however, are more active with insoluble substrates, but often have a negligible Michaelis-Menten-type activity toward soluble substrates. Desnuelle observed that the lipases, when acting on emulsions of triglycerides, are adsorbed at the substrate–water interface, following Langmuir isotherms (12). A strong correlation was found between the amount of lipase adsorbed and the enzymic activity. Further studies revealed that the size of the droplets is important. The smaller the droplets were, the larger was the area available to the lipase for binding. A relationship between reaction rate and substrate concentration, expressed as surface area, could be found and was shown to be analogous to Michaelis-Menten kinetics. The problem with an interfacial K_m, in which concentration is expressed as moles per unit area instead of per unit volume, is that it is difficult to interpret. It may be more an estimate of the relatively unspecific protein–lipid interaction than an actual measure of substrate specificity. An important part of any model describing lipase kinetics is that it should distinguish between adsorption on the one hand and substrate binding, leading to catalysis, on the other.

A kinetic model for the hydrolysis of insoluble lipids has been proposed by Verger *et al.* (13–15). The model proposes an adsorption process of the lipase to the interface, described as being in equilibrium. The adsorption is followed by binding of the substrate to the enzyme at the surface and then by catalytic turnover in a Michaelis-Menten fashion. Accumulation of products at the interface may cause problems, such as induced changes of the physico-chemical properties of *the interface and product inhibition. The kinetic model takes into account a lipase* inactivation step caused by the interface. This inactivation depends upon the surface pressure, with low pressures (corresponding to high surface energy) being the most strongly denaturing. The production rate depends therefore both on the adsorption and the inactivation, and becomes constant in time when the adsorption is fast compared with the inactivation, or when the two processes have equal fluxes. The reversibility of the adsorption step seems to depend on the surface pressure.

1.1 Conformational changes in lipase catalysis

Conformational changes in enzymes are often an important part of their action, either in regulation and/or in the catalytic process. For example, triose phosphate isomerase has a lid that covers the active site during the catalytic event (16). The lid has to open to let the substrate and the product in and out. The open-lid and closed-lid conformers are stabilised by a number of hydrogen bonds between the lid and the rest of the protein. The lid of *Rhizomucor miehei* lipase (6, 7) has several features similar to those of triose phosphate isomerase. Both lids are helical and move as rigid bodies, and distinct hinge regions are present. The closed-lid conformer of the lipase is stabilised by hydrophobic interactions, while the open form is stabilised by hydrogen bonds. In the open form, the lipase has around the active site a non-polar surface, which can serve as a binding site to a hydrophobic interface (14).

In their model of lipase action, Verger and de Haas (15) assume that an active enzyme conformer is formed concomitantly with, or after, the binding of the lipase to the surface of the insoluble substrate. No active enzyme is thought to occur free in solution. An alternative model, which can include soluble substrates, would be that the lipase exists in two conformations in solution, by analogy with allosteric proteins. The tight state would correspond to the inactive, closed-lid conformation, while the relaxed state would correspond to the active, open-lid conformation. There would be a free equilibrium between the two forms, lying well over towards the closed-lid form in aqueous solution. When a non-polar interface is introduced into the system, enzyme in the relaxed form will bind to it, and the equilibrium will shift towards the active open bound conformer, as in allosteric activation. The low activity towards soluble substrates is then explained by the small fraction of relaxed lipase in the absence of an interface. Verger has reviewed the problems of orientation and conformational changes of lipases when they bind to interfaces (14).

A further question of detail is whether lipases with lids can have different conformations when active, with the lid more or less open, or whether they have only one active form, with the lid in a well-defined position. The relaxed form of *Rhizomucor miehei* lipase is stabilised by, among a number of other interactions, hydrogen bonds between Arg86 in the lid and backbone carbonyl oxygen atoms in a polar cavity at the protein surface. If this interaction is disturbed, by modification of the arginine residue or by competitive binding of guanidine, the activity towards tributyrin decreases (17). The degree of inhibition depends on the modification agent and the guanidine concentration, suggesting that the position of the lid is important for the activity of the lipase. Guanidine exhibits different degrees of inhibition for different enantiomers of the same substrate, which may be explained by enantiospecific interactions between the substrate and the lid (H.M. Holmquist *et al.*, unpublished results).

1.2 Adsorption

Most, if not all, enzymic lipolytic activity takes place at interfaces. The adsorption process of lipases is therefore of fundamental interest in the understanding of lipase activity. As mentioned above, the kinetic model consists of two steps: first, the binding to the surface and, secondly, the binding of the substrate in the surface.

Brockman *et al.* (18–22) have investigated the adsorption of pancreatic carboxyl-ester lipase to monomolecular films at the air–water interface using films compressed near to the point of collapse, making it a well-defined thermodynamic system. These films were made of lipids based on the same hydrophobic tail, but with different polar head groups. The adsorption of native and catalytically inactivated lipase (modified with di-*iso*-propyl fluorophosphate, DFP) to 13,16-docosadienoic acid (DA), and 1,3-diolein (DO) followed Langmuir adsorption isotherms with saturation levels of 3.3 and 4.2 pmol/cm^2, respectively. The dissociation constant ranged from 10 to 50 nM, depending on ionic strength and pH. A monolayer of a globular protein of the size in question would correspond to 4.0 pmol/cm^2. A DFP-inactivated enzyme was also used with other films, such as triolein, methyl oleate, oleyl alcohol and oleonitrile, or air. These surfaces were saturated at slightly lower levels in the range 2.3–3.1 pmol/cm^2. These results indicate a relative lack of chemical specificity in the adsorption step of this lipase. However, a comparison of the apparent rate constants of adsorption to DA and to DO as a function of pH indicated a dependence on the state of ionisation of the fatty acid. The rate constant for binding decreased to near zero at pH values above 7. Furthermore, the concentration of lipase adsorbed at equilibrium to 1-palmitoyl-2-oleoyl-phosphatidylcholine (POPC) was about 5% of that adsorbed to DA or DO. The conclusion is that anionic and zwitterionic lipids inhibits the lipase–lipid interaction. Studies have been performed using mixed films of POPC and the non-phospholipids mentioned above. In these films, POPC and the lipid form complexes of 'preferred packing array' that behave as pseudo-species with their own area and hydration level. At a certain mole fraction, the lipid also exists in uncomplexed form. In mixed films containing POPC-DA or POPC-DO, the binding affinities of native and DFP-modified lipase were reduced to 5–10% of the values for pure DA or DO films, and the amount of surface-bound lipase increased non-linearly with the mole fraction or area fraction of non-phospholipid. The lipase probably needs a certain area of minimal size for binding and, as its binding to POPC is very low, clusters of non-phospholipids above a certain minimal size are needed. The result is that only a part of the surface will be available for binding. As phospholipids are abundant *in vivo*, their presence may have a regulatory effect on the lipolytic process. As these studies were restricted to only one lipase, the conclusions may not be general.

Brockman also found that the degree of hydrolysis in mixed surface films is strongly dependent upon the balance between POPC and the non-phospholipid substrate. At a particular mole fraction of substrate, the conversion after a certain time increased from 5% to 95%. The results were the same for triolein, 1,3-diolein and methyl oleate. They could not be explained by the fact that less enzyme was bound, in comparison with pure substrate films. Instead, they were explained as being a consequence of substrate connectivity. When substrate is present at a low mole fraction it behaves as pools in a sea of POPC, but above a specific fraction this behaviour changes and a connected substrate film appears.

The adsorption of lipase at interfaces can be inhibited by a number of other surface-active molecules. Verger *et al.* (23, 24) have shown that proteins, such as bovine serum albumin and β-lactoglobulin A, inhibit the action of lipase from pancreas and from *Rhizopus delemar*. The inhibition correlated well with a decrease of lipase binding to the protein–lipid films. The inhibitory proteins were also able to cause dissociation of the lipases from the interface when added afterwards. The lipases were inhibited to 50% when the inhibitory protein covered only 5–10% of the surface of a dicaproin monolayer. This disproportionately strong inhibition is explained by Verger *et al.* as being due to long-range effects of the inhibitory proteins. Bile salts and other surfactants have a dual effect on lipase catalysis (25, 26). At low concentrations they seem to prevent lipase denaturation at interfaces, but at high concentrations they inhibit binding.

Protein denaturation, based on retained enzymic activity, differed between different surfaces (13, 20). This was shown to be an effect of surface pressure, with an increase of denaturation at reduced surface pressure. The effect has also been shown by others as inactivation at water–hydrocarbon or water–triglyceride interfaces (25).

2. KINETICS WITH WATER-SOLUBLE SUBSTRATES

In water, lipases have a marginal activity towards dissolved substrates. Even if the activity is low, studies with these substrates can give us important information about catalysis by lipase. For example, porcine pancreatic lipase can catalyse the hydrolysis of monomeric p-nitrophenyl acetate (PNPA) in water in the presence of 4% acetonitrile (27). The initial phase follows 'burst' kinetics, typical for serine hydrolases, with the rate-limiting step being deacylation (28). The presence of acetyl-lipase as an intermediate in the catalysis was also demonstrated (27). The study shows clearly that porcine pancreatic lipase is active with soluble substrates, even if the reaction velocity is very low compared with insoluble substrates. Another example with the same lipase is the hydrolysis of monomeric tripropionin (5). In this case, the reaction velocity was 0.3% of that for hydrolysis of emulsified tripropionin. The reaction rate with monomeric tripropionin was increased 10–20 times by the addition of water-miscible solvents such as dioxane, acetonitrile, formamide and *tert*-butanol (5). It was further shown that the increased activity was caused by an increase in V, while the affinity for the substrate decreased, with K_m increasing from 10 mM to 80 mM by an addition of 6 mol % acetonitrile (Table 1).

Brockman *et al.* (2) studied the binding of porcine pancreatic lipase to siliconised glass beads. They found a reversible, diffusion-controlled binding, with a dissociation constant of 1.3×10^{-8} M. They also found a 1000-fold increase in the specific activity, when the lipase was bound. The authors attributed the increased activity exclusively to an increase in the local concentration of the substrate, tripropionin, at the hydrophobic surface. Today, the knowledge that pancreatic lipase has to undergo a conformational change to expose its active site to the substrate makes it plausible that at least part of the activation is due to stabilisation of the open, active conformer of the lipase at the surface.

A lipase from *Chromobacterium* was shown to be active on monomeric PNPA (29). In contrast to the pancreatic lipase, the bacterial lipase was shown to exhibit its rate-limiting step in acylation and not deacylation of the enzyme. This conclusion was based on the absence of burst kinetics. The soluble substrate (PNPA) showed 5% of the activity of emulsified olive oil. A K_m of 0.5 mM was determined for PNPA, and it was shown that p-nitrophenol and phenol were non-competitive inhibitors, with respective K_i values of 2 mM and 25 mM. An interesting observation was that the reaction rate could be increased by the addition of glass beads with hydrophobic coatings. While the reaction rate increased, no change in K_m was seen (Table 1).

Table 1. Examples of K_m values for lipase hydrolysis of esters dissolved in monomeric form in water-based solutions.

Enzyme source and additive	Substrate	K_m (mM)	V (%)	Reference
Porcine pancreas				
None	Tripropionin	10	100	5
5 mol% acetonitrile	Tripropionin	80	900	5
Chromobacterium species				
None	PNP acetate	0.5	100	29
Silicone-oil-coated beads	PNP acetate	0.5	270	29
Paraffin-coated beads	PNP acetate	0.5	220	29
Rhizomucor miehei				
17 vol % HMPT	PNP octanoate	0.03	100	30
17 vol % HMPT	PNP 2-MO	0.05	12	30
Candida cylindracea				
17 vol % HMPT	PNP octanoate	0.03	100	30
17 vol % HMPT	PNP 2-MO	0.21	38	30

Values of V are given only for comparison of different substrates and different additions, for the same lipase. PNP, *p*-nitrophenyl; HMPT, hexamethylphosporic triamide; 2 MO, 2 methyloctanate. Data were taken directly from the references given or calculated from data therein.

Sonnet and Baillergen (30) determined kinetic constants for a number of substrates dissolved in a buffer system containing 17% hexamethylphosphoric triamide. They studied a number of lipases, but focused on two lipases from *Rhizomucor miehei* and *Candida cylindracea*. The substrates used were esters of *n*-octanoic and methyl-*n*-octanoic acids. The ethyl thioesters did not show saturation kinetics within their range of solubility. The corresponding *p*-nitrophenyl esters were much better substrates, and saturation kinetics were observed. The values for K_m fall between 0.03 and 0.2 mM and did not differ significantly between the two lipases. The *p*-nitrophenyl ester of *n*-octanoic acid was the best substrate, assessed by V/K_m, for both enzymes. For the *Candida cylindracea* lipase, V was 2–7 times higher for the *n*-octanoic ester than for the branched analogues, except for the 3-methyloctanoic ester, which showed no activity (Table 1). The lipases from *Rhizomucor miehei* and *Candida cylindracea* had opposite enantioselectivity for *p*-nitrophenyl 2-methyloctanoate. It was further shown that in both cases the main part of the enantioselectivity was caused by differences in V and not K_m (Table 2).

Table 2. Calculated enantioselectivities of *Rhizomucor miehei* and *Candida cylindracea* lipases in the hydrolysis of *p*-nitrophenyl 2-methyl decanoate in aqueous solutions.

Enzyme source	K_m (mM)		V (mM·min^{-1}·mg^{-1})		V/K_m (min^{-1}·mg^{-1})		E
	(*R*)-	(*S*)-	(*R*)-	(*S*)-	(*R*)-	(*S*)-	
R. miehei	0.040	0.054	5.78	0.26	145	4.75	30.4
C. cylindracea	0.028	0.040	68.6	217	2650	5830	0.45

E is the ratio $(V/K_m)_R/(V/K_m)_S$, also known as the enantiomeric ratio (67). Data from reference 30.

A central question in connection with lipase action on monomeric substrates in aqueous solutions, is what conformation the enzyme adopts. The possibility, at least in the cases of *Rhizomucor* (6, 7), *Geotrichum* (9) and pancreatic (8) lipases, that the enzyme is active when the lid is closed over the active site can be ruled out because of inaccessibility of the active site. The question is rather whether the lid can adopt conformations intermediate between fully open and closed, and what specific activities such intermediate forms would have. Addition of water-miscible solvents that increase the hydrophobicity of the solution, should be able to shift the equilibrium between closed and open conformers in favour of the open ones. This might explain the increased reaction rate observed, but it fails to reveal whether a totally open form is needed for activity.

The low values of K_m reported from studies with monomeric substrates in aqueous solutions are not surprising. It must be energetically favourable for the hydrophobic substrates to move from the aqueous solution to the hydrophobic binding site of the lipase.

3. LIPASE KINETICS IN CONTINUOUS ORGANIC PHASE

3.1 Lipases entrapped in reversed micelles

Lipases entrapped in reversed micelles catalyse hydrolysis (31–33), synthesis (33, 34) and glycerolysis (35) of fats and other esters. The benefit of reversed micelles lies in the easy preparation of a two-phase system with a very large and stable interfacial area. The micelles have diameters of a few nanometres, large enough to accommodate enzyme molecules. With the huge surface area in these systems it may be supposed that all enzyme molecules can be adsorbed at the interface.

One of the most important parameters for optimal enzyme activity in reversed micelles is the molar ratio of water to surfactant, often called R. A distinct maximum in enzymic activity is usually found, when it is measured as a function of R (31, 32, 34). The size of the micelles is related to this parameter. At low ratios of water to surfactant, the micelles may be too small to accommodate lipase molecules, and at high ratios the large amount of water present may interfere with the enzymic reaction.

The choice of surfactant is important in attaining high enzymic activity. For example, *Chromobacterium viscosum* lipase was five times more active in systems based on the anionic surfactant Aerosol-OT (AOT) than in systems based on various zwitterionic phosphatidylcholines. The non-ionic surfactant polyethyleneglycol 6000 and the cationic surfactant cetyltrimethylammonium bromide afforded enzyme activities only about one-tenth of those found with AOT (35). Both physical differences of the reversed micelles and different interactions between the lipase and the surfactants may be reasons for the different activities seen. As will be discussed below, the choice of surfactant even can change the kinetics completely. In one case a direct interaction between the surfactant and the lipase was detected. In the hydrolysis of palm oil, catalysed by *Rhizomucor miehei* lipase in microemulsions based on either AOT or pentaethylene glycol monododecyl ether (C12AO5), a great difference in reaction rate and yield of free fatty acid was observed (36). It could be shown that the low rate of reaction in the C12EO5-system was caused by competitive binding of the surfactant to the lipase. Furthermore, the yield of free fatty acids was low because the surfactant served as acyl acceptor, allowing formation of an ester between the fatty acid and C12EO5.

The continuous oil phase is an important part of the system. Chang and Rhee (35) and Han and Rhee (31) tried a number of solvents. They showed that *iso*-octane was the solvent that gave the highest reaction rate. In both publications a large difference between octane and *iso*-octane is reported. The reason for the three- to fivefold higher reaction rate in *iso*-octane is not clear. It might reflect differences in the molecular arrangement at the interface, or have a more direct kinetic explanation in terms of competitive binding of octane to the active site, which could be less pronounced with the more bulky *iso*-octane.

The lipases from *Rhizopus arrhizus* (32) and *Candida rugosa* (31) were more unstable in the reversed micelles than in solutions without surfactants. The loss of enzyme activity was about 70% in an hour. In the case of *C. rugosa* lipase the stability depended on the ratio R. At low ratios of water to surfactant, the stability was greater than at high ratios. This effect might parallel that seen in monolayers with different surface tensions. However, Chang and Rhee showed an increased stability of the *C. viscosum* lipase in reversed micelles (35). The ratio R was in this case very low, 0.65, and a high concentration of glycerol was used.

In several reversed-micelle systems lipase catalysis follows simple Michaelis-Menten kinetics, and K_m and V can be determined. Table 3 shows examples of

Table 3. Examples of kinetic constants determined for lipases in reversed micelles with iso-octane as the continuous phase.

Enzyme source	Surfactant	R	Substrate	K_m (mM)	V (U·mg^{-1})	Reference
Candida rugosa	AOT	10.5	Olive oil	10	1020	31
Rhizopus arrhizus	AOT	13	Palm kernel oil	136	5714	32
Rhizopus delemar	AOT	11	Tricaprylin	32	87	33
	Lecithin	2.2	Tricaprylin	∞	–	33
Chr. viscosum	AOT	0.65	Triolein (glycerol)	0.87	0.029	35
Chr. viscosum (Lipase A)	AOT	6.7	Propionic acid	188	2.4	34
	AOT	6.7	Oleyl alcohol	394	2.6	34
Chr. viscosum (Lipase B)	AOT	6.7	Propionic acid	232	0.89	34
	AOT	6.7	Oleyl alcohol	389	1.0	34

Chr., Chromobacterium.

such kinetic constants. It is not evident how much the K_m values reflect the actual affinity of the enzymes towards their substrates and how much they depend on the micellar system used. One reason to be aware of the problem raised by the change from saturation to non-saturation kinetics that occurs when lecithin is substituted for AOT in the hydrolysis of tricaprylin catalysed by Rhizopus delemar (33). The authors suggest that the absence of saturation kinetics is caused by diffusion limitation of tricaprylin in reversed micelles of lecithin. The same authors demonstrate esterification of caprylic acid with glycerol, although the reaction rate was extremely low. Aires-Barros and Cabral found saturation kinetics both in respect of the acyl-donor (propionic acid) and the acyl-acceptor (oleyl alcohol) in esterifications catalysed by two lipases from C. viscosum (34).

Substrate inhibition by alcohols or esters in reversed micelles has not been reported. Aires-Barros and Cabral observed inhibition by propionic acid at concentrations above 130 mM (34). They explained this by physical changes in the media and not by substrate inhibition. Schmidli and Luisi found no inhibition by the less water-soluble octanoic acid at concentrations below 500 mM (33).

The evaluation of lipase kinetics in reversed micelles is difficult, for several reasons. We may be sure that the available interfacial area is unrestricted, or, put differently, the enzyme is saturated by surface. What is not easily understood is how different surfactants give different enzymic activity. Does the surfactant interact directly with the enzyme, or does it create an environment in which the enzyme fails to adopt an optimal conformation for catalysis? In at least one case, it is known that the surfactant interacts directly, as a substrate, with the enzyme (see above and reference 36). The surfactant may change the availability of the substrate by restricting its diffusion. The low stability of the lipases in the systems is disturbing and must be taken account of if long incubation times are used. In several cases, the lipases show clear saturation kinetics, but it is not clear whether the kinetic constants derived reflect the properties of the lipases or of the assay systems.

3.2 Polyethyleneglycol-modified lipases

Enzymes such as lipases, chymotrypsin, catalase and peroxidase have been made soluble and active in organic solvents by covalent modification with polyethylene glycol (PEG) (37). The technique has been used by Inada's group for studies of lipase kinetics in organic solvents and for lipase-catalysed synthesis.

PEG was linked to 58% of the free amino groups of lipoprotein lipase from *Pseudomonas fluorescens* (38). The modified lipase retained 75% of its activity in hydrolysis of olive-oil emulsions in water. The PEG-lipase exhibited saturation kinetics in respect of both acids and alcohols, when catalysing esterifications in benzene. It was shown that V increased with the chain length of the carbon chain of the fatty acid and of the alcohol. K_m values did not show any dependence on chain length and lay for both acids and alcohols in the range 100–300 mM. Fatty acids branched at C-2 or C-3 were not substrates, but showed inhibition with K_i values similar to the K_m of their unbranched analogues. Branched-alcohol substrates were almost as good as the unbranched ones, but the secondary alcohol 1-methylbutyl alcohol gave a value for V that was only 1/26th that of pentyl alcohol. The tertiary alcohol 1,1-dimethylpropyl alcohol was neither a substrate nor an inhibitor.

In a similar study, a lipase from *Pseudomonas fragi* was modified with PEG (39). With 49% of the amino groups bound to PEG, the hydrolytic activity was 43% of that of the natural enzyme. The PEG-lipase was most active in 1,1,1-trichloroethane; lower rates were found for benzene and chloroform. No activity was seen in acetone or dimethylsulphoxide. For the *P. fluorescens* PEG-lipase, the value of V increased with the length of the acid chain. The length of the alcohol, in contrast, did not affect the reaction rate. The K_m value for butyric acid was 370 mM, while longer acids all gave K_m values between 200 and 250 mM. V/K_m increased by a factor of five in going from butyric acid to dodecanoic acid. The enzyme did not accept methyl-substituted pentanoic acids or a double bond in the 2-position. 4-pentenoic acid showed normal kinetics, with parameters identical to those of hexanoic acid. Butanol had a K_m value of 50 mM, while the longer alcohols had K_m values close to 100 mM. Introduction of a methyl group or a

Table 4. Kinetic constants determined for *Pseudomonas fragi* PEG-lipase used in benzene for resolution of secondary alcohols by esterification with *n*-dodecanoic acid.

Alcohol	K_m (mM)		V (U $\cdot$ mg^{-1})		V/K_m (U $\cdot$ mg^{-1} $\cdot$ M^{-1})		E
	(*R*)-	(*S*)-	(*R*)-	(*S*)-	(*R*)-	(*S*)-	
2-Butanol	430		1.18	1.23	2.7	2.9	0.96
2-Pentanol	430	1340	1.25	0.45	2.9	0.3	8.7
2-Octanol	540	1500	1.68	0.42	3.1	0.3	11
2-Nonanol	500	1660	1.37	0.30	2.7	0.2	15

E is the ratio $(V/K_m)_R/(V/K_m)_S$, also known as the enantiomeric ratio (12). Data from reference 28.

double bond in pentanol did not affect the values of K_m, but the values of V were low for 3- and 4-methyl pentanols.

The *P. fragi* PEG-lipase was used for esterification of secondary alcohols by *n*-dodecanoic acid in benzene (40). The enzyme exhibited no enantioselectivity towards 2-butanol. For the homologues 2-pentanol, 2-octanol and 2-nonanol, the enantiomeric ratio increased from 8.7 to 15 in favour of the *R*-enantiomer. The increase in enantioselectivity from 2-butanol to 2-nonanol was caused by an increase in K_m concomitant with a decrease in the value of V for the *S*-enantiomers, while the kinetic constants of the *R*-enantiomers remained constant (Table 4). The lipase showed a very high enantioselectivity against 1-phenyl ethanol. The *R*-enantiomer reacted with kinetic parameters similar to those of the other *R*-alcohols tested, while the *S*-enantiomer did not react at all. The authors did not report any substrate inhibition similar to that seen for unmodified lipases in organic solvents, even though they studied the enzymes with up to 1 M alcohol or 0.5 M acid.

The activity of PEG-enzymes was shown to depend upon their being surrounded by a certain amount of water. This was, according to the authors, the reason why *Pseudomonas fluorescens* lipoprotein lipase was more active in water-immiscible solvents than in water-miscible solvents. In the immiscible solvents, the water remained associated with the enzyme, while the water-miscible solvents dispersed it (41, 42). The addition of water to PEG-lipase in benzene dried with molecular sieve increased the rate of pentyl pentanate synthesis. The highest rate of reaction was observed in water-saturated benzene (42).

Lipases modified with PEG and used in water-saturated benzene should experience only a marginal difference in their physico-chemical environment compared with lipases entrapped in reversed miscelles. These systems should have comparable diffusion barriers for substrates and products, and the activity of the water around the enzyme could be very similar. This may be the reason why the two systems show very similar kinetics.

3.3 Solid and immobilised lipases

There are many examples of enzymes that express catalytic activity in organic solvents. These include proteases, lipases, oxidoreductases and isomerases (43–48). Interest in the application of enzymes in organic solvents has developed very much in the last decade. Among these enzymes, lipases have shown a great potential in different types of synthetic reactions, *e.g.*, the synthesis of asymmetric esters. Concurrently, an interest in basic understanding of enzyme kinetics and action in organic solvents has emerged.

Water is essential for enzymic activity, because it participates in many essential non-covalent interactions within the protein molecule and plays an important rôle in protein dynamics (49). The proteins do not need to be totally dissolved in water to be fully hydrated, as long as the essential water within and at the surface of the molecule is present. Hydration studies have been conducted on lysozyme using calorimetric and a number of spectroscopic techniques, and the amount of water required for full hydration were found to be 0.38 g water per gramme protein (49). This is important for enzyme catalysis in organic media with low water content.

Klibanov *et al.* (50) showed some time ago that porcine pancreatic lipase and lipases from *Candida* and *Mucor* species function in nearly anhydrous media (< 0.01% water), catalysing a variety of reactions such as esterifications, alcoholysis, aminolysis, thiolysis and oximolysis. Lower activity was observed in hydrophilic solvents compared with hydrophobic solvents. This has been interpreted as a desorption of water from the protein by the solvents. This observation has been repeated for other enzymes (51–53). However, large differences have been observed between enzymes. In the case of lipases, porcine pancreatic lipase retained its activity much better in hydrophilic solvents than did lipases from *Mucor* and *Candida* (50). This probably means that porcine pancreatic lipase keeps its essential water more tightly bound than the other lipases do.

Isotherms of the adsorption of water to proteins in other solvents have been established by Yamane *et al.* (54, 55). The studies have revealed dramatic differences in the shapes of the isotherms, depending on the solvent. In water-miscible solvents, for example ethanol, the level of protein-bound water approached saturation. In water-immiscible solvents, for example benzene, no saturation was seen. Instead, the amount of protein-bound water increased with

Table 5. Examples of apparent kinetic constants for lipase catalysis in organic solvents.

Enzyme, Solvent	Substrates		K_m (mM)		V (U/mg)	Reference
	Acyl donor	Acyl acceptor	Acyl donor	Acyl acceptor		
Rhizomucor miehei lipase						
Hexane	Oleic acid	Ethanol	120	190 ($K_i = 40$)	5.7	64
Porcine pancreas lipase						
Hexane	Tributyrin	Methanol	42	33	0.17	50
		Heptanol	29	13	0.08	50
		Dodecanol	83	5.2	0.28	50
	Cl$_3$Et butyrate	Heptanol	39	4.5	0.028	50
	Me butyrate	Heptanol	No saturation	–	0.033	50
Candida cylindracea lipase						
Heptane	Octanoic acid	(±)-Menthol	130	90	–	63
Cyclohexane	Lauric acid	Dilaurin	17	160	0.10	73
Pseudomonas fluorescens lipoprotein lipase						
Acetonitrile	Vn butyrate	1-Butanol	–	42	0.051	74
Cyclohexane	Vn butyrate	1-Butanol	–	13	0.83	74

Data were taken directly from the references or calculated from data in the references. Me, methyl; Cl$_3$Et, trichloroethyl; Vn, vinyl. Concentrations of vinyl butyrate, octanoic acid and menthol were 200, 500 and 500 mM respectively when these were fixed for the determination of K_m for the other substrate.

the amount of free water in the solvent until the latter was saturated and a water phase started to form. These results were the same for four proteins tested, *Pseudomonas* species lipase, chymotrypsinogen, lactoglobulin and bovine serum albumin.

It has also been shown that the carrier on which enzymes are immobilised affects the water distribution in the same manner as the solvents do (56). Hydrophilic carriers cause the desorption of water from the enzyme, while hydrophobic carriers may expel interfering water from the enzyme's environment (57). The water in the system should thus be treated as being in an equilibrium distribution between the protein, the carrier and the solvent. Studies should be made at similar water activities, rather than similar water concentrations, as discussed by Halling (58). Attempts have been made to correlate enzyme activity with the parameters of the solvent and the carrier. Good correlations have been found between enzymic activity and the $\log(P)$ value of the solvent (59, 60) and $\log(Aq)$ of the carrier (56).

All enzymes in water have some kind of activity–pH profile. It has been shown, by using porcine pancreatic lipase, that it is important to have the enzyme in a buffer with the optimal pH for activity prior to dehydration, and to use organic solvents with low or no buffering capacity (50). The enzyme will retain the pH of the water phase unless substrates such as acids or amines change it. Halling *et al.* (61, 62) have measured the pH in the water phase localised around the enzyme in organic solvents, using hydrophobic esters of fluorescein as indicators.

Although the kinetics of hydrolases in water can often be interpreted by simple one-substrate kinetics, the reaction involves water as a second substrate and follows a so-called 'bi-bi-ping-pong' mechanism (28). When these enzymes are used in solvents that do not function as acyl acceptors, the kinetics will be more complex and will call for a two-substrate model. Klibanov *et al.* have shown that porcine pancreatic lipase acts by a bi-bi-ping-pong mechanism, with saturation kinetics shown by both acyl donors and acyl acceptors (50). The apparent values of K_m are listed in Table 5. The table includes some more examples of apparent kinetic constants found in organic media. An interesting observation is the absence of saturation kinetics when the acyl donor is methyl butyrate. No substrate inhibition was observed with porcine pancreatic lipase. In contrast, inhibition of both alcohols and acids were found in studies with *Candida cylindracea* lipase (63). The inhibition was explained by changes in the physical environment, caused by the substrates, rather than by specific substrate–lipase interactions. As in the case of porcine pancreatic lipase, no saturation was achieved with aliphatic esters. In a thorough study of the esterification of oleic acid by ethanol, catalysed by *Mucor (Rhizomucor) miehei* lipase (Lipozyme), Chulalaksananukul *et al.* showed a strong inhibition by ethanol (64). This was shown to be due to a direct substrate–lipase interaction. Their results were consistent with a bi-bi-ping-pong mechanism with a competing dead-end enzyme–ethanol complex (Figure 1). The inhibition was not caused by desorption of water, as the same results were obtained with a high content of water in the enzyme preparation, and it was so strong that the highest observed reaction rate was only 10% of the calculated maximal rate. It should be possible to overcome the inhibition by ethanol by raising the concentration of acid, but no such attempt was reported, in spite of the fact that no inhibition by oleic acid (up to 60 mM) was found. The same type of competitive inhibition was caused by octanol in the esterification of octanoic acid in heptane, catalysed by a lipase from *Candida antarctica* (authors' unpublished observations). The apparent values of K_i and K_m were more than one order of

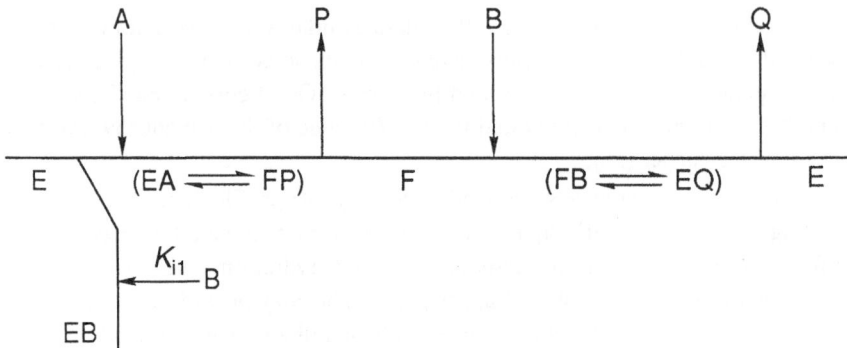

Figure 1. Schematic representation of competitive substrate inhibition in a ping-pong system. This kinetic system was shown to be the best model for esterifications catalysed by lipases from *Candida antarctica* and *Mucor miehei*. In the case of esterification, the symbols in the figure denote the following: A, acid; B, alcohol; P, water; Q, ester; E, enzyme; F, acyl enzyme; EA, EB, EQ, FP, FB, complexes between enzyme species and substrates; K_{i1}, dissociation constant for EB.

magnitude lower than in the case of *Mucor*. Here, the substrate inhibition by octanol could be overcome by high concentrations of the acyl donor, even when the enzyme was saturated with octanol. However, the use of different carriers, substrates and solvents makes comparison of the two enzymes difficult.

Enzyme kinetics in heterogeneous systems have always to be interpreted carefully, because of limitations due to diffusion. External diffusion can be overcome by good agitation. Internal diffusion is more difficult to control, but monolayer immobilisation should minimise this problem. However, the carrier will influence the kinetics to some extent. The importance of the distribution of water in the system has been discussed above, and the same detailed attention should be given to the substrates. The level of saturation at a given substrate concentration may vary, depending on the concentrations of solvent, carrier and water.

The conformations of the active sites of lipases used in organic solvents must be similar to those in aqueous solutions, for otherwise no activity would be seen. The position of the lid in lipases is more hypothetical. As mentioned in section 2, it has to be open at least to some extent, in order to allow the substrate to enter. The hydrophobic environment in organic solvents should be very similar to that experienced by the lipase around the active site when it is bound to an insoluble substrate in aqueous media. There is thus no reason why the lid should not be fully open, and stabilised in that conformation by electrostatic forces and hydrogen bonds between the lid and the rest of the protein.

4. COMPETITION METHODS

True kinetic constants are difficult to determine in most systems in which lipases are studied. As a consequence, a number of competition methods have emerged. These methods are useful because of their versatility in many systems. Ransac *et al.* (65) and de Haas *et al.* (66) have studied competitive inhibition of lipases at the water–lipid interface using water-insoluble inhibitors. They emphasize the importance of having inhibitors with the same physical properties as the substrate, if the result is to reflect interactions at the active site of the lipase. Several general models for studying enantioselectivity in water or organic phases by competition methods have been developed (67–70). In studies with enantiomers, there is no problem with physical discrimination between the competing substrates. The specificity for acyl donors and acceptors of lipases has been studied in organic solvents by competition methods (70–72). These studies of similar substrates will reflect the properties of the catalytic site better than if the kinetic parameters are determined for each substrate individually.

Competition studies with enantiomers should help in comparisons of the properties of the active site of a given lipase used in different assay systems. Such studies could show whether the lipase adopts different conformations in different systems, with consequences for the catalysis.

5. CONCLUSION

In our concluding discussion we refer to the kinetic model depicted in Figure 2. The model involves a conformational change of the lipase and is an extension of the model of Verger *et al.* (14). The only additional feature is an active form of the lipase in solution (E*), which is active with soluble substrates.

The evaluation of lipase kinetics by measuring reaction rates in two-phase emulsion systems is very complex, as many steps are involved in the catalysis. A change in one parameter may influence more than one part of the scheme, making it practically impossible to draw unequivocal conclusions. A second complication is that almost all systems in which lipases can be studied have some kind of boundary between the enzyme and the substrate. This boundary often restricts the diffusion of the substrates and the products.

It is uncertain whether the active form of the enzyme in solution (E*) and the active form adsorbed to the surface (E_s*) have the same conformation, or even whether there are several conformations of each. If we assume that all possible conformations are in equilibrium with each other and have similar properties, then their number will not change the model. Unfortunately, differences in k_{cat} of different conformers, if present, are not easily determined.

Studies of the kinetics with water-soluble substrates would in our model involve E, E* and E*S_w. Such studies thus open the possibility of studying the

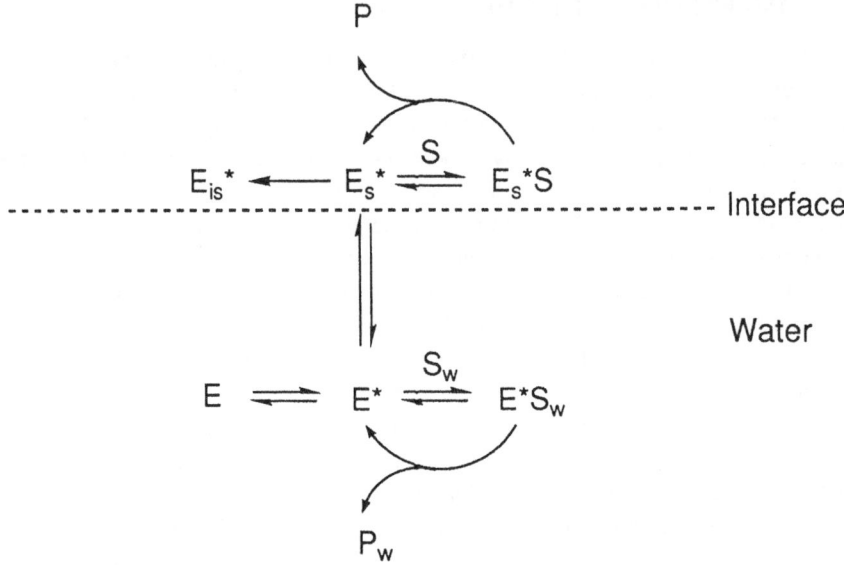

Figure 2. Model for lipase action on soluble and insoluble substrates, by lipases that undergo conformational changes upon activation. The model is an extension of the model proposed by Verger *et al.* (14). The symbols denote: E, inactive dissolved lipase; E*, active dissolved lipase; E_s*, adsorbed active lipase; E_{is}*, adsorbed inactive lipase; S_w, water-soluble substrate; S, water-insoluble substrate; E*S_w and E_s*S, lipase–substrate complexes.

equilibrium between E and E* indirectly. The catalytic properties of E* can also be studied in such systems. Direct studies of E* have not been reported.

Adsorption studies involve E, E_s* and E_{is}*. If there is an equilibrium between E, E* and E_s*, the species E* will not be noticed, if it is not looked for specifically. The interface-induced inhibition of E_s* to E_{is}* cannot be studied without keeping a check on the adsorption step.

Kinetics observed in systems composed of one water phase and one substrate phase will always involve E, E_s*, E_s*S and E_{is}*. Variations in the substrate phase can affect both the adsorption and the catalysis. The result is that exclusive studies of the substrate specificity of the catalytic step are not possible without the knowledge that the adsorption is unchanged. Competitive studies with enantiomers should be possible in emulsion systems. The same enzyme species are involved when lipases are entrapped in reversed micelles. The influence of the surfactants on the lipase complicates the situation, but diffusion restrictions would be small in this system. In the case of PEG-modified lipases, the enzyme species are all changed, and conclusions from kinetic studies should be restricted to the active site itself.

It is not known what type of boundary exists between the lipase and the substrate when the enzyme is used directly in organic solvents. At least in organic solvents with low water activity, only the catalytic part of the model, with E_s^* and E_s^*S, is involved. How similar the active enzyme in organic solvent is to E_s^* remains unknown. A good method to compare these two enzyme species would be to study competition between enantiomers in emulsions and organic solvents.

The main problem in the studies of lipase kinetics is the presence of several phases. The reason for investigation of the kinetics must be kept in mind. For optimisation of industrial processes, an understanding of parts of the model may be enough, but for an understanding of lipase action *in vivo* all steps must be accounted for.

REFERENCES

1. Brockerhoff, H. (1968). Substrate specificity of pancreatic lipase. *Biochim. Biophys. Acta*, 159, 296–303.
2. Brockman, H, Law, J.H. and Kézdy, F.J. (1973). Catalysis by adsorbed enzymes. *J. Biol. Chem.*, 248, 4965–4970.
3. Wells, M.A. (1974). Mechanism of interfacial activation of phospholipase A_2. *Biochemistry*, 13, 2248–2257.
4. Sarda, L. and Desnuelle, P. (1958). Action de la lipase pancréatique sur les esters en émulsion. *Biochim. Biophys. Acta*, 30, 513–521.
5. Entressangles, B. and Desnuelle, P. (1974). Action of pancreatic lipase on monomeric tripropionin in the presence of water-miscible organic compounds. *Biochim. Biophys. Acta*, 341, 437–446.
6. Brzozowski, A.M., Derewenda, U., Derewenda Z.S., Dodson, G.G., Lawson, D.M., Turkenburg, J.P., Björkling, F., Huge-Jensen, B., Patkar, S.A. and Thim, L. (1991). A model for interfacial activation in lipases from the structure of a fungal lipase–inhibitor complex. *Nature*, 351, 491–494.
7. Brady, L., Brzozowski, A.M., Derewenda, Z.S., Dodson, E., Dodson, G., Tolley, S., Turkenburg, J.P., Christiansen, L., Huge-Jensen, B., Norskov, L., Thim, L. and Menge, U. (1990). A serine protease triad forms the catalytic centre of a triglycerol lipase. *Nature*, 343, 767–770.
8. Winkler, F.K., D'Arcy, A. and Hunziker, W. (1990). Structure of human pancreatic lipase. *Nature*, 343, 771–774.
9. Schrag, J.D., Li, Y., Wu, S. and Cygler M. (1991). Ser-His-Glu triad forms the catalytic site of the lipase from *Geotrichum candidum*. *Nature*, 351, 761–764.
10. Thunnissen, M.M.G.M, Ab, E., Kalk, K.H., Drenth, J., Dijkstra, B.W., Kuipers, O.P., Dijkman, R., de Haas, G.H. and Verheij, H.M. (1990). X-ray structure of phospholipases A_2 complexed with a substrate derived inhibitor. *Nature*, 347, 689–691.
11. Cygler, M. (1992). Insight into the mechanism of interfacial activation of lipases provided by the structures of *Geotrichum candidum* and *Candida rugosa* lipases. Communicated to the 3rd. Nordic conference on Protein Engineering, Korpilampi, Finland.

12. Benzonana, G. and Desnuelle, P. (1965). Etude cinétique de l'action de la lipase pancrétique sur des triglycerides en émulsion. Essai d'une enzymologie en milieu heterogène. *Biochim. Biophys. Acta*, 105, 121–136.
13. Rietsch, J., Pattus, F., Desnuelle, P. and Verger, R. (1977). Further studies of mode of action of lipolytic enzymes. *J. Biol. Chem.*, 252, 4313–4318.
14. Verger, R. (1980). Enzyme kinetics in lipolysis. *Methods Enzymol.* 64, 340–392.
15. Verger, R. and de Haas, G.H. (1976). Interfacial enzyme kinetics of lipolysis. *Annu. Rev. Biophys. Bioengin.*, 5, 77–117.
16. Joseph, D., Petsko, G.A. and Karplus, M. (1990). Anatomy of a conformational change: hinged 'lid' motion of the triose phosphate isomerase loop. *Science*, 249, 1425–1428.
17. Holmquist, M., Norin, M. and Hult, K. (1993). The role of arginines in stabilising the active open-lid conformation of *Rhizomucor miehei* lipase. *Lipids* (in the press).
18. Brockman, H.L. and Muderhwa, J.M. (1991). Phospholipid–glyceride interactions as regulators of carboxylester lipase adsorption and catalysis. In *Lipases: Structure, Mechanism and Genetic Engineering*, GBF monographs, vol. 16, ed. Alberghina, L., Schmid, R.D. & Verger, R., pp. 95–104. Weinheim: VCH.
19. Muderhwa, J.M. and Brockman, H.L. (1990). Binding of pancreatic carboxylester lipase to mixed lipid films. Implications for surface organisation. *J. Biol. Chem.*, 265, 19644–19651.
20. Tsujita, T. and Brockman, H.L. (1987). Regulation of carboxylester lipase adsorption to surfaces. 1. Chemical specificity. *Biochemistry*, 26, 8423–8429.
21. Tsujita, T., Muderhwa, J.M. and Brockman, H.L. (1989). Lipid–lipid interactions as regulators of carboxylester lipase activity. *J. Biol. Chem.*, 264, 8612–8618.
22. Tsujita, T., Smaby, J.M. and Brockman, H.L. (1987). Regulation of carboxylester lipase adsorption to surfaces. 2. Physical state specificity. *Biochemistry*, 26, 8430–8434.
23. Gargouri, Y., Piéroni, G., Rivière, C., Sarda, L. and Verger, R. (1986). Inhibition of lipases by proteins: a binding study using dicaprin monolayers. *Biochemistry*, 25, 1733–1738.
24. Verger, R., Rivière, C., Moreau, H., Gargouri, Y., Rogalska, E., Moulin, A., Ransac, S., Carrière, F., Cudrey, C. and Trétout, N. (1991). Enzyme kinetics of lipolysis. Lipase inhibition by proteins. In *Lipases: Structure, Mechanism and Genetic Engineering*, GBF monographs, vol. 16, ed. Alberghina, L., Schmid, R.D. & Verger, R., pp. 105–116. Weinheim: VCH.
25. Brockerhoff, H. (1971). On the function of bile salts and proteins as cofactors of lipase. *J. Biol. Chem.*, 246, 5828–5831.
26. Momsen, W.E. and Brockman, H.L. (1976). Effects of colipase and taurodeoxycholate on the catalytic and physical properties of pancreatic lipase B at an oil–water interface. *J. Biol. Chem.*, 251, 378–383.
27. Sémériva, M., Chapus, C., Bovier-Lapierre, C. and Desnuelle, P. (1974). On the transient formation of an acyl enzyme intermediate during the hydrolysis of *p*-nitrophenyl acetate by pancreatic lipase. *Biochem. Biophys. Res. Commun.*, 58, 808–813.
28. Fersht, A. (1985). *Enzyme Structure and Mechanism*, 2nd edn, New York: W H Freeman.
29. Sugiura, M. and Isobe, M. (1976). Studies on the mechanism of lipase reaction. IV. Action of the lipase from *Chromobacterium* on monomeric *p*-nitrophenyl acetate. *Chem. Pharm. Bull.*, 24, 1822–1828.
30. Sonnet, P.E, Baillargeon, M.W. (1991). Methyl-branched octanoic acids as substrates for lipase-catalyzed reactions. *Lipids*, 26, 295–299.
31. Han, D. and Rhee, J.S. (1986). Characteristics of lipase-catalyzed hydrolysis of olive oil in AOT-isooctane reversed micelles. *Biotech. Bioengin.*, 28, 1250–1255.
32. Kim, T. and Chung, K. (1989). Some characteristics of palm kernel olein hydrolysis by *Rhizopus arrhizus* lipase in reversed micelle of AOT in *iso*-octane, and additive effects. *Enzyme Microb. Technol.*, 11, 528–532.

33. Schmidli, P.K. and Luisi, P.L. (1990). Lipase-catalyzed reactions in reversed micelles formed by soybean lecithin. *Biocatalysis*, 3, 367–376.
34. Aires-Barros, M.R. and Cabral, J.M.S. (1991). Purification and kinetic studies of lipases using reversed micellar systems. In *Lipases: Structure, Mechanism and Genetic Engineering*, GBF monographs, vol. 16, ed. Alberghina, L., Schmid, R.D. & Verger, R., pp. 407–416. Weinheim: VCH.
35. Chang, P.S. and Rhee, J.S. (1990). Characteristics of lipase catalyzed glycerolysis of triglyceride in AOT-isooctane reversed micelles. *Biocatalysis*, 3, 343–355.
36. Skagerlind, P., Jansson, M. and Hult, K. (1992). Surfactant interference in lipase-catalysed reactions in microemulsions. *J. Chem. Biotechnol.*, 54, 277–282.
37. Takahashi, K., Saito, Y. and Inada, Y. (1988). Lipase made active in hydrophobic media. *J. Am. Oil Chem. Soc.*, 65, 911–916.
38. Takahashi, K., Yoshimoto, T., Ajima, A. and Inada, Y. (1984). Modified lipoprotein lipase catalyzes ester synthesis in benzene. *Enzyme*, 32, 235–240.
39. Nishio, T., Takahashi, T., Tsuzuki, T., Yashimoto, T., Kodera, Y., Matsushima, A., Saito, Y. and Inada, Y. (1988). Ester synthesis in benzene by polyethylene glycol-modified lipase from *Pseudomonas fragi* 22.39B. *J. Biotech.*, 8, 39–44.
40. Kikkawa, S., Takahashi, K., Katada, T. and Inada, Y. (1989). Esterification of chiral secondary alcohols with fatty acid in organic solvents by polyethylene glycol-modified lipase. *Biochem. Internat.* 19, 1125–1131.
41. Inada, Y., Nishimura, H., Takahashi, K., Yoshimoto, T., Saha, A.R. and Saito, Y. (1984). Ester synthesis catalyzed by polyethylene glycol-modified lipase in benzene. *Biochem. Biophys. Res. Commun.*, 2, 845–854.
42. Takahashi, K., Nishimura, H., Yoshimoto, Y., Okada, M., Ajima, A., Matsushima, A., Tamaura, Y., Saito, Y. and Inada, Y. (1984). Polyethylene glycol-modified enzymes trap water on their surface and exert enzymic activity in organic solvents. *Biotech. Letters*, 6, 765–770.
43. Chen, C.-S. and Sih, C.J. (1989). General aspects and optimisation of enantioselective biocatalysis in organic solvents: the use of lipases. *Angew. Chem. Int. Ed.*, 28, 695–707.
44. Dordick, J. (1989). Enzymatic catalysis in monophasic organic solvents. *Enzyme Microb. Technol.*, 11, 194–211.
45. Klibanov, A.M. (1989). Enzymatic catalysis in anhydrous organic solvents. *Trends Biochem. Sci.*, 14, 141–144.
46. Klibanov, A.M. (1990). Asymmetric transformations catalyzed by enzymes in organic solvents. *Accts Chem. Res.*, 23, 114–120.
47. Wong, C.-H. (1989). Enzymatic catalysis in organic synthesis. *Science*, 244, 1145–1152.
48. Zaks, A. (1991). Enzymes in organic solvents. In *Biocatalysts for industry*, ed. Dordick, J.S., pp. 161–180. New York and London: Plenum Press.
49. Rupley, J.A., Gratton, E. and Careri, G. (1983). Water and globular proteins. *Trends Biochem. Sci.*, 8, 18–22.
50. Zaks, A. and Klibanov, A.M. (1985). Enzyme catalyzed processes in organic solvents. *Proc. Natl. Acad. Sci. USA*, 82, 3192–3196.
51. Parida, S. and Dordick, J.S. (1991). Substrate structure and solvent hydrophobicity control lipase catalysis and enantioselectivity in organic media. *J. Am. Chem. Soc.* 113, 2253–2259.
52. Zaks, A. and Klibanov, A.M. (1988). Enzyme catalysis in nonaqueous solvents. *J. Biol. Chem.*, 263, 3194–3201.
53. Zaks, A. and Klibanov, A.M. (1988). The effect of water on enzyme action in organic media. *J. Biol. Chem.*, 263, 8017–8021.
54. Yamane, T. (1988). Importance of moisture content control for enzymatic reactions in organic solvents: A novel concept of 'microaqueous'. *Biocatalysis*, 2, 1–9.
55. Yamane, T., Kojima, Y., Ichiryu, T. and Shimizu, S. (1988). Biocatalysis in a

microaqueous organic solvent. *Enzyme Engineering*, 9, 282–293.

56. Reslow, M., Adlercreutz, P. and Mattiasson, B. (1988). On the importance of the support material for bio-organic synthesis. Influence of water partition between solvent, enzyme and solid support in water-poor reaction media. *Eur. J. Biochem.*, 172, 573–578.

57. Norin, M., Boutelje, J., Holmberg, E. and Hult, K. (1988). Lipase immobilised by adsorption. Effect of support hydrophobicity on the reaction rate of ester synthesis in cyclohexane. *Appl. Microbiol. Biotechnol.*, 28, 527–530.

58. Halling, P.J. (1989). Organic liquids and biocatalysts: theory and practice. *Trends Biotechnol.*, 7, 50–52.

59. Laane, C., Boeren, S., Hilhorst, R. and Veeger, C. (1987). Optimisation of biocatalysis in organic media. In *Biocatalysis in Organic Media*, ed. Laane, C., Tramper, J. and Lilly, M.D., pp. 65–84. Amsterdam: Elsevier Science Publishers.

60. Laane, C., Boeren, S., Vos, K. and Veeger, C. (1987). Rules for optimisation of biocatalysis in organic solvents. *Biotechnol. Bioengin.* 30, 81–87.

61. Brown, L., Halling, P.J., Johnston, G.A., Suckling, C.J. and Valivety, R.H. (1990). Water insoluble indicators for the measurement of pH in water immiscible solvents. *Tetrahedron Letters*, 31, 5799–5802.

62. Valivety, R.H,. Rakels, J.L.L., Blanco, R.M., Johnston, G.A., Brown, L., Suckling, C.J. and Halling, P. (1990). Measurement of pH changes in an inaccessible aqueous phase during biocatalysis in organic media. *Biotech. Letters*, 12, 475–480.

63. Langrand, G.A. (1986). Reactions catalysées par les lipases en emulsion et en milieu organique. Competition et resolution. Application a la seperation de (–)menthol. Thesis Université de Droit, d'Economie et des Sciences d'Aix-Marseille.

64. Chulalaksananukul, W., Condoret, J.S., Delorme, P. and Willemont, R.M. (1990). Kinetic study of esterification by immobilised lipase in *n*-hexane. *FEBS Letters*, 276, 181–184.

65. Ransac, S., Rivière, C., Soulié, J.M., Gancet, C., Verger, R. and de Haas, G.H. (1990). Competitive inhibition of lipolytic enzymes. I. A kinetic model applicable to water-insoluble competitive inhibitors. *Biochim. Biophys. Acta*, 1043, 57–66.

66. de Haas G.H., Dijkman, R., van Oort, M.G. and Verger, R. (1990). Competitive inhibition of lipolytic enzymes. III. Some acylamino analogues of phospholipids are potent competitive inhibitors of porcine pancreatic phospholipase A_2. *Biochim. Biophys. Acta*, 1043, 75–82.

67. Chen, C.-S., Fujimoto, Y., Girdaukas, G and Sih, C.J. (1982). Quantitative analysis of biochemical kinetic resolutions of enantiomers. *J. Am. Chem. Soc.*, 104, 7294–7299.

68. Chen, C.-S., Wu, S.-H., Girdaukas, G. and Sih, C.J. (1987). Quantitative analysis of biochemical kinetic resolution of enantiomers. 2. Enzyme catalyzed esterifications in water–organic solvent biphasic systems. *J. Am. Chem. Soc.*, 109, 2812–2817.

69. Jongejan, J.A., van Tol, J.B.A., Geerlof, A. and Duine, J.A. (1991). Enantioselective enzymatic catalysis. 1. A novel method to determine the enantiomeric ratio. *Rec. Trav. Chim. Pays-Bas*, 110, 247–254.

70. Langrand, G., Baratti, J., Buono, G. and Triantaphylides, C. (1988). Enzymatic separation and resolution of nucleophiles: a predictive kinetic model. *Biocatalysis*, 1, 231–248.

71. Deleuze, H., Langrand, G., Millet, H., Baratti, J., Buono, G. and Triantaphylides, C. (1987). Lipase-catalyzed reactions in organic media: competition and applications. *Biochim. Biophys. Acta*, 91, 117–120.

72. Rangheard, M.-S., Langrand, G., Triantaphylides, C. and Baratti, J. (1989). Multi-competitive enzymatic reactions in organic media: a simple test for the determination of lipase fatty acid specificity. *Biochim. Biophys. Acta*, 1004, 20–28.

73. Miller, D.A., Prausnitz, J.M. and Blanch, H.W. (1991). Kinetics of lipase-catalysed interesterfication of triglycerides in cyclohexane. *Enzyme Microb. Technol.*, 13, 98–103.

74. Adams, K.A.H., Chung, S.-H. and Klibanov, A.M. (1990). Kinetic isotope effect investigation of enzyme mechanism in organic solvents. *J. Am. Chem. Soc.*, 112, 9418–9419.

9.

Gastric lipases: cellular, biochemical and kinetic aspects

FRÉDÉRIC CARRIÈRE, YOUSSEF GARGOURI, HERVÉ MOREAU, STEPHANE RANSAC, EWA ROGALSKA and ROBERT VERGER

Until recently, the hydrolysis of dietary triglycerides was thought to begin in the intestinal lumen and to be catalysed entirely by pancreatic lipase. The stomach was thought of as a transient storage organ, the rôle of which was limited to mixing and dispersing lipids with the other nutriments. Although many authors observed the occurrence of a pre-duodenal lipolysis in humans and several other species, the gastric phase of lipolysis was assumed to be negligible and without significance in comparison with the intestinal step. Gastric lipolysis was even attributed to pancreatic contamination after a duodeno-gastric reflux. In the middle of the last century, however, it was observed that gastric juice could hydrolyse fat. In 1901, Volhard (1) stated that gastric lipase is the 'ferment' present in gastric juice and that it is responsible for fat hydrolysis. Finally, as early as 1917, the gastric origin of the lipase present in dog gastric juice was demonstrated by Hull and Keaton (2) in dogs with Pavlov stomachs under conditions precluding the possibility of any pancreatic contamination.

These findings were neglected for many years, until the physiological importance of pre-duodenal lipolysis was demonstrated indirectly, both in cases of pancreatic lipase deficiency, such as cystic fibrosis and chronic pancreatitis, and in premature infants, where the pancreatic enzyme and bile secretion levels are very low. In all these cases, about 70% of the ingested dietary fat is absorbed (3–5). An earlier study on the mechanism of fat absorption in patients with congenital isolated pancreatic lipase deficiency (6) showed that in the complete absence of pancreatic lipase more than 50% of the ingested dietary fat was absorbed. This demonstrated that a second lipolytic enzyme is involved in triglyceride digestion. The exact physiological contribution of the pre-duodenal lipase to overall lipolysis has not yet been clearly established, however. Some data suggest that this enzyme may act preferentially under duodenal conditions (7), co-operating with pancreatic lipase (8).

In order to purify and characterise pre-duodenal lipases, screening assays were carried out along the upper digestive tract (from the tongue to the pylorus) of

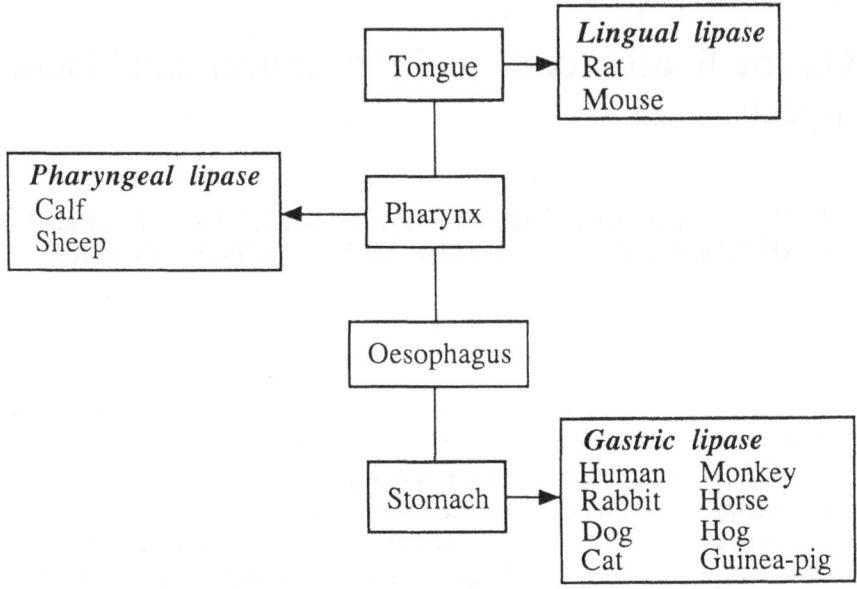

Figure 1. Anatomical localisation of pre-duodenal lipases in several mammals.

several mammals (9, 10). In each species tested, the pre-duodenal lipase activity was associated mainly with a single tissue, which was located either in the lingual, the pharyngeal or the gastric area (Figure 1). Unlike pancreatic lipase, all these lipases were characterised by their high stability and high levels of enzymic activity under acidic conditions. To date, rat lingual (11–13), human gastric (HGL; reference 14), calf pharyngeal (15), rabbit gastric (RGL; 16) and dog gastric (DGL; 17) lipases have been purified and characterised biochemically. In this review, we deal mostly with gastric lipases and illustrate our discussion with results obtained in our laboratory.

1. CELLULAR LOCATION AND SECRETION OF GASTRIC LIPASES

The choice of the enzyme source is of prime importance if a satisfactory degree of purification is to be achieved. A good knowledge of the anatomical distribution of gastric lipases is therefore needed when this enzyme is to be purified directly from gastric tissue, as in the case of RGL (16) or DGL (17). On the other hand, when the enzyme is to be purified from gastric juice collected *in vivo*, as in the case of HGL, it is important to know which cells secrete the enzyme and which drugs stimulate its secretion. These are some of the reasons why we first investigated the tissular and cellular localisation of gastric lipases.

In the three different species studied up to now (human, rabbit and dog), the tissular and cellular distributions of gastric lipase have been found to be different. HGL and RGL are present in restricted areas of the stomach: HGL is found only in the fundic mucosa (18, 19), *i.e.*, from the cardia to approximately the middle of the human stomach, whereas RGL is localised in the small cardiac mucosa close to the œsophagus (9, 20). In the case of the dog, gastric lipase activity was found to occur in the whole gastric mucosa, with a decreasing concentration gradient from the cardia to the pylorus (17). As regards the cellular localisation (for a general scheme of a gastric gland, see Figure 2(*d*)), HGL was found to be co-localised with pepsinogen in chief cells of fundic glands (21), RGL was located in different chief cells from those producing pepsinogen (20), and DGL was detected in mucous pit cells, while dog pepsinogen was present in chief cells (Figure 2(*a,c*); reference 22).

HGL was purified from human gastric juice (14). The secretion of HGL is stimulated by intravenous injection of pentagastrin (23, 24), which was previously known to stimulate the secretion of pepsinogen. This is in good agreement with the cellular co-localisation of these two enzymes. Pentagastrin increases the lipase output but not the enzyme concentration in the gastric juice. Other gastro-intestinal hormones, such as secretin and CCK-PZ, had no effect on the secretion of HGL (23).

Using dogs with chronic gastric fistulae, we first observed the absence or low level of DGL activity in basal or stimulated gastric secretion, despite the fact that the gastric mucosa contains substantial amounts of lipolytic activity (17). This seeming paradox was explained when we observed that DGL was inactivated irreversibly by gastric acid below pH 1.5, such as is found in the stomach under fasting conditions. By buffering the acid secretion *in vivo*, or by using an anti-acid secretion drug such as Omeprazole during stimulation, we determined the effects of several gastric secretagogues on the secretion of DGL (Figure 3; reference 22). Unlike HGL secretion, DGL secretion was poorly stimulated by pentagastrin, while Urecholine, 16,16-dimethyl prostaglandin E2 and secretin were potent secretagogues. These secretagogues were also known to stimulate gastric mucus secretion. As predicted from the cellular localisation of DGL, a clear-cut parallel was observed between the secretion of DGL and that of gastric mucus.

2. GASTRIC LIPASES ARE GLYCOPROTEINS

HGL was cloned and expressed in *E. coli* or yeast (25). The amino-acid sequence derived from cDNA showed that HGL consisted of a single 379-amino-acid

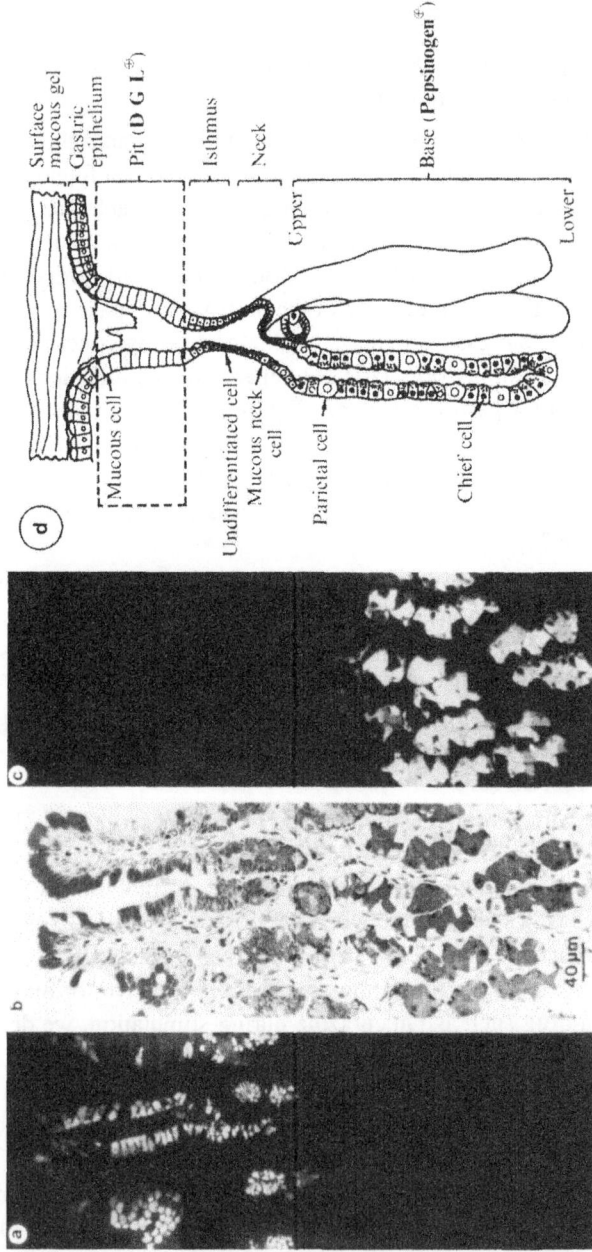

Figure 2. Dog fundic gland immunolabelling and scheme of fundic gastric gland. Biopsy sections were incubated with guinea-pig antibodies against (*a*) DGL and (*c*) dog pepsinogen, and the binding locations were revealed by fluorescein-labelled anti-guinea-pig immunoglobulins. (*b*) Another section was stained with Azur II for general identification of the cells in the fundic gland. (*d*) Scheme of fundic gastric gland with localisation of the different cellular types. (From reference 22.)

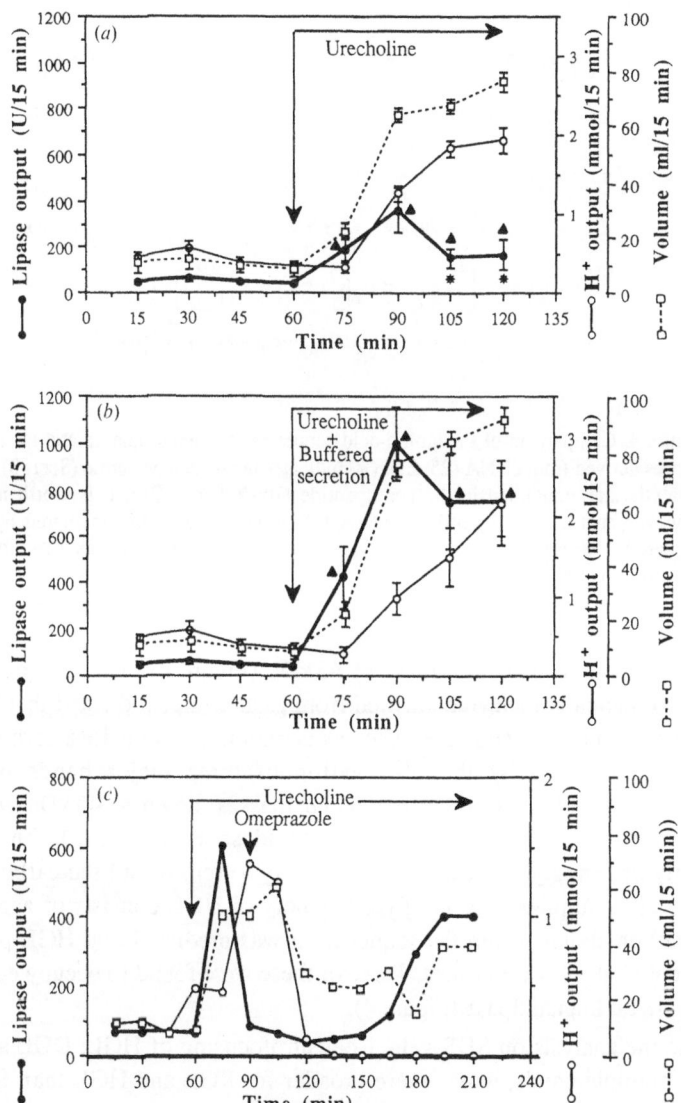

Figure 3. Effect of intravenous infusion of urecholine (200 µg·kg⁻¹h⁻¹) on DCL secretion. (*a*) Infusion of urecholine alone after a 60-min basal period. (*b*) Infusion of urecholine associated with 1 M glycine.HCl buffer, pH 6.0, sprayed regularly into the stomach after a 60-min basal period. (*c*) Infusion of urecholine associated with Omeprazole (10 mg). The urecholine stimulation is launched after a 60-min basal period and Omeprazole is administrated intraduodenally as a bolus dose 30 min after the start of stimulation. By buffering gastric secretion or by using Omeprazole, secreted DGL is preserved from inactivation by gastric acid secretion. (From reference 22.)

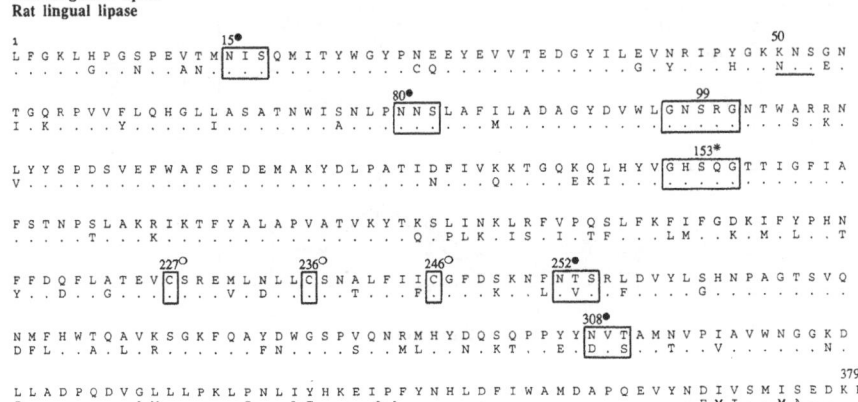

Figure 4. Comparison of the amino-acid sequences of human gastric and rat lingual lipases derived from cDNA (25, 27). (✱) indicates the presumed serine (Ser153) of the catalytic site, contained in the pentapeptide Gly-X-Ser-X-Gly. It is worth noting, however, that a second similar pentapeptide (containing Ser99) is shared by both sequences. (●) represents potentially glycosylated asparagine residues. (○) indicates the position of cysteine residues.

polypeptide with a molecular mass of 43 kDa (Figure 4). The molecular mass (50 kDa) of the native enzyme purified from gastric juice indicated that HGL, as well as RGL and DGL, are glycoproteins containing around 15% carbohydrate. This was confirmed by the observation of three major bands on SDS-polyacrylamide gels with lower molecular masses, down to 43 kDa, when the HGL was first incubated with endoglycosidase F (Figure 5). The known specificity of endoglycosidase F and its ability to remove all the carbohydrates linked to HGL implies that the glycan moiety of HGL consists of asparagine-linked carbohydrates. From the sequence, it was predicted that HGL possesses four potential glycosylation sites. Three of these were found to occupy equivalent positions in rat lingual lipase (Figure 4).

Unlike the analysis on SDS gels, isoelectrofocusing of HGL, DGL and RGL showed multiple bands, which were broader for RGL and HGL than for DGL. The observed isoelectric points ranged between 5.8 and 7.2 with RGL, between 6.8 and 7.8 with HGL (four isoforms) and between 6.3 and 6.5 with DGL.

Several isoforms of RGL and HGL have been purified by FPLC. All the purified isoforms were found to have the same specific lipase activity (around 1200 units per mg of protein, measured with tributyrin as substrate). The isoforms of DGL are more closely related, and could not be separated (26).

Several crystallisation trials on purified native preparations of RGL and HGL were unsuccessful, whereas good-quality crystals were obtained from native DGL

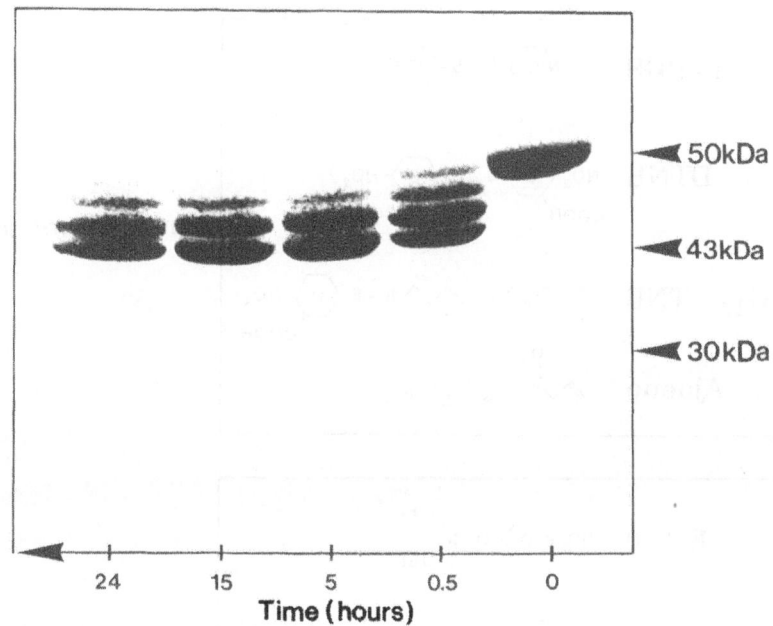

Figure 5. SDS-polyacrylamide gel electrophoresis of HGL after treatment with EndoF and various times of incubation.

and all the purified isoforms of RGL and HGL. Charge homogeneity seemed to be essential to obtain single crystals. The best crystals were obtained with DGL and were found to be orthorhombic, and to belong to the space group $P2_12_12$, with the cell dimensions $a = 182.8\,\text{Å}$, $b = 211.2\,\text{Å}$ and $c = 97.9\,\text{Å}$. They diffracted to a resolution of $4.5\,\text{Å}$ (26).

All of the crystals obtained had large cell dimensions, some extremely large, probably with four to eight molecules in the asymmetric unit in the case of DGL. The tendancy to crystallise in this way is probably connected with the fact that large amounts of carbohydrates are present in the protein.

3. GASTRIC LIPASES ARE SERINE ESTERASES

A high degree of sequence identity (up to 80%; Figure 4) was found between HGL (25) and rat lingual lipase (27). The recent molecular cloning of cDNA for human lysosomal acid lipase (HLAL) demonstrated that pre-duodenal and lysosomal lipases belong to a new lipase gene family (28). The HLAL amino-acid sequence was found to be 58% and 57% identical with those of HGL and rat lingual lipases, respectively. No sequence homology between these lipases and other mammalian or microbial lipases was detected, although they all possessed

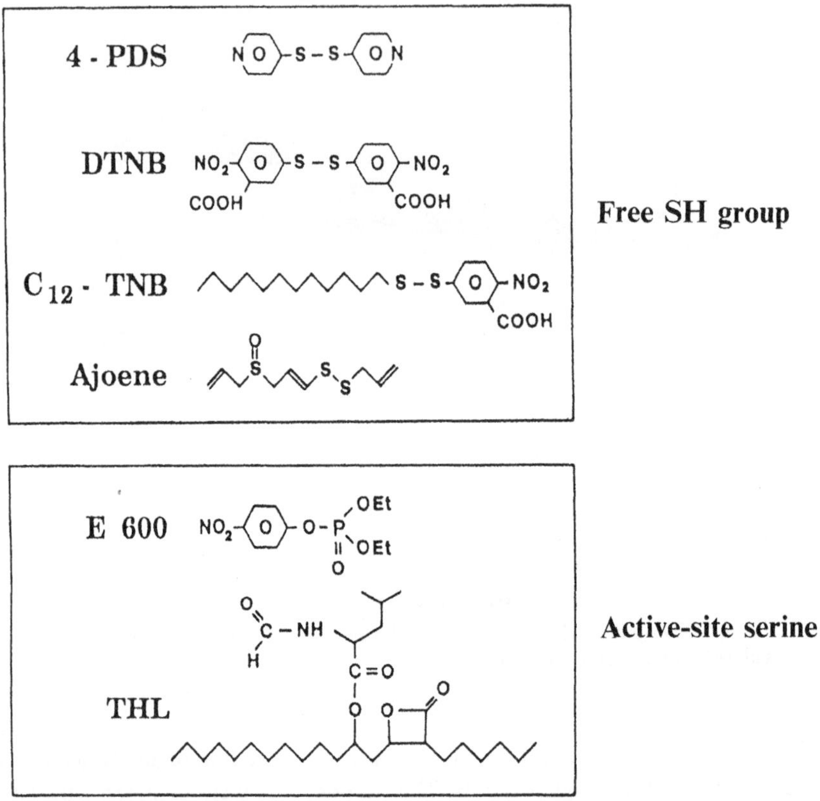

Figure 6. Chemical structures of some inhibitors of gastric lipases.

the pentapeptide Gly-X-Ser-X-Gly, a motif shared by all known lipases (29–31) as well as serine proteases such as trypsin. The conserved serine within this pentapeptide (Ser153 in HGL) corresponds to Ser152 in porcine and human pancreatic lipases (32, 33) and to Ser144 in *Mucor miehei* lipases (34). As shown by the three-dimensional structures of human pancreatic (33) and *M. miehei* (34) lipases, this essential serine is part of the classical Asp...His...Ser triad, and probably constitutes the nucleophilic residue essential for catalysis.

Gastric lipases are inhibited specifically by micellar diethyl *p*-nitrophenyl phosphate (E600, Figures 6 and 7(*a*); reference 35), which is a well-known inhibitor of serine esterases. Reaction of gastric or pancreatic lipases with mixed E600/bile-salt micelles resulted in a stoichiometric inactivation of these enzymes, as tested with emulsified tributyrin and triolein as substrates. Gastric lipases (HGL and RGL) inactivated by E600 were also inactive towards the water-soluble substrate *p*-nitrophenyl acetate, whereas the modified pancreatic lipase was still

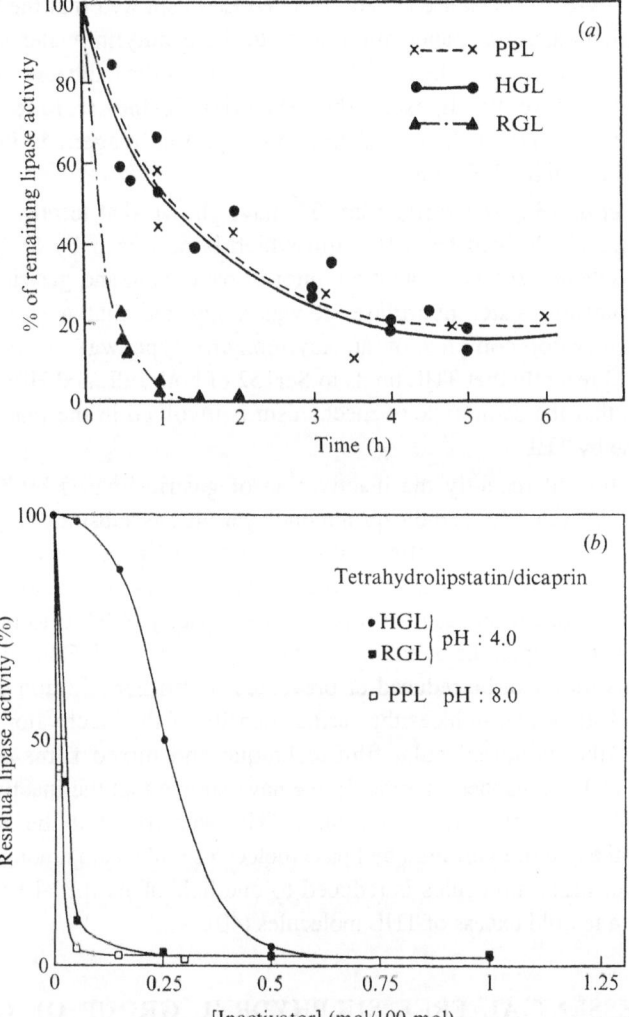

Figure 7. Inactivation of gastric lipases by serine reagents. (*a*) Time-course of inactivation of HGL, RGL and PPL during incubation with radiolabelled E600 (molar ratio E600:lipase = 84:1). Lipases (25 nmol) were incubated with 2.1 μmol E600 (200 μCi) at 25 °C in 50 mM acetate buffer, pH 6.0, 50 mM NaCl, 25 mM CaCl$_2$, and 3 mM NaTDC. Residual lipase activity was measured on tributyrin as substrate (35). (*b*) Effect of increasing surface concentration of THL on the rate of hydrolysis of dicaprin monolayers at a constant surface pressure (25 mN/m) by HGL (final concentration 2 nM; ●) or RGL (final concentration 0.4 nM; ■) or PPL (final concentration 0.2 nM; □). The subphase was 50 mM glycine.HCl, pH 4.0, 100 mM NaCl and 10 mM CaCl$_2$ in the HGL and RGL assays, and 10 mM Tris.HCl, pH 8.0, 100 mM NaCl, 21 mM CaCl$_2$ and 1 mM EDTA in the PPL assays. (From reference 40.)

able to hydrolyse this. The use of radiolabelled E600 showed that the binding of E600-modified gastric or pancreatic lipases to the tributyrin–water interface is comparable to that of native lipases. All in all, these results indicated that, in both gastric and pancreatic lipases, the essential serine residue that was stoichiometrically labelled by this organophosphorus reagent is involved in catalysis and not in lipid binding.

Hadvàry *et al.* (36) and Borgström (37) have shown that tetrahydrolipstatin (THL, Figure 6), derived from lipstatin which is produced by *Streptomyces toxytricini,* acts *in vitro* as a potent inhibitor of pancreatic and gastric lipases as well as cholesterol-ester hydrolase. It was suggested that a stoichiometric enzyme–inactivator complex of an acyl-enzyme type was formed. It was demonstrated recently that THL binds to Ser152 of both PPL and HPL (38). One can assume that the same type of mechanism is involved in the inactivation of gastric lipase by THL.

We investigated recently the inactivation of gastric lipases by THL using emulsified tributyrin (39) and dicaprin monolayer (40) as substrate (Figure 7(*b*)). In the presence of an emulsified substrate, gastric lipases were completely inactivated when THL was injected before or after the enzyme addition, and this is in agreement with earlier results (36, 37). The capacity of THL to inactivate is directly dependent upon the interfacial area of the system used. Consequently, the lipase inactivation can be reduced or prevented by further addition of a water-insoluble substrate that reduces the surface density of the inactivator molecules (40). Using the monomolecular-film technique and mixed films of dicaprin containing THL ('poisoned interface'), we have shown that the inactivation was independent upon the surface pressure. THL was found to be a powerful inactivator: the rate of hydrolysis by lipase molecules embedded among a 10^5-fold excess of substrate molecules is reduced to one-half of its initial value by the presence of a tenfold excess of THL molecules (40).

4. THE ESSENTIAL FREE SULPHYDRYL GROUP OF GASTRIC LIPASES

The amino-acid composition reveals three cysteine residues per molecule of HGL (25), RGL (17) and DGL (17). According to the amino-acid sequence of HGL (25), these three cysteines are clustered between residues 227 and 244 (Figure 4).

Titration of HGL (41), RGL (42) or DGL (17) by incubation with classical sulphydryl reagents, such as dithio-nitrobenzoic acid, DTNB, or di(thiopyridine), 4-PDS (Figure 6), revealed one titrable sulphydryl group per enzyme molecule (Figure 8). Further addition of a denaturing agent (SDS or urea) did not increase the number of modified SH groups. As only one free sulphydryl group was titrated, it can be concluded that gastric lipases contain a single disulphide bridge.

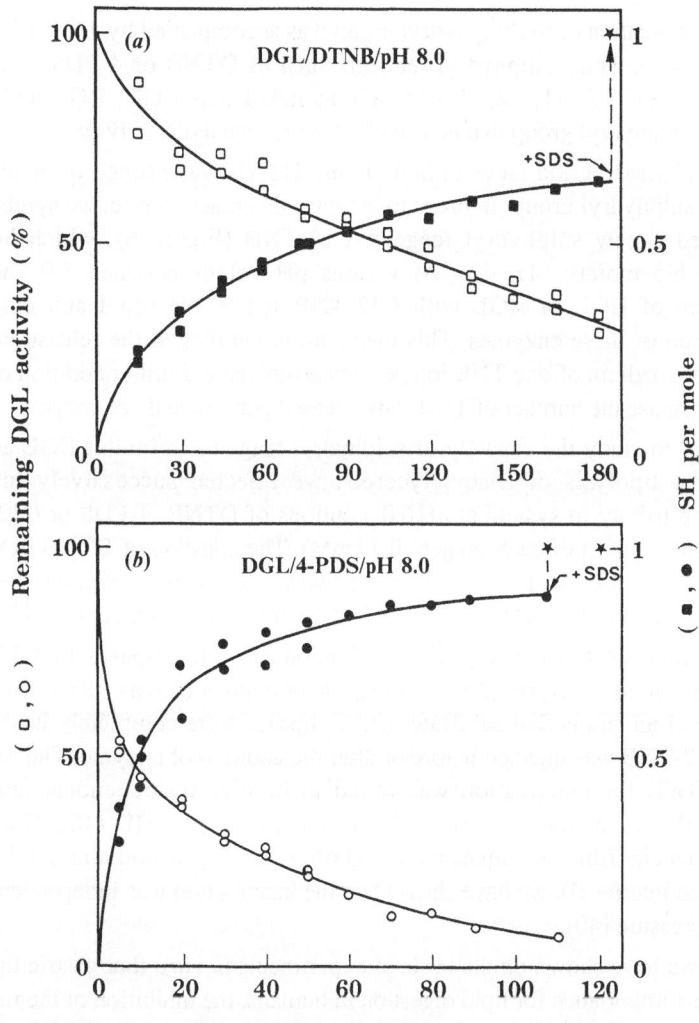

Time (min)

Figure 8. Variation in DGL activity and sulphydryl-group modification during incubation with sulphydryl reagents. (*a*) DGL (13 nmol) was incubated at pH 8.0 and 25 °C with DTNB (1200 nmol). The 2-nitro-5-thiobenzoic acid (TNB) released was measured spectrophotometrically at 412 nm ($A_{412} = 13\,600$). The number of sulphydryl groups modified per DGL molecule was calculated at various incubation times (■). In parallel, the residual DGL activity was measured by tributyrin hydrolysis (□). (*b*) DGL (13 nmol) was incubated at pH 8.0 and 25 °C with 4-PDS (1200 nmol). The 4-thiopyridone released was measured spectrophotometrically at 324 nm ($A_{324} = 19\,800$). The number of sulphydryl groups modified per human gastric lipase molecule was calculated at various incubation times (●). Residual DGL activity was measured on tributyrin as substrate (○). The arrows indicate the injection of 3% SDS, and ★ indicates the value reached on addition of SDS. (From reference 17.)

The modification of the sulphydryl group was accompanied by a loss of gastric lipase activity. Thus, sulphydryl reagents such as DTNB or 4-PDS inactivate gastric lipases (17, 41, 42). It can be concluded that HGL, RGL and DGL possess a sulphydryl group that is essential for their catalytic activity.

An incubation period (several hours) with DTNB was needed to modify the essential sulphydryl group. In order to increase this reaction rate, we synthesized and tested a new sulphydryl reagent, C12-TNB (Figure 6), which bears a hydrophobic moiety (41, 42). At various pH values between 3.0 and 8.0, incubation of HGL or RGL with C12-TNB led to the rapid and complete inactivation of these enzymes. This inactivation parallelled the release into the incubation medium of one TNB ion per lipase molecule. Further addition of SDS did not increase the number of TNB ions released per molecule of enzyme.

In order to study the capacity of sulphydryl reagents to inhibit HGL activity during the lipolysis of triacylglycerols, we injected successively into the emulsified tributyrin system at pH 6.0 solutions of DTNB, 4-PDS or C12-TNB (final concentration of each reagent 0.12 mM). The addition of DTNB or 4-PDS after HGL injection had no effect on the hydrolysis of tributyrin. On the other hand, C12-TNB abolished the activity of HGL almost immediately (41, 42).

We re-investigated recently the inactivation of gastric lipases by C12-TNB using emulsified tributyrin (39) and dicaprin monolayer (40) as substrates. In the presence of an emulsified substrate, gastric lipases were completely inactivated when C12-TNB was injected before or after the addition of enzyme. The capacity of C12-TNB for inactivation was found to be directly dependent upon the interfacial area of the system used, as in the case of THL (40). Using the monomolecular film technique and mixed films of dicaprin containing C12-TNB ('poisoned interface'), we have shown that the inactivation was independent upon surface pressure (40).

Since we have shown on the basis of experiments *in vitro* that gastric lipolysis is of prime importance for lipid digestion in humans, the inhibition of the first step in gastrointestinal lipolysis could induce a decrease in the overall dietary fat absorption. In studies *in vitro*, Gargouri *et al.* (43) selectively inactivated HGL and RGL after pre-incubation with ajoene (a disulphide derived from garlic extracts, Figure 6), whereas diallyldisulphide was found to be much less effective. On the other hand, pancreatic lipase remains fully active after these disulphide treatments.

A priori, the loss of activity after chemical modification of a lipolytic enzyme seems to be attributable either to a lack of lipid binding or to blockage of the catalytic site. In order to determine what catalytic step was affected by the modification of the enzyme, we prepared [^{14}C]-TNB gastric lipases and studied their interactions with lipid–water interfaces (44). It was established that, after modification of one sulphydryl group, gastric lipases continue to bind to monomolecular films of phosphatidylcholine or dicaprin.

We can thus conclude that sulphydryl modification affects the hydrolysis step and not the interfacial binding step, that is, that TNB-lipases are still capable of adsorbtion to penetration into lipid monolayers while, unlike native lipases, they are unable to catalyse the hydrolysis of the ester bond. These results, which were quite unexpected, provide positive evidence that the lipid-binding domain in gastric lipases is functionally and topologically distinct from the catalytic site.

On the basis of the inactivation studies described here, we have shown that gastric lipases are 'sulphydryl enzymes'. This has since been interpreted as indicating that the sulphydryl group is part of the catalytic site. However, there is no direct evidence for this. Moreover, gastric lipases are probably serine enzymes, like other lipases. In conclusion, it is tentatively proposed that the modification of the sulphydryl group prevents the access of the substrate to the catalytic site.

5. EFFECTS OF AMPHIPHILES AND INTERFACIAL TENSION ON GASTRIC LIPASE ACTIVITY

Using pure HGL *in vitro* under conditions similar to those prevailing in the stomach and the upper small intestine, we investigated the effects of human bile salts and non-steroid detergents on HGL activity, with tributyrin as substrate. The effects of phosphatidylcholine (PC) and proteins (including some that are part of the normal diet: β-lactoglobulin, ovalbumin, protein-inhibiting lipase from soya) were also determined (7).

Bile salts

It has been established clearly that bile salts activate HGL, whereas they inhibit the activity of pure pancreatic lipase *in vitro*.

HGL activity was measured in the presence of various concentrations of the individual bile salts present in human bile, and also in a mixture of pure bile salts prepared in the same proportions as that found in human bile.

In the absence of bile salt, no HGL activity was detected. Conversely, in the presence of glycocholate, bile-salt mixture, taurocholate or taurodeoxycholate (TDC), HGL activity increased with increasing concentrations of bile salt up to a plateau; glycodeoxycholate, glycochenodeoxycholate and taurochenodeoxycholate activated HGL less strongly. With these bile salts, a specific activity of only 340 units/mg was reached at a bile-salt concentration of 8 mM, as compared with 600 units/mg obtained with TDC at the same concentration. In the presence of mixed bile salts at a concentration of 8 mM, the specific activity of HGL was 450 units/mg.

Synthetic detergents

Non-steroid detergents, both charged and neutral, were added to the assay system in order to ascertain whether the activating effects of bile salt on HGL activity were a general phenomenon, common to all detergents.

Each detergent was added to the tributyrin emulsion one minute before injection of HGL. In the presence of 2 mM of SDS, benzalkonium chloride, Tween 80 or Triton X-100, no HGL activity was detected. In the presence of 2 mM TM3-14 (a zwitterionic detergent), only 12.5% of the maximal HGL activity was found. When SDS or benzalkonium chloride was present in the assay system, a further addition of 5 mM TDC, five minutes after HGL injection, failed to activate the enzyme, whereas in the presence of TM3-14, Tween 80 or Triton X-100, 37%, 36% and 34% (respectively) of the maximal HGL activity was measured after the addition of TDC (5 mM).

The effects of synthetic detergents added during the course of tributyrin hydrolysis by HGL in the presence of 5 mM TDC were also studied. Triton X-100 and benzalkonium chloride reduced HGL activity to less than 5% of the value recorded in the presence of TDC prior to detergent addition. TM3-14, SDS and Tween-80 reduced the HGL activity to 37%, 22.5% and 20% of its initial value, respectively. These results indicate that all synthetic detergents inhibit HGL, even though bile salt is present in the assay system.

Phosphatidylcholine

Egg PC was added to tributyrin emulsion before the addition of enzyme. HGL activity increased with increasing concentration of egg PC in the assay system. At a PC concentration of 5 mM, the specific activity of HGL was 345 units/mg. Further addition of TDC (5 mM) or mixed bile salts had no effect on the rate of hydrolysis, whereas the simultaneous addition of bile salts (5 mM) and PC, one minute before the enzyme, increased the HGL specific activity up to 545 units/mg.

Proteins

The rate of hydrolysis of emulsified tributyrin was determined as a function of the protein concentration. Various proteins were injected two minutes before HGL addition. The activity of HGL increased with the concentration of β-lactoglobulin, bovine serum albumin (BSA) or ovalbumin. Maximal specific activities of 340, 280 and 240 units/mg, respectively, were obtained at a protein concentration of 8 μM. Injection of 5 mM TDC, five minutes after addition of enzyme in the presence of β-lactoglobulin or BSA, increased HGL activity to a value equal to that measured with bile salts alone. When myoglobin, mellitin and protein-inhibiting lipase from soya were injected prior to the enzyme, no activity was detected, even after further addition of 5 mM bile salts.

In order to investigate the lack of HGL activity in the absence of amphiphiles, the influence of the consecutive addition of BSA, bile salt and enzyme on the kinetics of hydrolysis of emulsified tributyrin was studied. No activity was detected when HGL (8 units) was injected into the system containing only tributyrin emulsion. Addition of BSA or mixed bile salts four minutes after HGL injection failed to induce enzyme activity. However, if the enzyme (8 units) was injected in the presence of BSA or bile salts, all the expected HGL activity (8 units) was exhibited. Addition of β-lactoglobulin, ovalbumin or PC after enzyme injection also failed to induce HGL activity in the tributyrin assay in the absence of bile salts. In all cases, the addition of pure horse pancreatic colipase (colipase:HGL = 20:1 mol/mol) had no effect on the rate of hydrolysis of tributyrin by HGL, either in the presence or in the absence of amphiphile (7).

Amphiphiles

The relationship between the activity of HGL, measured in the presence of an amphiphile, and the lowering of surface tension at the triacylglycerol–water interface brought about by the amphiphile was determined. Bile salts (TDC), which activate HGL, decreased the interfacial tension from 15 to 9 dyne/cm. This contrasted with synthetic detergents, all of which inactivated HGL and reduced the interfacial tension from 15 to less than 3 dyne/cm (Figure 9(*a*)). It is thus tempting to speculate that the low interfacial tension value reached in the presence of the synthetic detergents might be responsible for the inactivation of HGL. Figure 9(*b*) shows that there is a correlation between HGL activity and the interfacial tension of tributyrin in the presence of various amphiphiles. It appears from this correlation curve that, among the detergents tested, bile salts have the unique property of allowing HGL to be active. Among the proteins tested in these studies, only mellitin, myoglobin and protein-inhibiting lipase from soya, which strongly decreased interfacial tension, were found to inhibit HGL (7).

It can be concluded that, in order to measure HGL activity with tributyrin as substrate, bile salts, proteins or PC should be added prior to HGL. Addition of amphiphile decreases the interfacial energy of the tributyrin–water interface. The existence of a correlation between HGL activity and surface tension at the substrate–water interface has been established (45, 46). Figure 9(*b*) shows clearly that HGL activity is restricted to an interfacial tension ranging from 8 to 13 dyne/cm. A pure triacylglycerol–water interfacial tension of 15 dyne/cm seems rapidly and irreversibly to inactivate HGL. A tension of less than 6 dyne/cm, which was obtained in the presence of all synthetic detergents and some highly tensioactive proteins, mellitin, myoglobin and protein-inhibiting lipase from soya, results in no HGL activity. The only amphiphiles compatible with HGL activity are most alimentary proteins and naturally occurring bile salts; both of these decreased the triacylglycerol–water interfacial tension to values between 8 and 13 dyne/cm.

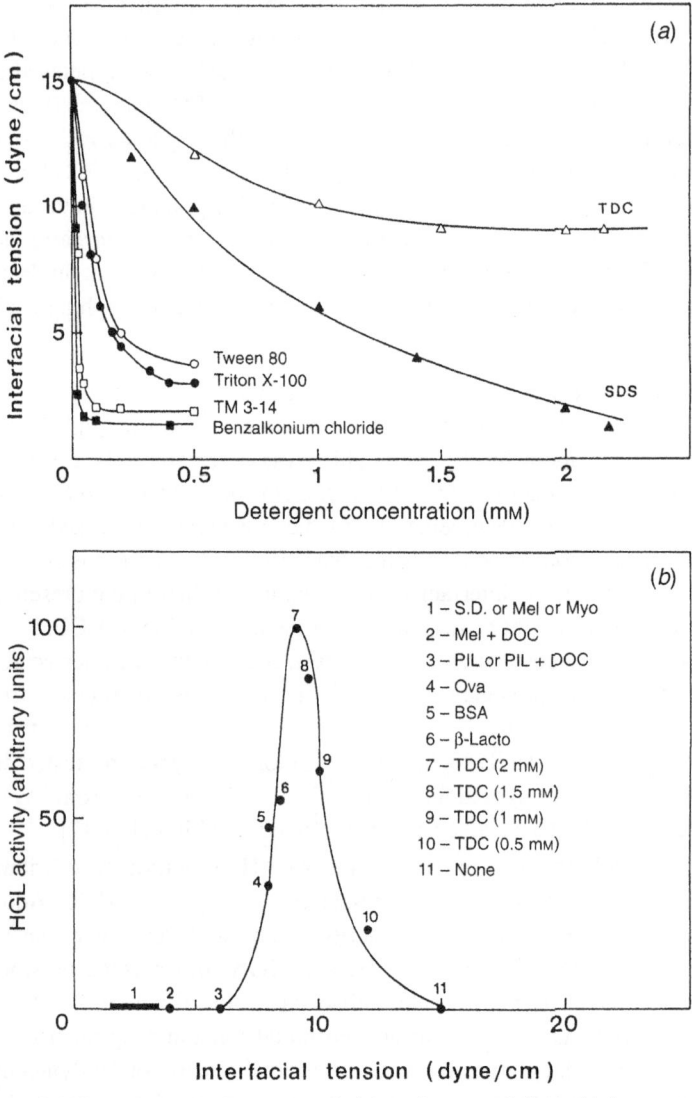

Figure 9. Effect of amphiphiles on interfacial tension: correlation with human gastric lipase activity. (*a*) Effect of increasing concentrations of various synthetic detergents and TDC on the interfacial tension at the triacylglycerol–water interface. (*b*) HGL activity as a function of the interfacial tension as modified by synthetic detergents or TDC (Figure 9(*a*)) or by proteins (7). The initial rate of HGL hydrolysis is measured in the presence of amphiphile, added one minute before enzyme injection. All proteins (melittin (Mel), ovalbumin (Ova), β-lactoglobulin (β-Lacto), protein-inhibiting lipase from soya (PIL), and myoglobin (Myo)) were used at a final concentration of 2μM. Synthetic detergent (S.D.) and deoxycholate (DOC) were used at final concentrations of 2mM.

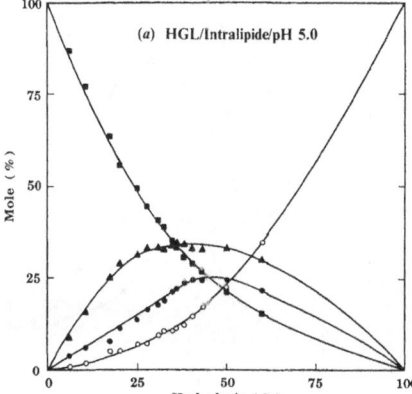 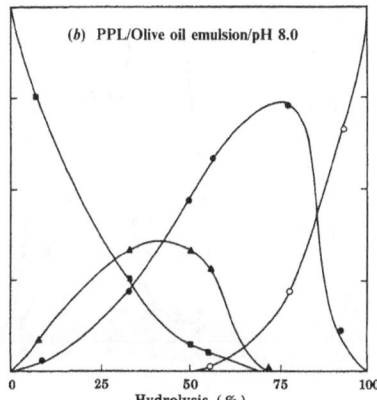

Figure 10. Specific activities of DGL on triacylglycerols of different chain length. (*a*) Influence of pH on the specific activity of DGL toward short-chain (tributyrin, ●), medium-chain (trioctanoin, ▲) and long-chain (30% Intralipide emulsion, ○) triacylglycerols, measured at 37 °C. (*b*) Activity–pH profile of pure DGL (○), HGL (●) and RGL (▲) on long-chain triacylglycerols (30% Intralipide emulsion). 1 unit corresponds to 1 µmol fatty acid released per minute. See reference 17 for assay conditions.

6. CHAIN-LENGTH SPECIFICITY OF GASTRIC LIPASES FOR VARIOUS TRIACYLGLYCEROLS

6.1 Effects of pH and fatty-acid chain length on gastric lipase activity

Many authors have reported that gastric lipases display a high specificity *in vitro* towards short-chain triacylglycerols. Low enzymic rates are observed when long-chain triacylglycerols are used as substrates (3, 14). These data obtained *in vitro* were in apparent contradiction with the observations showing that gastric lipase *in vivo* was able to catalyse the hydrolysis of alimentary long-chain triacylglycerols (3). The problem of the chain-length specificity of gastric lipases was re-investigated. This involved establishing optimal assay conditions with short-chain

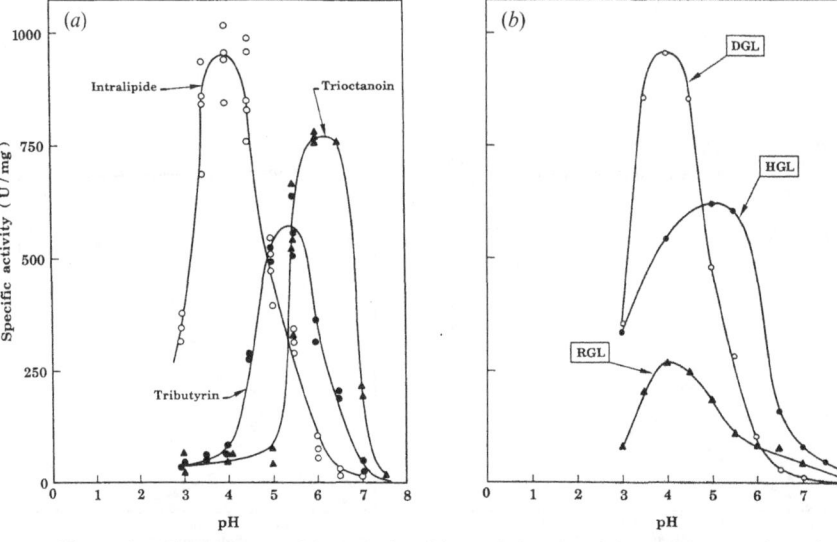

Figure 11. Time-course of hydrolysis of long-chain triacylglycerol by gastric and pancreatic lipases. (*a*) Hydrolysis of Intralipide emulsion (17). (*b*) Hydrolysis of olive-oil emulsion by PPL (48). Triglycerides (■), 1,2(2,3)-diglycerides (▲), 1(3)-monoglycerides (●) and glycerol (○) are expressed as a percentage of the total number of triglyceride molecules (mol %) present initially and are plotted as a function of the percent hydrolysis. (From reference 17.)

triacylglycerols (tributyrin) and long-chain triacylglycerols (Intralipide: purified soybean oil emulsified with egg phosphatidylcholine). With these substrates, HGL activity was determined as a function of pH (47). The maximum specific activity of 1160 units/mg was reached at pH 6.0 with tributyrin (1 unit = 1 μmol fatty acid liberated per min). With Intralipide emulsion, the optimum pH of lipolysis was 600 units/mg and lay in the range 4.5–5.5.

The influence of pH on the specific activities of RGL (16) and DGL (17) was also tested, by using emulsions of short-chain (tributyrin), medium-chain (trioctanoin) and long-chain (Intralipide) triacylglycerols as substrates. DGL was found to have a higher specific activity when long-chain triacylglycerol emulsion was used (Figure 10(*a*)). The specific activity of DGL on Intralipide reached 950 U/mg at an optimal pH value of 4.0, which approached the highest hydrolysis rates (1160 U/mg at pH 6.0 and 1200 U/mg at pH 5.5) obtained on short-chain triacylglycerol with HGL (47) and RGL (16), respectively. The specific activities obtained as a function of pH on long-chain triacylglycerols in the cases of DGL, HGL and RGL are compared in Figure 10(*b*).

HGL, DGL and RGL also hydrolyse medium-chain triacylglycerols (trioctanoin or Liprocil). These are hydrolysed at similar rates, with a maximum at pH 6.

These observations support the conclusion that gastric lipases show no intrinsic specificity for short-chain triacylglycerols, and that their apparent specificity is a consequence of variations in pH and the presence of amphiphiles in the incubation medium. Our observations support the view that gastric lipolysis may play an important rôle in long-chain fat digestion (47).

6.2 Time-course of long-chain triacylglycerol hydrolysis by gastric lipases

Hydrolysis was carried out at constant pH on Intralipide emulsion using gastric lipases. In order to determine the time-course of the concentrations of partial glycerides and fatty acids during hydrolysis, sampling was carried out at regular time intervals and the lipolysis products were analysed by HPLC (17).

The hydrolysis profiles obtained with HGL are presented in Figure 11(*a*). Partial glycerides and glycerol are expressed as a molar percentage of the initial triglycerides (TG) and are plotted as a function of the hydrolysis levels (100% hydrolysis corresponds to 3 mol fatty acid released per mol triglyceride). One can see that the TG levels decreased continuously and that an early production of glycerol occurred, indicating that all three ester groups initially present on the TG molecule can be hydrolysed. On the other hand, when using porcine pancreatic lipase (PPL), which is known to be strictly specific for primary ester-bond hydrolysis, Constantin *et al.* (48) reported a late glycerol production as resulting from the cleavage of 1(3)-MG (monoglycerides) after isomerisation from 2-MG (Figure 11(*b*)). With HGL and RGL, a transient accumulation of both DG (diglycerides) and MG occurred, reaching a maximum value at around 50% hydrolysis. This finding contrasts with the case of PPL, where the DG disappear quite early whereas MG accumulate until chemical isomerisation renders them subject to lipase hydrolysis. With DGL, a transient accumulation of DG occurs, whereas no MG were ever observed. These results may be attributable to the fact that the rate of production of DG is greater than the rate of their hydrolysis, whereas MG are hydrolysed as fast as they are produced from DG.

7. STEREOSELECTIVITY OF GASTRIC LIPASES

7.1 Stereoselective hydrolysis of triglycerides by gastric lipases

The stereospecificity of pre-duodenal lipases has been investigated by several authors using various substrates. It was observed that gastric lipase hydrolysed preferentially the *sn*-3 ester of triglycerides in comparison with the *sn*-1 ester. The ester at the position *sn*-2 was also hydrolysed to some extent, contrary to what was observed with pancreatic lipase. However, in all these previous experiments, triglycerides containing chemically non-equivalent acyl chains at the 3-position of glycerol were used as substrate, and the fatty-acid specificity could not be

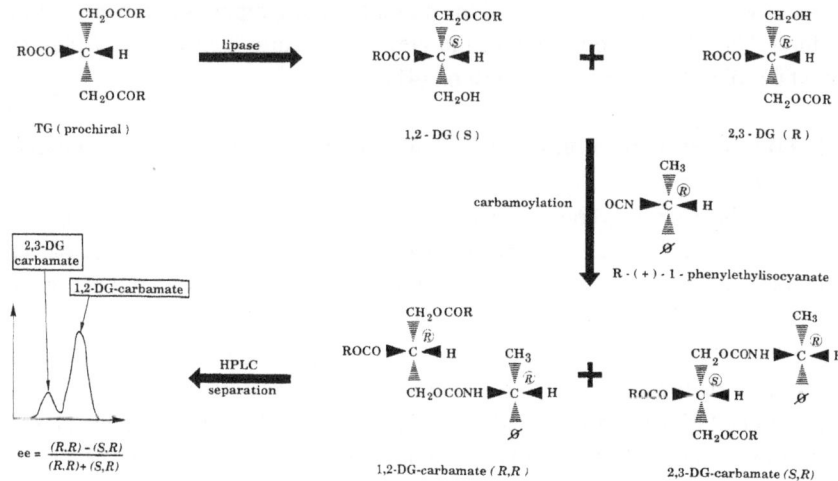

Figure 12. Principle of the method used to study the stereoselective hydrolysis of triglycerides by lipases. DG, diglyceride; TG, triglycerides. (From reference 55.)

dissociated from the stereospecificity of the lipase. For example, milk fat is the dietary fat that has generally been used for studying gastric lipase specificity (3, 49). It has been shown (50) that the predominant enantiomers of the diacylglycerols obtained had the 1,2-*sn* configuration. It should be recalled that the 4:0, 6:0 and 8:0 fatty acids are mostly esterified at the *sn*-3 position of milk triglycerides, while unsaturated fatty acids are mostly found at the position *sn*-2 (51–53). Thus, the reported (54, 55) preferential release of short-chain and medium-chain fatty acids by gastric lipase may in fact be partly attributable to its stereoselectivity, and only to a lesser extent to a specificity for short-chain fatty acids.

The stereoselectivity of gastric lipases towards chemically similar, but sterically non-equivalent ester groups within one single triglyceride molecule has been studied (56). Lipolysis reactions were carried out on synthetic trioctanoin or triolein, which are homogeneous, prochiral triglycerides, chosen as models for lipase physiological substrates. Diglyceride mixtures resulting from lipolysis were converted to their optically active *R*-(+)-1-phenylethylisocyanate derivative, in order to obtain diastereoisomeric carbamate mixtures, which were further separated by HPLC. Resolution of diastereoisomeric carbamates gave enantiomeric excess values, which reflect the stereochemical bias of the lipases (Figure 12).

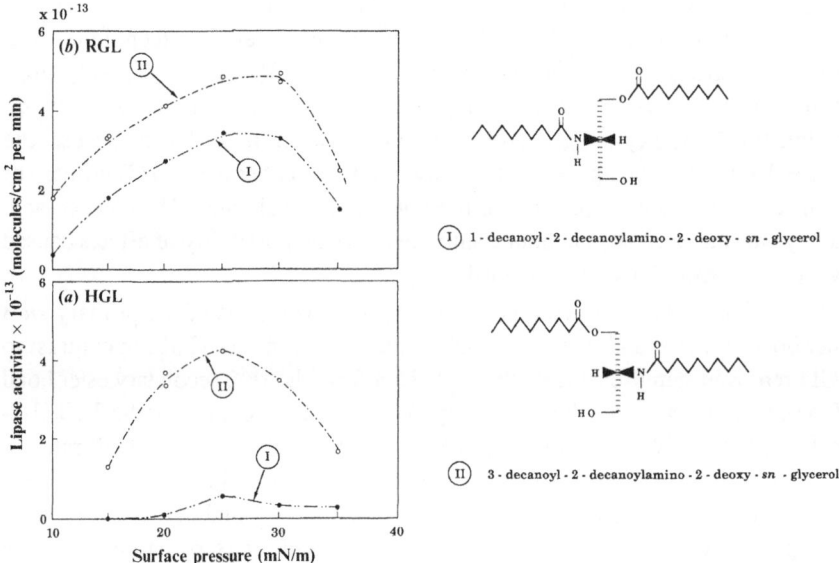

Figure 13. Gastric lipase activities on monomolecular films of diglyceride analogues. Surface pressure profiles of (*a*) RGL and (*b*) HGL activities at pH 5.0 and 25° C using 1-decanoyl-2-decanoylamino-2-deoxy-*sn*-glycerol (I, ●) and 3-decanoyl-2-decanoyl amino-2-deoxy-*sn*-glycerol (II, ○) as substrates. Surface pressure was kept constant by the barostat technique. (From reference 57.)

The results of the separation by HPLC of the diastereoisomeric carbamate derivatives of the diglycerides released from trioctanoin and triolein hydrolysis HGL and RGL demonstrated clearly the existence of a stereochemical preference by both gastric lipases for the *sn*-3 position. The respective stereoselectivity values of HGL and RGL, expressed as the enantiomeric excess percentage, were 54% and 70% for trioctanoin and 74% and 47% for triolein. In comparison, the corresponding values for PPL were 3% in the case of trioctanoin and 8% in that of triolein. It is worth noting that RGL, unlike HGL, became more stereoselective for the triglyceride with shorter acyl chains (trioctanoin). This is one of the most striking catalytic differences between these two gastric lipases yet observed. For the present, in the absence of any three-dimensional structure of these lipases, it is difficult to interpret these data in terms of selective enzyme–substrate interactions with different triacylglycerides.

7.2 Hydrolysis of enantiomeric glyceride analogues by gastric lipases

The stereospecificity of HGL and RGL toward enantiomeric glyceride derivatives has been investigated kinetically by the monomolecular film technique (57). Pseudoglycerides such as enantiomeric 1(3)-alkyl-2,3(1,2)-diacyl-sn-glycerol, enantiomeric 1(3)-alkyl-2-acyl-sn-glycerol and enantiomeric 1(3)-acyl-2-acylamino-2-deoxy-sn-glycerol were synthesised in order to assess the stereoselectivity of the lipase during the hydrolysis of either the sn-1(3) primary or the sn-2 secondary ester position of these glyceride analogues. The cleaved acyl moiety was the same (C_8) in both enantiomers, so the possibility of effects caused by fatty-acid specificity was excluded.

The gastric lipases catalysed preferentially the hydrolysis of the primary sn-3 ester bond of the enantiomeric monoalkyl-diacyl pair tested. HGL, in contrast to RGL, removed stereoselectively the acyl chain from the sn-2 secondary ester bond of the enantiomeric monoalkyl-monoacyl pair, with a preference for the 3-alkyl-2-acyl-sn-glycerol. When this enantiomeric substrate pair was used, high rates of hydrolysis and no chiral discrimination were observed in the case of RGL, whereas low rates and a clear chiral discrimination were found in the case of HGL during the hydrolysis of the secondary ester bond. It is possible that high stereoselectivity may be generally associated with low catalytic rates.

Using diglyceride analogues such as the enantiomeric 2-acylamino derivatives (Figure 13), the same preference for the sn-3 ester bond was shown by both HGL and RGL. In the case of HGL, decreasing the lipid packing increased the lipase's preference for the sn-3 substrate.

In conclusion, gastric lipases possess a preference for the sn-3 site of both triglycerides and diglyceride analogues.

REFERENCES

1. Volhard, F. (1901). Über das Fettspaltende Ferment des Magen. Z. Klin. Med., 42, 414–429.
2. Hull, M. & Keaton, R.W. (1917). The existence of a gastric lipase. J. Biol. Chem., 32, 127–140.
3. Hamosh, M. (1984). Lingual lipase. In Lipases, ed. Borgström, B. & Brockman, H.L., pp. 49–81. Amsterdam: Elsevier.
4. Ross, C.A.C. & Sammons, H.C. (1955). Non-pancreatic lipase in children with pancreatic fibrosis. Arch. Dis. Child., 30, 428–431.
5. Roulet, M., Weber, A.M., Paradis, Y., Roy, C.C., Chartraud, L., Lasalle, R. & Morin, C.L. (1980). Gastric emptying and lingual lipase activity in cystic fibrosis. Pediatr. Res., 14, 1360–1362.
6. Muller, D.P.R., McCollum, J.P.K., Trompeter, R.S. & Harries, J.T. (1975). Studies on

the mechanism of fat absorption in congenital isolated lipase deficiency. *Gut*, 16, (Abstr.) 838.

7. Gargouri, Y., Piéroni, G., Lowe, P.A., Sarda, L. & Verger, R. (1986). Human gastric lipase. The effect of amphiphiles. *Eur. J. Biochem.*, 156, 305–310.

8. Gargouri, Y., Piéroni, G., Rivière, C., Lowe, P.A., Saunière, J.F., Sarda, L. & Verger, R. (1986). Importance of human gastric lipase for intestinal lipolysis: an in vitro study. *Biochim. Biophys. Acta*, 879, 419–423.

9. Moreau, H., Gargouri, Y., Lecat, D., Junien, J.L. & Verger, R. (1988). Screening of pre-duodenal lipases in several mammals. *Biochim. Biophys. Acta*, 959, 247–252.

10. DeNigris, S.J., Hamosh, M., Kasbekar, D.K., Lee, T.C. & Hamosh, P. (1988) Lingual and gastric lipases: species differences in the origin of pre-pancreatic digestive lipases and in the localisation of gastric lipase. *Biochim. Biophys. Acta*, 959, 38–45.

11. Hamosh, M., Ganot, D. & Hamosh, P. (1979). Rat lingual lipase: characteristics of enzyme activity. *J. Biol. Chem.*, 254, 12121–12125.

12. Field, R. & Scow, R.O. (1983). Purification and characterization of rat lingual lipase. *J. Biol. Chem.*, 258, 14563–14569.

13. Roberts, I.M., Montgomery, R.K. & Carey, M.C. (1984). Lingual lipase: partial purification, hydrolytic properties and comparison with pancreatic lipase. *Am. J. Physiol.*, 247, G385–G393.

14. Tirruppathi, C. & Balasubramanian, K.A. (1982). Purification and properties of an acid lipase from human gastric juice. *Biochim. Biophys. Acta*, 712, 692–697.

15. Bernback, S., Hernell, O. & Bläckberg, L. (1985). Purification and molecular characterization of bovine pre-gastric lipase. *Eur. J. Biochem.*, 148, 233.

16. Moreau, H., Gargouri, Y., Lecat, D., Junien, J.L. & Verger, R. (1988). Purification, characterization and kinetic properties of the rabbit gastric lipase. *Biochim. Biophys. Acta*, 960, 286–293.

17. Carrière, F., Moreau, H., Raphel, V., Laugier, R., Bénicourt, C., Junien, J.L. & Verger, R. (1991). Purification and biochemical characterization of dog gastric lipase. *Eur. J. Biochem.*, 202, 75–83.

18. Moreau, H., Laugier, R., Gargouri, Y., Ferrato, F. & Verger, R. (1988). Human pre-duodenal lipase is entirely of gastric fundic origin. *Gastroenterology*, 95, 1221–1226.

19. Abrams, C.K., Hamosh, M., Lee, T.C., Ansher, A.F., Collen, M.J., Lewis, J.H., Benjamin, S.B. & Hamosh, P. (1988). Gastric lipase: localisation in the human stomach. *Gastroenterology*, 95, 1460–1464.

20. Moreau, H., Bernadac, A., Tretout, N., Gargouri, Y., Ferrato, F. & Verger, R. (1990). Immunocytochemical localisation of rabbit gastric lipase and pepsinogen. *Eur. J. Cell Biol.*, 51, 165–172.

21. Moreau, H., Bernadac, A., Gargouri, Y., Pieroni, G. & Verger, R. (1989). Immunocytolocalisation of human gastric lipase in chief cells of the fundic mucosa. *Histochemistry*, 91, 419–423.

22. Carrière, F., Raphel, V., Moreau, H., Bernadac, A., Devaux, M.A., Grimaud, R., Barrowman, J.A., Bénicourt, C., Junien, J.L., Laugier, R. & Verger, R. (1992). Dog gastric lipase: stimulation of its secretion *in vivo* and cytolocalisation in mucous pit cells. *Gastroenterology*, 102, 1535–1545.

23. Szafran, Z., Szafran, H., Popiela, T. & Trompeter, G. (1978). Coupled secretion of gastric lipase and pepsin in man following pentagastrin stimulation. *Digestion*, 18, 310–318.

24. Moreau, H., Saunière, J.F., Gargouri, Y., Piéroni, G., Verger, R. & Sarles, H. (1988). Human gastric lipase: variations induced by gastrointestinal hormones and by pathology. *Scand. J. Gastroenterol.* 23, 1044–1048.

25. Bodmer, M.W., Angal, S., Yarranton, G.T., Harris, T.J.R., Lyons, A., King, D.J., Piéroni, G., Rivière, C., Verger, R. & Lowe, P.A. (1987). Molecular cloning of human

gastric lipase and expression of the enzyme in yeast. *Biochim. Biophys. Acta*, 909, 237–244.

26. Moreau, H., Abergel, C., Carrière, F., Ferrato, F., Fontecilla-Camps, J.C., Cambillau, C. & Verger, R. (1992). Isoform purification of gastric lipases: towards crystallisation. *J. Mol. Biol.*, 225, 147–153.

27. Docherty, A.J.P., Bodmer, M.W., Angal, S., Verger, R., Rivière, C., Lowe, P.A., Lyons, A., Emtage, J.S. & Harris, T.J.R. (1985). Molecular cloning and nucleotide sequence of rat lingual lipase cDNA. *Nucl. Acids Res.*, 13, 1891–1903.

28. Anderson, R.A. & Sando, G.N. (1991). Cloning and expression of cDNA encoding human lysosomal acid lipase/cholesteryl ester hydrolase. Similarities to gastric and lingual lipases. *J. Biol. Chem.*, 266, 22479–22484.

29. Komaromy, M.C. & Schotz, M.C. (1987). Cloning of rat hepatic lipase cDNA: evidence for a lipase gene family. *Proc. Natl. Acad. Sci. USA*, 84, 1526–1530.

30. Lowe, M.E., Rosenblum, J.L. & Strauss, A.W. (1989). Cloning and characterization of human pancreatic lipase cDNA. *J. Biol. Chem.*, 264, 20042–20048.

31. Shimida, Y., Sugihara, A., Tominaga, Y. & Tsunasawa, S. (1989). cDNA molecular cloning of *Geotrichum candidum* lipase. *J. Biochem.*, 106, 383–388.

32. De Caro, J., Boudouard, M., Bonicel, J., Guidoni, A., Desnuelle, P. & Rovery, M. (1981). Porcine pancreatic lipase: completion of the primary structure. *Biochim. Biophys. Acta*, 671, 129–138.

33. Winkler, F.K., D'Ary, A. & Hunziker, W. (1990). Structure of human pancreatic lipase. *Nature*, 343, 767–770.

34. Brady, L., Brzozowski, A.M., Derewenda, Z.S., Dodson, E., Dodson, G., Tolley, S. & Turkenburg, J. (1990). A serine protease triad forms the catalytic centre of a triacylglycerol lipase. *Nature*, 343, 767–770.

35. Moreau, H., Moulin, A., Gargouri, Y., Noël, J.P., Verger, R. (1991). Inactivation of gastric and pancreatic lipases by diethyl *p*-nitrophenyl phosphate. *Biochemistry*, 30, 1037–1041.

36. Hadvàry, P., Lengsfeld, H. & Wolfer, H. (1988). Inhibition of pancreatic lipase *in vitro* by the covalent inhibitor tetrahydrolipstatin. *Biochem. J.*, 256, 357–361.

37. Borgström, B. (1988). Mode of action of tetrahydrolipstatin: a derivative of the naturally occurring lipase inhibitor lipstatin. *Biochim. Biophys. Acta*, 962, 308–316.

38. Peng, Q., Hadvàry, P. & Maerki, H.P. (1991). Identification of the THL binding site on human pancreatic lipase. In *Lipases: Structure, Mechanism and Genetic Engineering*, GBF monographs, vol. 16, ed. Alberghina, L., Schmid, R.D. & Verger, R., pp. 129–133. Weinheim: VCH.

39. Gargouri, Y., Chahinian, H., Moreau, H., Ransac, S. & Verger, R. (1991). Inactivation of pancreatic and gastric lipases by THL and C12:0-TNB: a kinetic study with emulsified tributyrin. *Biochim. Biophys. Acta*, 1085, 322–328.

40. Ransac, S., Gargouri, Y., Moreau, H. & Verger, R. (1991). Inactivation of pancreatic and gastric lipases by tetrahydrolipstatin and acyldithio-5-(2-nitrobenzoic acid): a kinetic study with dicaprin monolayers. *Eur. J. Biochem.*, 202, 395–400.

41. Gargouri, Y., Moreau, H., Pieroni, G. & Verger, R.(1988). Human gastric lipase: a sulfhydryl enzyme. *J. Biol. Chem.*, 263, 2159–2162.

42. Moreau, H., Gargouri, Y., Pieroni, G. & Verger, R. (1988). Importance of sulfhydryl group for rabbit gastric lipase activity. *FEBS Letters* 236, 383–387.

43. Gargouri, Y., Moreau, H., Jain, M.K., de Haas, G.H. & Verger, R. (1989). Ajoene prevents fat digestion by human gastric lipase in vitro. *Biochim. Biophys. Acta*, 1006, 137–139.

44. Gargouri, Y., Moreau, H., Piéroni, G. & Verger, R. (1989). Role of a sulfhydryl group in gastric lipases. A binding study using the monomolecular-film technique. *Eur. J. Biochem.*, 180, 367–371.

45. Gargouri, Y, Julien, R., Piéroni, G., Verger, R. & Sarda, L. (1984). Studies on the inhibition of pancreatic and microbial lipases by soybean proteins. *J. Lipid Res.*, 25, 1214–1221.

46. Gargouri, Y., Julien, R., Sugihara, A., Verger, R. & Sarda, L. (1984). Inhibition of pancreatic and microbial lipases by proteins. *Biochim. Biophys. Acta*, 795, 326–331.

47. Gargouri, Y., Piéroni, G., Rivière, C., Saunière, J.F., Lowe, P.A., Sarda, L. & Verger, R. (1986). Kinetic assay of human gastric lipase on short- and long-chain triacylglycerol emulsions. *Gastroenterology*, 86, 919–925.

48. Constantin, M.J., Pasero, L. & Desnuelle, P. (1960). Quelques remarques complémentaires sur l'hydrolyse des triglycerides par la lipase pancréatique. *Biochim. Biophys. Acta*, 43, 103–109.

49. Bernbäck, S., Blackberg, L. & Hernell, O. (1990). The complete digestion of human milk triacylglycerol *in vitro* requires gastric lipase, pancreatic colipase-dependent lipase and bile salt stimulated lipase. *J. Clin. Invest.*, 85, 1221–1226.

50. Lok, C.L. (1979). Identification of chiral 1,2-diacylglycerols in fresh milk fat. *Recueil des Travaux Chimiques*, 98, 92–95.

51. Breckenridge, W.C. (1978). Stereospecific analysis of triacylglycerols. In *Handbook of Lipid Research*, vol. 1, *Fatty Acids and Glycerides*, ed. Kuksis, A., pp.197–232. New York: Plenum Press.

52. Staggers, J.E., Fernando-Warnakulasuriya, G.J.P. & Wells, M.A. (1981). Studies on fat digestion, absorption, and transport in the suckling rat. II. Triacylglycerols, molecular species, stereospecific analysis and specificity of hydrolysis by lingual lipase. *J. Lipid Res.*, 22, 675–679.

53. Parodi, P.W. (1982). Positional distribution of fatty acids in the triglyceride classes of milk fat. *Lipids*, 17, 437–442.

54. Clark, S.B., Brause B. & Holt, P.R. (1969). Lipolysis and absorption of fat in the rat stomach. *Gastroenterology*, 56, 214–222.

55. Barrowman, J.A. & Darnton, S.J. (1970). The lipase of rat gastric mucosa. A histochemical demonstration of the enzyme activity against a medium-chain triglyceride. *Gastroenterology*, 59, 13–21.

56. Rogalska, E., Ransac, S. & Verger, R. (1990). Stereoselectivity of lipases. II. Stereoselective hydrolysis of triglycerides by gastric and pancreatic lipases. *J. Biol. Chem.*, 265, 20271–20276.

57. Ransac, S., Rogalska, E., Gargouri, Y., Deveer, A., Paltauf, F., de Haas, G.H. & Verger R. (1990). Stereoselectivity of lipases. 1. Hydrolysis of enantiomeric glyceride analogues by gastric and pancreatic lipases, a kinetic study using the monomolecular film technique. *J. Biol. Chem.*, 265, 20271–20276.

10.

Lipase inhibitors

SHAMKANT PATKAR and FREDRIK BJÖRKLING

Lipase inhibitors have been used in the study of structural and mechanistic properties of lipases. Most recently, the crystal structure of a *Rhizomucor miehei* lipase was determined without (1) and with (2) an inhibitor, and the analysis of these structures has afforded important insight in the selectivity and mode of action of lipases. In addition, the search for inhibitors of pharmacological interest, *e.g.*, inhibitors of the lipases active in fat metabolism, which could lead to drugs for the treatment of obesity, have guided development in this field. We discuss here some of the important aspects of the wide variety of compounds found to act as inhibitors of lipases.

1. REVERSIBLE INHIBITORS

In general, inhibitors of enzymes are classified as reversible or irreversible. Reversible enzyme inhibitors are further divided into competitive, non-competitive and uncompetitive. This classification is based on the effect of the inhibitor concentration upon K_m and/or V_{max}. One of the basic requirements for this classification is that the enzyme fulfil Michaelis-Menten saturation kinetics.

However, in the case of lipases normal Michaelis-Menten kinetics do not apply, because the lipase substrates, lipids, are not soluble in water, and form a two-phase system with the lipase adsorbed at the interface. A working model for lipolysis has been suggested (3). This model takes into account parameters such as surface area, enzyme adsorption and interfacial enzyme activation, which all are of great importance in the understanding of lipase kinetics.

1.1 Kinetics of lipase inhibition

Some of the inhibitors of lipolytic enzymes act at the lipid–water interface and prevent enzyme–substrate complex formation rather than inhibiting at the active site. Therefore, interpretation of inhibition data has to take into account the kinetics of lipase action at the lipid–water interface. A kinetic model applicable to water-insoluble competitive inhibitors of lipolytic enzymes has been described (4). In order to simplify the kinetic treatment of inhibition at the interface, several assumptions are made. (a) The water-insoluble substrate (S), inhibitor (I) and detergent (D) are located exclusively at the interface. (b) Water-soluble reaction products are desorbed and diffuse rapidly away from the interface. (c) The concentrations of substrate, inhibitor and detergent are constant during the reaction, and the concentration of product formed is negligible compared with that of the substrate. (d) The volume occupied by the molecules constituting the interface is negligible compared with the total volume. (e) The area occupied by the various enzyme species penetrating the lipid surface (E*, E*S, E*I, E*D) is negligible compared with the total interfacial area of the penetrating enzyme in molecules per surface unit. (f) The substrate, inhibitor and detergent behave ideally, with no specific interactions between them. By using these assumptions, a kinetic interpretation could be made.

When the relative velocity is plotted against the inhibitor concentration (mole fraction), three different curves can in principle be obtained. If the inhibitor binds to the enzyme with the same affinity as the substrate, a straight line is obtained. If the inhibitor has a lower or a higher affinity than the substrate, a concave or convex curve is obtained. This determination of the relative enzyme velocity in presence of different inhibitors has been used to calculate the efficiencies of various competitive inhibitors. For example, this model was applied to study the inhibition of phospholipase A_2, with acylamino analogues of phospholipids as inhibitors (5).

1.2 Reversible non-specific inhibitors

Compounds that do not act directly at the active site, but inhibit lipase activity by changing the conformation of lipase or interfacial surface properties, are defined as non-specific inhibitors.

Surfactants

Surface-active materials influence lipase activity either by a direct effect of the surfactant on the lipase, leading to unfolding and denaturing, or by a change in the interfacial properties. In general, a surfactant concentration below the critical micelle concentration (CMC) weakly activates lipases, whereas high concentrations inhibit them.

Anionic surfactants such as sodium dodecyl sulphate (SDS) and linear alkylbenzene sulphonates (LAS) are known denaturing agents of proteins. The effect of increasing surfactant concentration on polylysine has been shown to change its conformation from the coil to the β form (6).

Pancreatic lipase was found to be inactivated by unfolding and denaturation when the SDS concentration was increased above the CMC (7). When bile salt was also present during incubation of pancreatic lipase, the lipase was protected against denaturation and irreversible inactivation (8).

Non-ionic surfactants such as alcohol ethoxylate are commonly used in detergents. Alcohol ethoxylate monoalkyl deca-(oxyethylene) ether was found to enhance the rate of hydrolysis by *Mucor* lipase below a concentration of half its CMC. Higher concentrations of the ethoxylate inhibited the lipase (9).

Cationic surfactants, such as quaternary ammonium salts, have been shown to enhance the rate of hydrolysis of olive oil by *Mucor* lipase. This increased hydrolysis rate was related to the oil:surfactant ratio, and an increase in adsorption of lipase to the interface. Optimal concentration of the cationic surfactant to achieve this activation varied with different detergent compositions. High concentrations of cationic surfactant were generally inhibitory (9).

There is still some uncertainty surrounding these reported effects of surfactants on lipases, because of the complex combination of effects on the substrate, the interfacial surface, and the lipase.

Bile salts

Bile salts, choleic acid derivatives of cholesterol, are known to influence lipase activity, either as activators or inhibitors. As in the case of surfactants, their effect is activating at low concentrations and inhibiting at high. Bile salts have also been shown to protect against the denaturing effect of SDS on pancreatic lipase by preventing the binding of SDS to the lipase (10).

The inhibitory effect of bile salt was explained by two hypotheses (11). (A) Bile salts accumulate at the substrate surface and inhibit lipase adsorption to the interface (12). (B) Bile salts form a complex with the lipase, which has low affinity towards the bile-salt-covered surface (13).

The interactions of bile salts with pancreatic lipase have been studied thoroughly. It was found that bile salts at low concentrations stabilise pancreatic lipase and at high concentrations inhibit the lipase. Furthermore, the inhibition of pancreatic lipase caused by bile salts is reversed by colipase, which seemingly prevents the interaction between lipase and bile salt (13). Explanations suggested for the inhibition (11) were the prevention of the adsorption of lipase to its substrate, and a change in the emulsion area, which could influence the amount of adsorbed bile salts.

The effect of bile salts on carboxyl-ester lipase was found to be more direct, as the initial rate of hydrolysis of tributyrin by carboxyl-ester lipase was not decreased by the addition of bile salts above the CMC. To explain this, binding sites for bile salt on human pancreatic carboxyl ester lipase were suggested (14). One site was non-specific, but the other was specific for the $3\alpha,7\alpha$-hydroxylated position of primary bile salt. This site was postulated to be involved in dimerisation of the enzyme and thus in its ability to act on water-insoluble substrates. This indicated a probable conformational change of the enzyme (15, 16).

The milk from human and higher primates contains a bile-salt-stimulated lipase. Its optimum bile-salt concentration is just right for its biological function in the intestine (where bile salts are normally above 2 mM), as compared with the low concentrations found in the blood, where the lipase consequently has a very low activity.

The effect of bile salts on the regulation of lipase activity is complex (17). However, specific binding sites for bile salts may in part explain the effect (14–16).

Proteins as inhibitors of lipases

There are several reports on the isolation of lipase-inhibiting proteins from various sources such as fungus *Rhizopus microsporus* (18), *Mycobacteria* (19), sunflower seeds (20) oil seeds (21) and a melanoma cell line, associated with the cancer Cachexia (22). The enzyme–protein interaction is reversible, and the inhibitory proteins may have the physiological rôle of regulating lipases intracellularly. Similarly, there are many proteins that inhibit proteases, and the mechanism of this inhibition is well studied (23). However, although several reports have been published on the inhibition of lipases by proteins, these investigations show no direct similarity between the mechanisms of inhibition of the lipases and the proteases.

The mechanism of inhibition of lipases by proteins and polypeptides is still not clearly understood. To explain the inhibitory action, two separate hypotheses have been suggested: (A) the inhibitor protein is adsorbed at the interface and induces changes in interfacial properties, and (B) the protein interacts directly with the lipase and competes with the substrate.

Horse pancreatic lipase and lipases from *Rhizopus arrhizus* and *R. delemar* have been investigated for inhibition by a series of common proteins including myoglobin, serum albumin, ovalbumin, mellitin, and β-lactoglobulin. All these proteins inhibit horse pancreatic lipase in the absence of colipase and bile salt. Interestingly, the lipase from *R. delemar* was inhibited by the proteins, while lipase from *R. arrhizus* was not affected by the same proteins in the same concentration range (24).

It was shown that the interfacial tension at the triolein–water interface decreases with increasing concentration of proteins, but there was no correlation between enzyme activity and interfacial tension. Increasing the concentration of lipase did not increase the lipase activity. These results were interpreted as inhibition caused by desorption of lipase from substrate due to a change in interfacial quality.

As lipases act at interfaces, it is preferable to perform the experiments with monolayer techniques, rather than in a bulk emulsion. The main difference between monolayer and bulk systems lies in the ratios of interfacial area to volume, which differ by several orders of magnitude. In the inhibition of lipase from *R. delemar* by proteins, using a dicaprin monolayer, a good correlation was found between the adsorption of radioactively labelled lipase to mixed protein–dicaprin films and the activity of the lipase (25).

One of the most interesting observations was the inhibition of pancreatic lipase at constant surface pressure in the hydrolysis of dicaprin monolayers. When mixed films of protein–dicaprin were used at constant surface pressure, the lipase was inhibited, suggesting that inhibition is not simply related to surface tension, but is rather associated with the adsorption of the lipase, which is in turn influenced by mixed films of protein and dicaprin (26).

Human gastric lipase differs from pancreatic lipase in its activity on tributyrin and its pattern of inhibition by proteins (27). By using monolayers of dicaprin, it was shown that inhibitory properties of proteins on gastric lipase are dependent on the penetrating power of the proteins in the lipid interface (28). Inhibition studies with the monolayer technique have so far been limited by availability of the apparatus and by the extrapolation of results to mixed-lipid layers.

The second hypothesis of direct protein-lipase interaction was favoured as result of an investigation of the activity of lipase towards water-soluble or -insoluble substrates. Studies with homo-poly-aminoacids as inhibitors of lipase from *Candida cylindracea*, acting on soluble substrates, showed that inhibition occurred in the absence of an interface. The efficiency of the proteins as inhibitors depended on the hydrophobicity of the proteins (29). A protein equal in hydrophobicity to the lipase inhibited was found to be the most effective inhibitor.

Metal ions as inhibitors of lipases

Some metal ions bind to the enzyme, altering its activity by stabilisation or destabilisation of the protein conformation. The metal ions may also act as scavengers of free fatty acids at the interface, minimising product inhibition from the free fatty acids, giving the metal ions a stimulatory effect.

The influence on lipase activity of a series of metal ions has been reported, but only a few metal ions, including calcium, iron and mercury, have direct or indirect effect on lipase activity.

The activating effect of calcium on a lipase derived from *Aspergillus niger* and a lipase derived from *Humicola lanuginosa* was explained by the removal of free fatty acids from the interface (30, 31). This effect is of great importance in the use of lipases for washing, where the calcium content of the water will affect the enzymes' activity and their performance in detergents.

Salts of Fe (II and III) were found to inhibit lipase from *Aspergillus niger* (32) and lipases from fungi of the genus *Geotrichum* (33). The mechanism of this interaction has not yet been clarified.

Mercury derivatives of proteins are used commonly in X-ray crystallographic investigations, in which the formation of heavy-atom derivatives is essential in the process of structure elucidation. Mercury is known to bind to thiol groups of proteins, forming stable complexes.

Mercury as Hg(II) has been shown to inhibit the hormone-sensitive lipase of rat adipose tissue (34). This loss of activity could be reversed by the addition of excess dithioerythritol. The inhibition involved only one specific cysteine residue out of five present in the lipase, which indicated the importance of this thiol group for lipase activity. However, to our knowledge, there is no lipase in which a cysteine residue replaces serine in the active site, as is known for proteases.

Platinum chloride has also been shown slowly to inhibit lipase from *Geotricum candidum* (35).

1.3 Specific reversible inhibitors

Boronic acids as reversible inhibitors

Boronic acids (R-B(OH)$_2$) are commonly described as transition-state analogues in the inhibition of proteases. Transition states have a short lifetime and can be studied only with difficulty. An important tool in the investigation of substrate–enzyme interactions at the active site is therefore the presence in the active site of long-lived analogues of substrates.

Boronic-acid derivatives have been shown to form reversible but long-lived complexes with the active-site serine of proteases, thereby revealing some of the specific interactions with the substrate (36). Similarly, boronic acids inhibit lipases, as exemplified by the inhibition of pancreatic lipase (37), lipoprotein lipase (38, 39) and bile-salt-stimulated lipases (40). The reversible inhibition of lipases with boronic acids is highly dependent on the inhibition conditions used. A change in substrate and/or pH gave a change in the inhibitory kinetics from non-competitive to mixed or competitive kinetics. The kinetics of boronic acid inhibition have been well studied, in particular for pancreatic lipase (37).

Phenylboronic acid inhibits competitively bile-salt-dependent lipase from human milk. The boronic acid binds near to or at the active-site serine, which was demonstrated by competitive inhibition on the serine residue by DFP (diisopropylphosphofluoridate). Phenylboronic acid does not interact with any bile-salt-binding site, since dissociation of the bile-salt–enzyme complex is not affected by the presence of inhibitor (40).

Tetrahedral Boronate RO-8-5263 (Figure 1, compound 1) is a competitive inhibitor of porcine pancreatic lipase (41).

Substrate analogues

The triglyceride analogue glycerol triether is also a competitive inhibitor of pancreatic lipase (41). However, the affinity of this compound for the enzyme is not high enough, compared with substrate, for the inhibitor to function as a

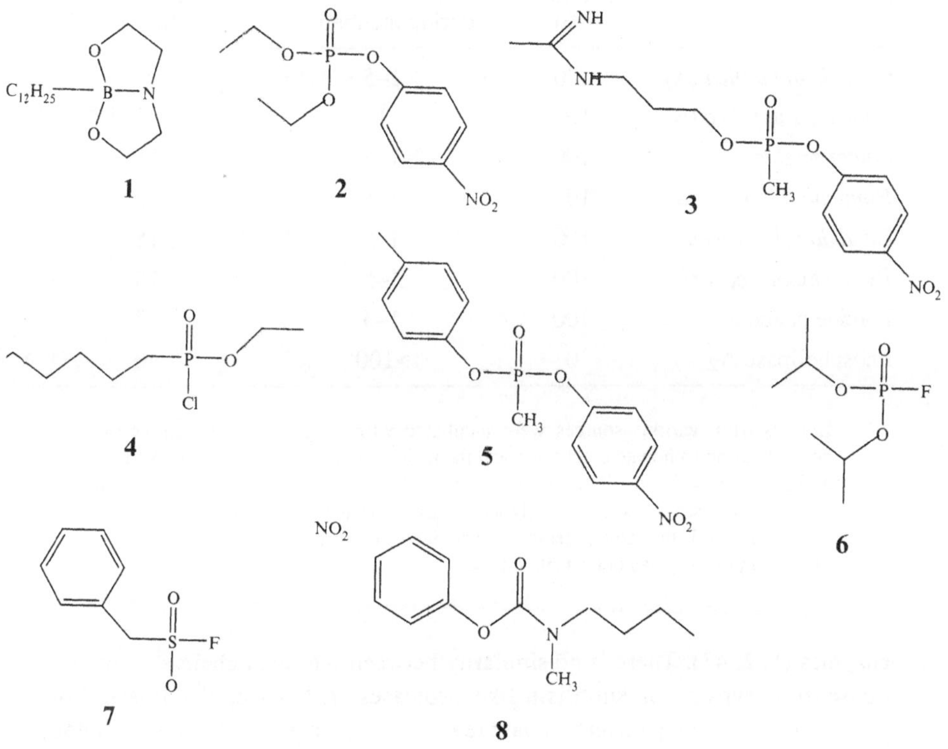

Figure 1. Structures of some lipase inhibitors referred to in this chapter.

transition-state analogue, which make it difficult to obtain useful information from these analogues.

2. ACTIVE-SITE-DIRECTED IRREVERSIBLE INHIBITORS

Recently, the crystal structures of the lipases from *Rhizomucor miehei* (1), *Geotricum candidum* (42) and human pancreatic lipase (HPL) (43) were reported. A catalytic triad similar to the Ser...His...Asp triad found in most proteases was also found in the *Rhizomucor miehei* and HPL lipases, while in *Geotricum candidum* Asp was replaced by Glu.

Interestingly, the catalytic triad in the lipases differs in geometry as compared with the proteases, which may have implications for the stereoselectivity of these

Table 1. Specificity of *n*-hexylchlorophosphonate ethyl ester as inhibitor.

Lipase source	Inhibition (%)	Molar ratio I/E during incubation	Time of incubation (min)
Candida antarctica (A)	100	1–5	5
Candida antarctica (B)	100	1	5
Mucor miehei	100	1–5	15
Humicola lanuginosa	100	1–5	15
Candida cylindracea	100	1–5	15
Pseudomonas cepacia	100	1–5	15
Porcine pancreas	100	1–5	15
Phospholipase A$_2$	0	>100	60

Lipases from various sources were incubated with varying concentrations of *n*-hexylchlorophosphonate ethyl ester, and the residual lipase activity was assayed using a pH-stat method and tributyrin as substrate. Phospholipase A$_2$ from snake venom (Sigma) was assayed with egg yolk as substrate and the pH-stat method. The Table shows that all the triglyceride lipases were inhibited by this inhibitor, but phospholipase A$_2$ was not sensitive to it.

enzymes (1, 2, 43). There is no similarity between the main-chain structures of lipase and trypsin- or subtilisin-like proteases. Likewise, the lipase from *R. miehei* and human pancreatic lipase reveal no sequence or structural homology apart from the consensus pentapeptide sequence around the essential serine.

In the crystal structure, access to the active site in the lipases was found to be blocked by a short helical segment or 'lid' which must move before the substrate can diffuse into the active-site pocket. To elucidate this, there was a need for an effective covalently-bound inhibitor to be used in X-ray crystallographic investigations (1, 2).

2.1 Phosphorus-containing inhibitors

Some of the most common active-site-directed inhibitors of proteases are reactive phosphorus compounds. They react with the active-site serine, forming a stable transition-state-like conformation. Similarly, phosphorus-containing compounds such as diethyl *p*-nitrophenyl phosphate, **2**, and 3-guanidopropyl *p*-nitrophenyl methylphosphonate, **3**, have been used as inhibitors for lipases (Figure 1).

With this background, *n*-hexyl chlorophosphonate ethyl ester, **4**, was prepared. This compound is a hydrophobic equivalent of the phosphorus and phosphonate substances mentioned above. The chlorophosphonate was found to be a very

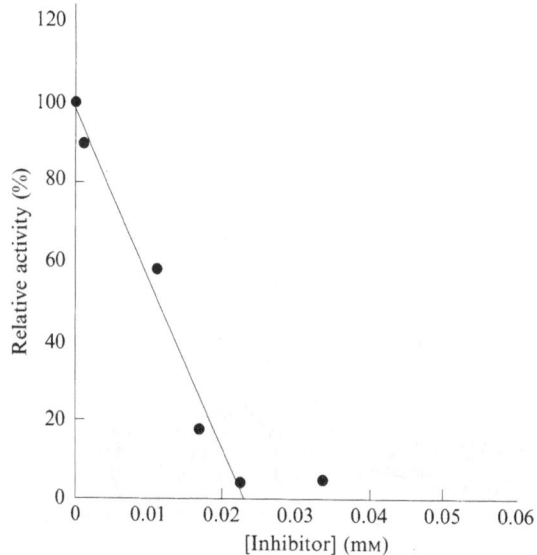

Figure 2. Inhibition of lipase B from *Candida antarctica* by $C_6H_{13}PO(OC_2H_5)Cl$ (compound **4** in Figure 1).

potent inhibitor, and a series of lipases was inhibited with only a small excess of inhibitor (Table 1). A typical curve for the inhibition of lipase B from *Candida antarctica* is given (Figure 2: taken from reference 44).

Very recently, the crystal structure of *Rhizomucor miehei* lipase inhibited by *n*-hexyl chlorophosphonate ethyl ester, **4**, was determined. It was found that the active site is exposed by movement of the helical lid, and that the inhibitor is bound covalently to the active-site serine (Figure 3). This movement also increased the hydrophobicity of the surface surrounding the catalytic site, strengthening the interactions with the lipid substrates. This structure of the enzyme–inhibitor complex was suggested to be equivalent to the activated state generated by the oil–water interface.

Much of today's knowledge of lipase mechanism and function has been acquired by using diethyl *p*-nitrophenyl phosphate, which has been shown to inhibit pancreatic (45–48) and gastric (49) as well as some microbial (50) lipases at high concentrations (Figure 4).

In the case of porcine pancreatic lipase (PPL), only the activity towards lipid substrates, and towards not water-soluble substrates, was inhibited, implying the presence of two separate active regions in this lipase. Limited proteolysis of PPL revealed two domains, one C-terminal domain with a molecular weight of 12 kDa, containing the esterase activity, and a domain containing lipase activity located at

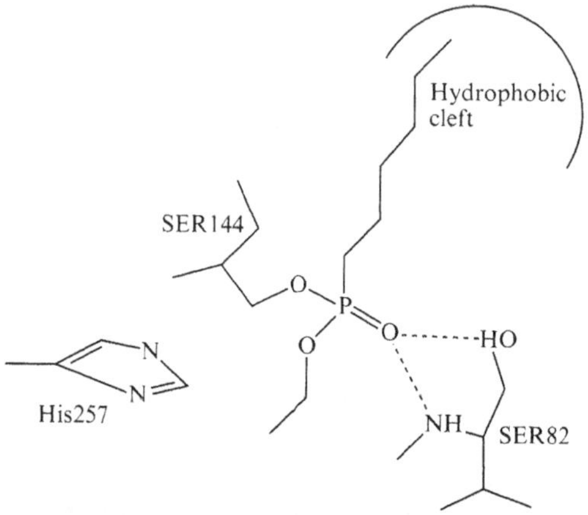

Figure 3. Schematic drawing of $C_6H_{13}PO(OC_2H_5)Cl$ (compound **4** in Figure 1) bound covalently to the active serine at the active site of lipase. See also Plate 6.

the N terminus (51). This second potential catalytic site in PPL is not inhibited by organophosphates, which is difficult to explain.

The water-soluble organophosphonate 3-guanidopropyl *p*-nitrophenyl methylphosphonate also inactivates PPL, in the absence of micelles, when tributyrin is used as a substrate. The modified serine residue was thus interpreted as being involved in the lipid-binding domain (52). Other phosphorus-based compounds used for the inhibition of lipases are bis-*p*-nitrophenyl phosphonate, **5** (53) and diisopropyl fluorophosphate, **6** (54).

2.2 Non-phosphorous compounds

Phenyl methyl sulphonyl fluoride

PMSF, **7**, which is a common inhibitor of serine proteases, has been used as an inhibitor of lipases (55). However, this compound needs a long incubation time, and it is not as effective as phosphorus-containing inhibitors. Recently, a new analogue of PMSF has been developed, a hexadecane sulphonyl fluoride. This compound has been shown to inhibit the activity of phospholipase from *E. coli*

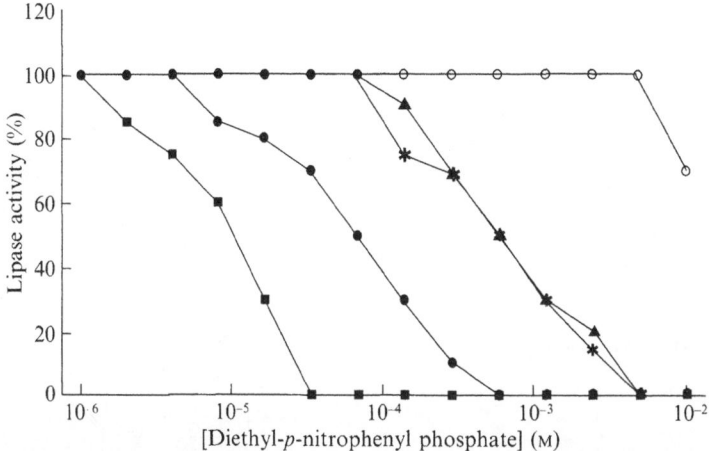

Figure 4. Inhibition of microbial lipases by diethyl-*p*-nitrophenyl phosphate. ●, *Candida antarctica* A lipase; ◆, *C. antarctica* B lipase; ▲, *Mucor miehei* lipase; ✳, *Pseudomonas cepacia* lipase; ○, *Humicola lanuginosa* lipase. The concentration of each enzyme was 2×10^{-5} M. Lipases A and B from *C. antarctica* needed only 5 min for inactivation, while the other lipases needed incubation for 1–20 h.

membrane. However, this finding does not correlate with the mechanism suggested for other phospholipases (56).

Carbamates

N-butyl-*N*-methyl-4-nitrophenyl carbamate, **8**, has been shown to be a specific active-site titrator of bile-salt-dependent lipases from human pancreas, human milk and dog pancreas. Other lipases, proteases and esterases were not influenced by this inhibitor (57).

Carbamates act at the active site, as shown by the competitive inhibition of bile-salt-dependent lipase with phenylboronic acid. Similarly, DFP-inactivated lipase did not react with the inhibitor, suggesting that both DFP and the carbamate act at the active site.

2.3 β-Lactone-containing inhibitors

Several β-lactones of microbial origin have been isolated and tested as lipase inhibitors. Among them are (1) esterastin, **9**, isolated from *Streptomyces lavendulae*, (2) valilactone, **10**, produced by *S. albolongus*, (3) lipstatin, **11**, produced by *S. toxytricini* and (4) ebalactone A and B, **12**, from *Actinomycetes* (Figure 5).

Although structurally similar, the above-mentioned β-lactones exhibit slightly different inhibition profiles towards lipases. For example, esterastin was found to

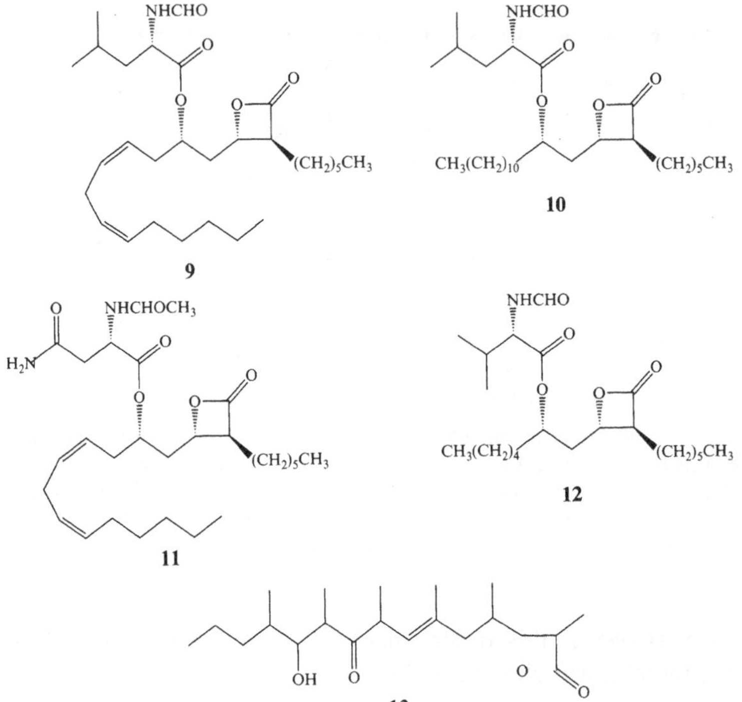

Figure 5. β-Lactone inhibitors of lipases.

be a potent and rather selective inhibitor of acidic lipases such as rabbit liver lysosomal acid lipase, whereas ebalactones A and B were not inhibitory for this lipase (58).

The most extensively studied β-lactone inhibitor is tetrahydrolipstatin, **13**, (THL), a saturated analogue of lipstatin (59). This compound was found to inhibit both human gastric and pancreatic lipases (60), and also microbial lipases from *S. aureus* and *Rhizophus ahrrizus* (61). THL has been shown to be bound covalently to the active-site serine in HPL (62). This was confirmed by peptide mapping and fast-atom-bombardment mass spectroscopy (63).

THL is a very hydrophobic compound and is active at the interface only. This could explain the low inhibition found for serine proteases such as trypsin, chymotrypsin, and pig liver esterase.

As THL inhibits pancreatic lipase (62), it may prove useful as a drug against obesity. At the time of writing, it is being evaluated in clinical trials.

3. CHEMICAL MODIFICATION

Chemical modification has often been used to determine active-site or other amino-acid residues involved in catalysis. Apart from the knowledge that the catalytic triad consists of His, Asp and Ser residues, two separate observations indicate that tryptophan residues may be involved in the catalysis by lipases.

3.1 Modification of tryptophan

Modification of tryptophan residues in *Rizophus delemar* lipase by dimethyl-(2-hydroxy-5-nitrobenzyl)-bromide (64) and the lipase of *Humicola lanuginosa* (65) by *N*-bromosuccinimide both led to a decreased lipase activity. The tryptophan residues may therefore play an important rôle in maintaining the conformation of the lipase.

3.2 Modification of cysteine

Modification of cysteine groups in lipases has been used to determine whether they belong to the group of sulphydryl enzymes, as found amongst proteases. Rabbit gastric lipase and human gastric lipase are inhibited by 5,5-dithiobis-(2-nitrobenzoic acid) and 4,4-dithiopyridine, respectively. In these cases, a direct correlation was found between the modification of one sulphydryl group per enzyme molecule and the loss of enzyme activity (66, 67).

The hydrophobic sulphydryl reagent dodecyldithio-5-(2-nitro benzoic acid) was found to be a very potent inhibitor of rabbit and human gastric lipases, and addition of the inhibitor during lipolysis stopped the hydrolysis almost instantaneously. These observations indicated that both the rabbit and the human gastric lipases need a sulphydryl group for optimal activity. This inhibition was not due to modification of the active site but rather to structural change in the enzymes.

4. PHARMACOLOGICALLY ACTIVE INHIBITORS

The main targets for the use of lipase inhibitors in medicine today is the possible treatment of obesity and hyperlipidæmia. Moreover, lipase inhibitors may find application in the treatment of acne.

Inhibition of the lipases involved in the metabolism of lipids in the intestine may lead to a less effective absorption of triglycerides. If the possible side-effects can be kept at an acceptable level, this may become an interesting alternative to existing treatments for obesity. Tetrahydrolipstatin, used for the inhibition of human pancreatic lipase, mentioned in section 2.3, is the best-characterised compound used for this purpose. The ongoing clinical trials will show if this approach for obesity treatment is feasible.

Derivatives of aminomethylene malonic acid have been suggested for the treatment of acne. These compounds have been shown to inhibit a lipase derived from *Rhizopus arrhizus* and could be used as additives in skin-cream formulations (68).

Several other compounds have been reported to inhibit a variety of important lipases *in vitro* and *in vivo*, as shown by the following examples.

The THL analogue esterastin, an acid lipase inhibitor, has been found to be a potent inhibitor of acid cholesterol esterase on cultured aortic smooth muscle cells (69).

1,6 di(O-carbamoyl cyclohexanone oxime) hexane RHC 80267 (Upjohn) has been found to be a potent inhibitor of diacylglycerol lipase from human platelets (70). This compound was not suitable for studying arachidonic acid metabolism in platelets, but *in vivo*, on cardiac myocytes, it was found to be a potent inhibitor (71).

Treatment of hypertension by propranolol and other β-blockers has been shown to increase the triglyceride level in plasma. This effect was attributed to inhibition of lipoprotein lipase *in vivo*. Propranolol has a cationic amphiphilic structure and was found to inhibit both phospholipase A_1 and lipoprotein lipase from bovine milk *in vitro* (72).

Amphetamine and other phenethylamines are known to interfere with fatty-acid metabolism. Various phenethylamines have been found to inhibit pancreatic lipase *in vitro*. The mechanism of inhibition is not clear, because K_i values were compared with diethyl-*p*-nitrophenyl phosphate as inhibitor; however, this is known to be an irreversible inhibitor. Further mechanistic investigation will be necessary to characterise inhibition by phenthylamines (73).

5. CONCLUSION

Our understanding of the mechanisms involved in lipase catalysis has gained greatly from the extensive studies of lipase–inhibitor interactions carried out in recent years. In the future, an even better understanding of the interfacial activation mechanism will surely be reached. This process is of course of profound importance, not only in the biological processing of lipids, but also in several new technical uses of lipases (74). Furthermore, investigation of the precise structural properties and substrate selectivity of lipases can now be a fruitful target of research. In particular, the combination of X-ray crystallographic investigations with substrate-like inhibitors, the exchange of specific amino-acid residues by protein engineering and the incorporation of molecular-modelling approaches may be expected to lead to a clear picture of lipase action.

REFERENCES

1. Brady, L., Brzozowaski, A. M., Derwenda, Z.S., Dodson, E., Dodson, G., Tolley, S., Turkenberg, J.P., Christansen, L., Huge-Jensen, B., Norskov, L., Thim, L. & Menge, U. (1990). A serine protease triad forms the catalytic centre of a triacylglycerol lipase. *Nature*, 343, 767–770.

2. Brazowaski, A.M., Derwenda, U., Derwenda, Z.S., Dodson, G.G., Lawson, D.M., Turkenberg, J.P., Björkling, F., Huge-Jensen, B., Patkar, S.A. & Thim, L. (1991). A model for interfacial activation in lipases from the structure of a fungal lipase–inhibitor complex. *Nature*, 351, 491–494.

3. Brockman, H.L. (1984). General features of lipolysis: reaction scheme, interfacial structure and experimental approaches. In *Lipases*, ed. Borgström, B. & Brockman, H.L., pp. 3–46. Amsterdam: Elsevier.

4. Ransac, S., Rivière, C., Soulie, C., Gancet, C., Verger, R. & de Haas, G.H. (1990). Competitive inhibition of lipolytic enzymes. I. A kinetic model applicable to water-insoluble competitive inhibitors. *Biochim. Biophys. Acta*, 1043, 57–66.

5. de Haas, G.H., Dijkman, R., Van Oort, M.G. & Verger, R. (1990). Competitive inhibition of lipolytic enzymes. III. Some acylamino analogues of phospholipids are potent competitive inhibitors of porcine pancreatic phospholipase A_2. *Biochim. Biophys. Acta*, 1043, 75–82.

6. Hayakawa, K., Ohara, K. & Satake, I. (1980). The stepwise conformational change of poly(L-lysine) in aqueous solution of sodium 1-octane sulphate. *Chemistry Letters* (Chem. Soc. Japan) 647–650.

7. Wills, E.D. (1954). The effect of anionic detergent and some related compounds on enzymes. *Biochem J.*, 57, 109–120.

8. Brockerhoff, H. (1971). On the function of bile salts and proteins as cofactor of lipase. *J. Biol. Chem.*, 246, 5828–5831.

9. Sugiura, M. & Ogiso, T. (1969). Studies on bile-sensitive lipase VII. Effect of surfactants on *Mucor* lipase (Studies on enzymes XLVIII). *Yakugaku Zasshi*, 89, 1284–1296.

10. Borgström, B. & Donner, J. (1976). Interactions of pancreatic lipase with bile salts and dodecylsulfate. *J. Lipid Res.*, 17, 491–497.

11. Verger, R. Pancreatic lipases (1984). In *Lipases*, ed. Borgström, B. & Brockman, H.L., pp. 83–150. Amsterdam: Elsevier.

12. Borgström, B. & Erlanson, C. (1973). Pancreatic lipase and colipase. Interaction and effects of bile salt and other detergents. *Eur. J. Biochem.*, 37, 60–68.

13. Momsen, W.E. & Brockman, H.L. (1976). Inhibition of pancreatic lipase B activity by taurodeoxy cholate and its reversal by colipase. *J. Biol. Chem.*, 251, 384–388.

14. Lombardo, D.D., Campese, L., Multigner, H., Lafont & Caro, A.D. (1983). On the probable involvement of arginine residues in bile-salt binding site of human carboxylester hydrolase. *Eur. J. Biochem.*, 133, 327–333.

15. Hernell, O. (1975). Human lipase. III. Physiological implication of the bile salt stimulated lipase. *Eur. J. Clin. Invest.*, 5, 267–272.

16. Hernell, O., Blackberg, L. & Olivecrona, T. (1981). Human milk lipase. In *Textbook of Gastroenterology and Nutrition in Infancy*, ed. Lebenthal, E., chapter 32, pp. 347–354. New York: Raven Press.

17. Tsujita, T. & Okuda H. (1990). Effect of bile salt on the interfacial activation of pancreatic carboxylester lipase. *J. Lipid Res.*, 31, 831–838.

18. Bezborodov, A.M., Davranov, K.D. & Akhmedova, (1985). Lipase inhibitor in *Rhizopus microsporus* cultures. In F.E.M.S. symposium 23, ed. by Kulaev, I.S., Dawes, E.A. & Tempest, D.W. (Environ. Regul. Microb. Metab.) pp. 145–149. London: Academic Press.

19. Kiyotani, K., Tasaka, H., Tsukiyama, F. & Matsuo, Y. (1983). Lipase activity in guinea-pig peritonal macrophages and mycobacterial lipase inhibitor. *Hiroshima J. Medical Sci.*,

32, 267–271.
20. Chapman, G.W. Jr (1987). A proteinaceous competitive inhibitor of lipase isolated from *Helianthus* seeds. *Phytochemistry*, 26, 3127–3131.
21. Wang, S. & Huang, A.H.C. (1984). Inhibitors of lipase activities in soyabean and other oil seeds. *Plant. Physiol.*, 76, 929–934.
22. Mori, M., Yamaguchi, K. & Abe, K. (1989). Purification of a lipoprotein lipase inhibiting protein produced by a melanoma cell line associated with cancer Cachexia. *Biochem. Biophys. Res. Commun.*, 160, 1085–1092.
23. Laskowski, M. Kato, I. (1980). Protein inhibitors of proteinases. *Annu. Rev. Biochem.*, 49, 593–626.
24. Gargouri, Y., Julian, R., Sugihara, A., Verger, R. & Sarda, L. (1984). Inhibition of pancreatic and microbial lipase by proteins. *Biochim. Biophys. Acta*, 795, 326–331.
25. Gargouri, Y., Julian, R., Pieroni, G., Rivière, C., Sarda, L. & Verger, R. (1986). Inhibition of lipases by proteins: a binding study using dicaprin monolayers. *Biochemistry*, 25, 1733–1738.
26. Verger, R., Rivère, C., Moreau, H., Gargouri, Y., Rogalska, E., Nury, S., Moulin, A., Ferrato, F., Ransac, S., Carriere, F., Cudrey, C. & Tretout, N. (1991). Enzyme kinetics of lipolysis. Lipase inhibition by proteins. In *Lipases: Structure, Mechanism and Genetic Engineering*, GBF monographs, vol. 16, ed. Alberghina, L., Schmid, R.D. & Verger, R., pp. 105–116. Weinheim: VCH.
27. Gargouri, Y., Pieroni, G., Lowe, P.A., Sarda, L. & Verger, R. (1986). Human gastric lipase. The effect of amphiphiles. *Eur. J. Biochem.*, 156, 305–310.
28. Gargouri, Y., Pieroni, G., Ferrato, F. & Verger, R. (1987). Human gastric lipase. A kinetic study with dicaprin monolayers. *Eur. J. Biochem.*, 169, 125–129.
29. Ishi, C., Endo, Y., Kimoto, H. & Taniguchi, K. (1988). Inhibition of lipases by proteins and their inhibitory mechanism. *Nippon Shokuhin Kogyo Gakkaishi*, 35, 430–439.
30. Iwai, M., Tsujisaka, Y. & Fukumoto, J. (1964). Studies on lipase. III. Effect of calcium ion on the action of the crystalline lipase from *Aspergillus niger. J. Gen. Appl. Microbiol.*, 10, 87–93.
31. Liu, W.H., Beppu, T. & Arima, K. (1973). Effects of various inhibitors on lipase action of thermophilic fungus *Humicola lanuginosa* S-38. *Agric. Biol. Chem.*, 37, 2487–2492.
32. Iwai, M., Tsujisaka, Y. & Fukumoto, J. (1970). Studies on lipase. V. Effect of iron ions on the *Aspergillus niger* lipase. *J. Gen. Appl. Microbiol.*, 16, 81–90.
33. Kgandopulo, G.B. & Ruben, E.L. (1974). Effect of various activators and inhibitors on the lipase activity of fungi of the genus *Geotrichum. Microbiologiya*, 43, 814–819.
34. Fredrikson, G., Strålfors, P., Nilsson, N.O. & Belfrage, P. (1981). Hormone sensitive lipase of rat adipose tissue: purification and some properties. *J. Biol. Chem.*, 256, 6311–6320.
35. Hata, Y., Matsuura, Y., Tanaka, N., Kakudo, M., Sugihara, A., Iwai, M. & Tsujisaka, Y. (1979). Low resolution crystal structure of lipase from *Geotrichum candidum* (ATCC 34614) *J. Biochem.*, 86, 1821–1827.
36. Lolis, E. & Petsko, G. (1990). Transition state analogues in protein crystallography probes of the structural source of enzyme catalysis. *Annu. Rev. Biochem.*, 59, 597–630.
37. Garner C.W. (1980). Boronic acid inhibitors of pancreatic lipase *J. Biol. Chem.*, 255, 5064–5068.
38. Vainio, P., Virtanen, J.A. & Kinnunen, P.K.J. (1982). Inhibition of lipoprotein lipase by benzene boronic acid: effect of apolipoprotein CII. *Biochim. Biophys. Acta*, 711, 386–390.
39. Vainio, P. (1983). N-(5-Dimethylaminonaphthalene-1-sulphonyl)-3-aminobenzene boronic acid as an active site directed fluorescent probe of lipoprotein lipase. *Biochim. Biophys. Acta*, 746, 217–219.
40. Aboukil, N. & Lombardo, D. (1989). Inhibition of human milk bile salt-dependent lipase

by boronic acids: implication to bile salt activator effect. *Biochim. Biophys. Acta*, 1004, 215–220.

41. Lengsfeld, D.H. & Wolfer, H. (1988). Inhibition of pancreatic lipase in vitro by the covalent inhibitor tetrahydrolipstatin. *Biochem. J.*, 256, 357–361.

42. Schrag, J.D., Wu, Y.S. & Cygler, M. (1991). Ser-His-Glu triad forms the catalytic site of the lipase from *Geotricum candidum. Nature*, 351, 761–764.

43. Winkler, F.K., D'Arcy, A. & Hunziker, W. (1990). Structure of human pancreatic lipase. *Nature*, 343, 771–774.

44. Authors' unpublished results.

45. Desnuelle, P., Sarda, L. & Ailhaud, G. (1960). Inhibition of pancreatic lipase by diethyl-*p*-nitrophenyl phosphate in emulsion. *Biochim. Biophys. Acta*, 37, 570–571.

46. Chapus, C. & Semeriva, M. (1976). Mechanism of pancreatic lipase action. 2. Catalytic properties of modified lipases. *Biochemistry*, 15, 4988–4991.

47. Maylie, M.F., Charles, M. & Desnuelle, P. (1972). Action of organophosphates and sulphonyl halides on pancreatic lipase. *Biochim. Biophys. Acta*, 276, 162–175.

48. Rouard, M., Sari, H., Nurit, S., Entressangles, B. & Desnuelle, P. (1978). Inhibition of pancreatic lipase by mixed micelles of diethyl *p*-nitrophenylphosphate and bile salts. *Biochim. Biophys. Acta*, 530, 227–235.

49. Moreau, H., Moulin, A., Gargouri, Y., Noël, J.P. & Verger, R. (1991). Inactivation of gastric and pancreatic lipases by diethyl *p*-nitrophenyl phosphate. *Biochemistry*, 30, 1037–1041.

50. Authors' unpublished results.

51. De Caro, J.D., Roumi, P. & Rovery, M. (1986). Hydrolysis of *p*-nitrophenyl acetate by the peptide chain fragment (336–449) of porcine pancreatic lipase. *Eur. J. Biochem.*, 158, 601–607.

52. Sikk, P.F., Oza, A.V. & Aaviksaar, A.A. (1979). A water-soluble organophosphorus inhibitor for pancreatic lipase. *Soviet J. Bioorg. Chem.*, 5, 706–708.

53. Sikk, P.F. & Aaviksaar, A.A. (1985). Irreversible inhibition of pancreatic lipase by bis-*p*-nitrophenyl methylphosphonate. *FEBS Letters*, 184, 193–196.

54. Reddy, N.M., Maraganore, J.M., Meredith, S.C., Heinrikson, R.L. & Kédzy, F.J. (1986). Isolation of an active site peptide of lipoprotein lipase from bovine milk and determination of its amino acid sequence. *J. Biol. Chem.*, 261, 9678–9683.

55. Olivecrona, T. & Bengtsson-Olivecrona, G. (1987). Lipoprotein lipase from milk – the model enzyme in lipoprotein lipase research. In *Lipoprotein Lipase*, ed. Borensztajn, J., pp. 15–58. Chicago, USA: Evener Publication.

56. Horrevoets A.J.G., Verheij, H.M. & de Hass, G.H. (1991). Inactivation of *Escherichia coli* outer-membrane phospholipase A by the affinity label hexadecanesulphonyl fluoride. Evidence for active-site serine. *Eur. J. Biochem.*, 198, 247–253.

57. Fourneron, J.D., Abouakil, N., Challan, C. & Lombardo, D. (1991). N-butyl-N-methyl-4-nitrophenyl carbamate as a specific active site titrator of bile salt dependent lipases. *Eur. J. Biochem.*, 196, 295–303.

58. Imanaka, T., Moriyama, Y., Ecsedi, G.G., Aoyagi, T., Amanuma-Muto, K., Ohkuma, S. & Takano, T. (1983), Esterastin, a potent inhibitor of lysosomal acid lipase. *J. Biochem.*, 94, 1017–1020.

59. Chadha, N.K., Batcho, A.D., Tang, P.C., Courtney, L.F., Cook, C.M., Wovkulich, P.M. & Uskokovic, M.R. (1991), Synthesis of tetrahydrolipstatin. *J. Org. Chem.*, 56, 4714–4718.

60. Borgström, B. (1988). Mode of action of tetrahydrolipstatin: a derivative of the naturally occurring lipase inhibitor lipstatin. *Biochim. Biophys. Acta*, 962, 308–316.

61. F. Winkler, personal communication.

62. Hadvarry, P., Lengsfeld, H. & Wolfer H. (1988). Inhibition of pancreatic lipase *in vitro* by the covalent inhibitor tetrahydrolipstatin. *Biochem. J.*, 256, 357–361.

63. Hadvary, P., Sidler, W., Meister, W., Vetter, W. & Wolfer, H. (1991). The lipase inhibitor tetrahydrolipstatin binds covalently to the putative active site serine of pancreatic lipase. *J. Biol. Chem.*, 266, 2021–2027.

64. Chiba, H., Histanke, M., Hirose, M. & Sugimoto, E. (1973). Role of tryptophan residues on the *Rhizopus delemar* lipase activity: chemical modification in a water olive oil emulsion. *Biochim. Biophys. Acta*, 327, 380–392.

65. Liu, W.H., Beppu, T. & Arima, K. (1977). The chemical modification of the lipase of *Humicola lanuginosa* by N-bromosuccinimide in urea solution. *Agric. Biol. Chem.*, 41, 131–135.

66. Moreau, H., Gargouri, Y., Pieroni, G. & Verger, R. (1988), Importance of sulfhydryl group for rabbit gastric lipase activity. *FEBS Letters*, 236, 383–387.

67. Gargouri, Y., Moreau, H., Pieroni, G. & Verger, R. (1988). Human gastric lipase: a sulfhydryl enzyme. *J. Biol. Chem.*, 263, 2159–2167.

68. Möller, H. (1981). German patent; European patent application no. 3018132.

69. Ecsedi, G.G., Amanuma, K., Imanaka, T., Aoyagi, T., Ohkuma, S. & Takano, T. (1985). Effect of esterastin, an acid lipase inhibitor, on the free and esterified cholesterol contents of cultured aortic smooth muscle cells treated with LDL and cholesterol ester liquid crystals. *Biochem. Internat.*, 10, 332–342.

70. Oglesby, T.D. & Gorman, R.R. (1984). The inhibition of arachidonic acid metabolism in human platelates by RHC 80267, a diacylglycerol lipase inhibitor. *Biochim. Biophys. Acta*, 793, 269–277.

71. Chuang, M. & Severson, (1990). Inhibition of diacylglycerol metabolism in isolated cardiac myocytes by U-57908 (RHC 80267), a diacylglycerol lipase inhibitor. *J. Mol. Cell Cardiol.*, 22, 1009–1016.

72. Kubo, M. & Hostetler, K. Y. (1987). Inhibition of purified bovine milk lipoprotein lipase by propranolol and other beta-adrenergic blockers *in vitro*. *Biochim. Biophys. Acta*, 918, 168–174.

73. Comani, K. & Sullivan, A.C. (1982). Phenethylamine inhibitors of partially purified rat and human pancreatic lipase. *J. Pharm. Sci.*, 171, 418–422.

74. Björkling, F., Godtfredsen, S.E. & Kirk, O. (1991). The future impact of industrial lipases. *Trends Biotechnol.*, 9, 360–363.

11.

Substrates for phospholipase C and sphingomyelinase from *Bacillus cereus*

Relation between chemical structure and enzyme hydrolysis

ULRICH MASSING and HANSJÖRG EIBL

Phospholipase C (1, 2) and sphingomyelinase from *Bacillus cereus* (3) are closely related enzymes in respect of their hydrolysis of the phosphate ester bond. They have nearly identical molecular weights, with values of 23 and 24 kDa for phospholipase C (1, 4) and sphingomyelinase (3) respectively. Divalent cations are required for enzyme activity, these being Zn^{2+} for phospholipase C (5) and Mg^{2+} for sphingomyelinase (3). Both enzymes preferentially hydrolyse naturally occurring lipids that contain phosphocholine. For instance, phosphatidylcholines are the best substrates for phospholipase C and sphingomyelins for sphingomyelinase. The chemical structures of these substrates are shown in Figure 1.

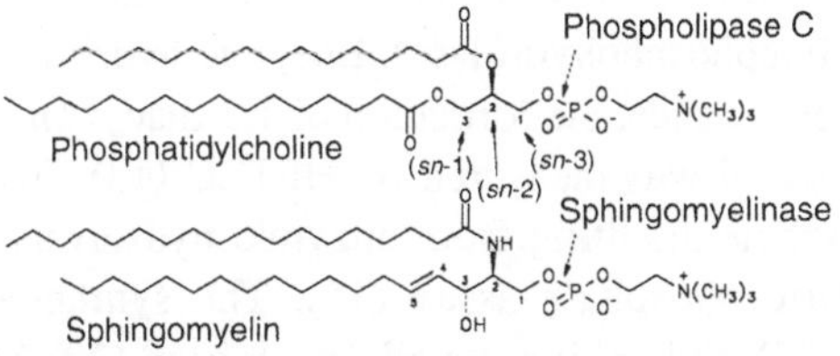

Figure 1. Typical structures of phosphatidylcholines and sphingomyelins in their natural configuration. The dotted arrows indicate the point of attack for phospholipase C or sphingomyelinase. Straight-line arrows indicate the positional distribution of substituents in the glycerol molecule according to Hirschmann (7).

Since the polar head-group phosphocholine is the same for both these lipid molecules, the pronounced differences in the substrate requirement of these enzymes must be due to strong involvement of the apolar region of the phospholipid molecules in the recognition process. The apolar region is dominated by two extended hydrocarbon chains, which can attain a perfect alignment in a bilayer arrangement. The existence of such preferred conformations has been demonstrated experimentally by crystal structure determination (6).

The products of hydrolysis by phospholipase C and sphingomyelinase are diacylglycerols and ceramides, respectively. Diacylglycerols are believed to act as secondary messenger molecules in the control of cell growth and differentiation (8). In particular, diacylglycerols have been reported to activate protein kinase C (9–11). According to recent publications, ceramides may also be mediators in cell differentiation (12).

The structural properties and biological importance of these molecules make it important to define precisely the specific structural elements in a substrate molecule that are necessary for hydrolysis by phospholipase C or sphingo-myelinase. Substrate recognition by phospholipase C or sphingomyelinase is not simply explained by the exchange of an ester bond for an amide bond at the 2-position of phosphatidylcholine. This has been demonstrated by the use of analogues of phosphatidylcholines and of sphingomyelins that bear structural modifications, mainly in their apolar region. In this chapter, we survey the results of such modifications on the rates of hydrolysis by the two enzymes, and draw conclusions concerning the detailed structural requirements made of their substrates.

In this survey, we draw extensively on our experiments concerning the substrate properties of newly designed phospholipid- or sphingomyelin-like molecules. The enzymes used were phospholipase C from *B. cereus* (Boehringer Mannheim, EC 3.1.4.3, phosphatidylcholine cholinephosphohydrolase) and sphingomyelinase from *B. cereus* (Boehringer Mannheim, EC 3.1.4.12, sphingomyelin cholinephosphohydrolase). Enzymic hydrolysis was followed by product analysis. For instance, the formation of 1,2-diacyl-*sn*-glycerol or of 1-O-alkyl-2-acyl-*sn*-glycerol was measured by HPTLC (13). In earlier work (14), water-soluble phosphate resulting from enzymic hydrolysis was measured by using a turbidimetric phosphate assay (15). The synthesis of the different phospholipids (16–19) and sphingomyelin analogues (20) has been described elsewhere.

A note on nomenclature

The nomenclature of phospholipids (21) used here follows that of Hirschmann (7). Naturally occurring phosphatidylcholines contain glycerol which is esterified at the 1- and 2-positions with fatty acids and at the 3-position with phospho-

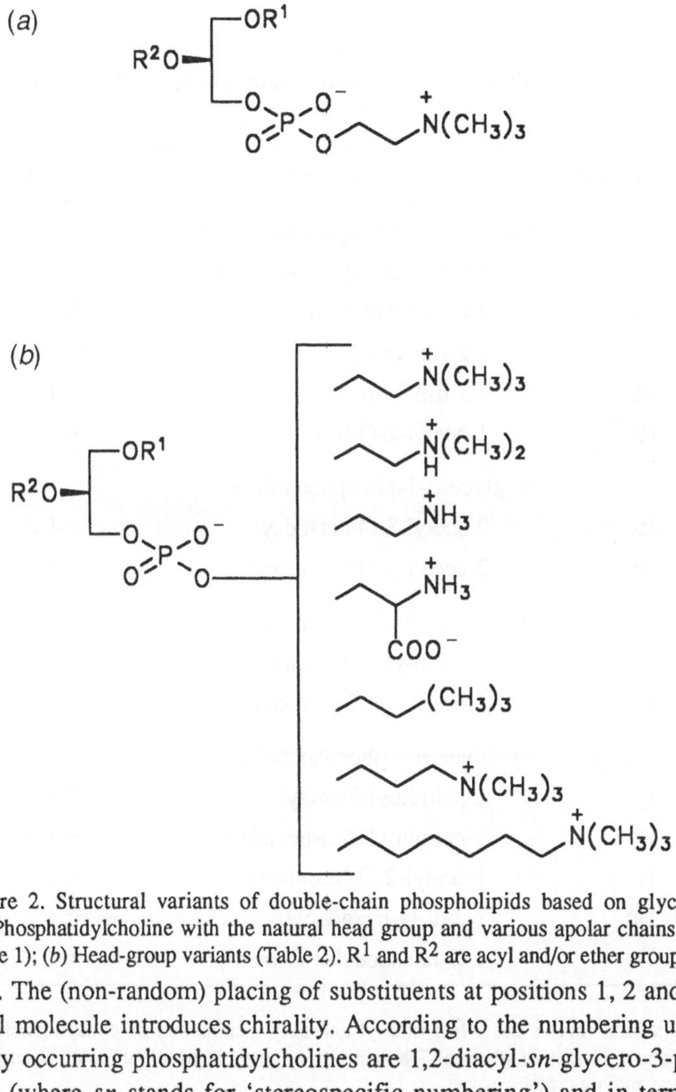

(a)

(b)

Figure 2. Structural variants of double-chain phospholipids based on glycerol. (a) Phosphatidylcholine with the natural head group and various apolar chains (see Table 1); (b) Head-group variants (Table 2). R^1 and R^2 are acyl and/or ether groups.

choline. The (non-random) placing of substituents at positions 1, 2 and 3 of the glycerol molecule introduces chirality. According to the numbering used here, naturally occurring phosphatidylcholines are 1,2-diacyl-*sn*-glycero-3-phospho-choline (where *sn* stands for 'stereospecific numbering') and in terms of the absolute configuration (*R*)-1,2-diacyl-3-phosphocholine-glycerol. Its opposite-handed enantiomer is 2,3-diacyl-*sn*-glycero-1-phosphocholine or (*S*)-1,2-diacyl-3-phosphocholine-glycerol. The structures and the numbering of these molecules are shown in Figure 1.

1. ESSENTIAL STRUCTURAL ELEMENTS FOR HYDROLYSIS BY PHOSPHOLIPASE C

Structures of double-chain phospholipids based on glycerol used as substrates for phospholipase C hydrolysis are shown in Figure 2. Their substrate properties are

Table 1. Rates of hydrolysis of double-chain phosphatidylcholines by phospholipase C.

Abbreviation	Compound	Activity
	sn-glycero-3-phosphocholine	
1a	1-O-octadecyl-2-palmitoyl	1360
1b	1-O-octadecyl-2-acetyl	1312
1c	1,2-dimyristoyl	905
1d	1,2-dipalmitoyl	371
1e	1-acetyl-2-O-hexadecyl	< 1
	sn-glycero-1-phosphocholine	
1f	2-acetyl-3-O-octadecyl	< 1
1g	2-acetyl-3-O-hexadecyl	< 1
	sn-glycero-2-phosphocholine	
1h	1-stearoyl-3-palmitoyl	< 1
1i	1-myristoyl-3-stearoyl	< 1
	rac-glycero-3-phosphocholine	
1j	1-palmitoyl-2-acetyl	411
1k	1-palmitoyl-2-O-hexadecyl	< 1
1l	1-acetyl-2-O-hexadecyl	< 1
1m	1,2-di-O-hexadecyl	< 1
1n	1-O-hexadecyl-2-O-methyl	< 1

Comparison reveals the effect of acyl and alkyl chains and of phosphocholine at different positions of glycerol. The activity is given in μmol substrate reacted per minute and mg protein.

summarised in Tables 1 and 2.

First we examine the substrate properties of various double-chain phosphatidylcholines (Table 1). Naturally occurring phosphatidylcholines such as 1,2-diacyl-sn-glycero-3-phosphocholines and 1-O-alkyl-2-acyl-sn-glycero-3-phosphocholines are excellent substrates for phospholipase C from *B. cereus*. In line with these observations, high rates of hydrolysis are observed for synthetic compounds that possess the same configuration, for instance 1-O-octadecyl-2-palmitoyl-sn-glycero-3-phosphocholine (compound 1a) or for the 2-acetyl

Table 2. Structural variations in the polar region of diacyl-*sn*-glycero-3-phosphate esters and their effect on phospholipase C hydrolysis.

Abbreviation	Compound	Activity
	sn-glycero-3-phosphate ester	
2a	1,2-dipalmitoyl-*sn*-G-3-PC	371
2b	1,2-dipalmitoyl-*sn*-G-3-P-(N,N-diMe)E	275
2c	1,2-dipalmitoyl-*sn*-G-3-PE	98
2d	1,2-dioleyl-*sn*-3-PS	61
2e	1,2-dipalmitoyl-*sn*-G-3-P-(3,3-diMe)butanol	<1
2f	1,2-dimyristoyl-*sn*-3-G-3-P-(N,N,N-triMe)propanolamine	<1
2g	1,2-dimyristoyl-*sn*-3-G-3-P-(N,N,N-triMe)hexanolamine	<1

G, glycerol; P, phosphate; PC, phosphocholine; Me, methyl; E, ethanolamine; PE, phosphoethanolamine; PS, phosphoserine; activity in μmol min^{-1} per mg protein.

analogue (PAF, compound 1b) with values of about 1350 U/mg protein (1 unit $\equiv$ 1 μmol min^{-1} substrate hydrolysis). For sphingomyelinase, the respective rate with PAF is only 3 U/mg under exactly the same experimental conditions as for phospholipase C. Thus, PAF is hydrolysed by sphingomyelinase at less than 0.25% of the rate with which it is hydrolysed by phospholipase C. The rate of hydrolysis by sphingomyelinase is probably in reality even lower, as trace amounts of phospholipase C are often present in the sphingomyelinase preparation. Thus, phospholipase C obviously hydrolyses preferentially substrates that have glycerol, rather than sphingosine, as their basic structural element. However, structurally well-defined synthetic analogues with the natural configuration are not invariably substrates for phospholipase C.

Surprisingly, an ether bond at the *sn*-2 position of glycerol, even in the natural configuration, completely abolishes the ability to act as a substrate for phospholipase C (compounds 1e and 1k). This is an important clue to the nature of substrate recognition by phospholipase C. These isomers have the natural configuration, but have never been detected in natural membranes. In addition, configurational isomers with phosphocholine attached to the *sn*-1 or the *sn*-2 position of glycerol are very poor substrates for phospholipase C.

Table 1 demonstrates that molecules amenable to hydrolysis by phospholipase C carry phosphate at the *sn*-3 position of glycerol, an acyl group at position 2 and acyl or alkyl at position 1, and Table 2 gives information on how variation of the chemical nature of the polar head group of the molecule affects the hydrolysis rate. The structures of the head-group variants are shown in Figure 2.

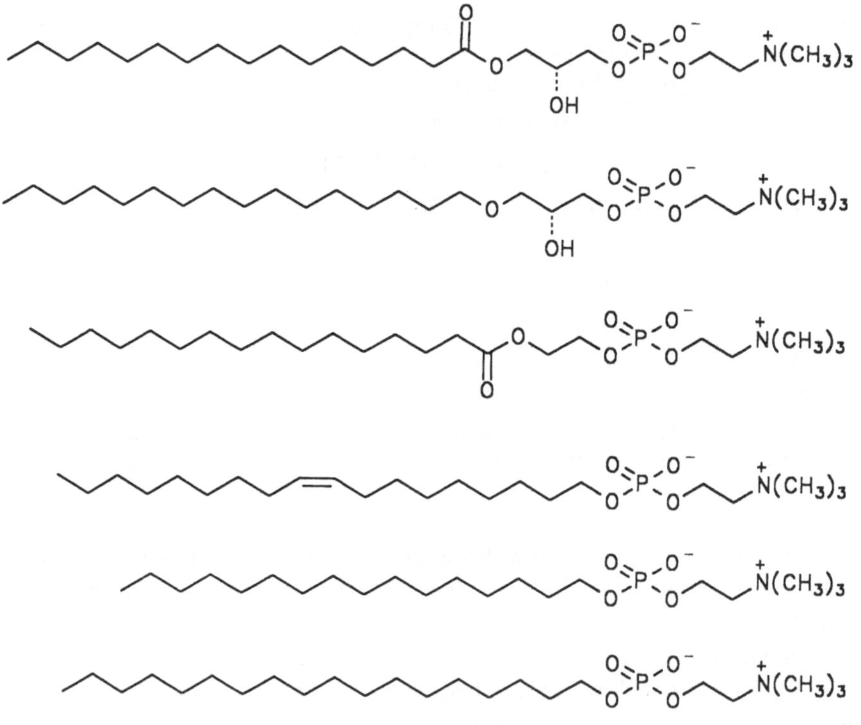

Figure 3. Structural variation in lysophosphatidylcholines (single-chain molecules). The structures from top to bottom correspond to compounds 3a–3f in Table 3.

As shown for compounds 2a, 2b and 2c, the rate of hydrolysis is strongly influenced by the degree of *N*-methylation. Thus, phosphatidylcholines are better substrates for phospholipase C from *B. cereus* than the respective phosphatidylethanolamines are. Note, however, that phosphatidylserine (compound 2d) is still a good substrate for phospholipase C. At pH 7, this molecule carries one nett negative charge, as the positive charge of the ammonium group is overcompensated by the negative charges of the phosphate and carboxyl groups.

However, replacement of the nitrogen by a carbon atom in phosphatidylcholines (compound 2e) results in a molecule with a negatively charged head group. This charge, no longer compensated by a positively charged ammonium group, completely prevents hydrolysis by phospholipase C. It demonstrates that the negative charge at the phosphate group must be balanced by a positively charged

Table 3. Single-chain phosphocholines as substrates for phospholipase C.

Abbreviation	Compound	Activity
	Phosphocholine	
3a	1-palmitoyl-*sn*-glycero-3-PC	0.70
3b	1-O-hexadecyl-*sn*-glycero-3-PC	0.03
3c	1-palmitoyl-2-PC-ethyleneglycol	8.5
3d	oleyl-PC	0.03
3e	hexadecyl-PC	0.02
3f	octadecyl-PC	0.02

PC, phosphocholine; activity in $\mu mol\,min^{-1}$ per mg protein.

ammonium group to fulfil the substrate requirements of the enzyme. But charge compensation alone is not sufficient. Compounds 2f and 2g contain both a negatively charged phosphate and a positively charged trimethylammonium group. However, their rates of hydrolysis are very low. In these molecules, the distance between the phosphate and the trimethylammonium group is greater than in phosphocholine: compound 2f contains one additional methylene group and 2g four additional methylene groups between the positive and negative centres. This results in a complete loss of the ability to act as substrate for phospholipase C, an effect that may be due either to increased charge separation or to steric interference with the formation of the enzyme–substrate complex.

A further conclusion drawn from the Tables 1 and 2 is that diacyl- or alkyl/acyl-glycerophosphocholines must bear phosphocholine at position *sn*-3 and acyl at position *sn*-2 of glycerol if they are to act as substrates for phospholipase C.

The importance of the acyl group at position 2 is also demonstrated by the properties of lysophospholipids as substrates (Figure 3 and Table 3). As shown in Table 3, the rates of hydrolysis are low if a free alcohol group is present at position *sn*-2 of glycerol, or, more important, if no ester is present at this position. For instance, 1-palmitoyl-*sn*-glycero-3-phosphocholine (compound 3a) is a poor substrate for phospholipase C. Only marginal activity is observed for the corresponding hexadecyl compound 3b. Compound 3c, in contrast, has an acyl group at the *sn*-2 of the glycerol. This results in a strong increase in the hydrolysis rate, and is in agreement with earlier observations that an ester bond at position 2 is important for hydrolysis by phospholipase C (14). This observation receives further confirmation from the properties of 2-acyl-*sn*-glycero-3-phosphocholines as substrates of phospholipase C. It has been demonstrated (22) that 2-acyl-*sn*-

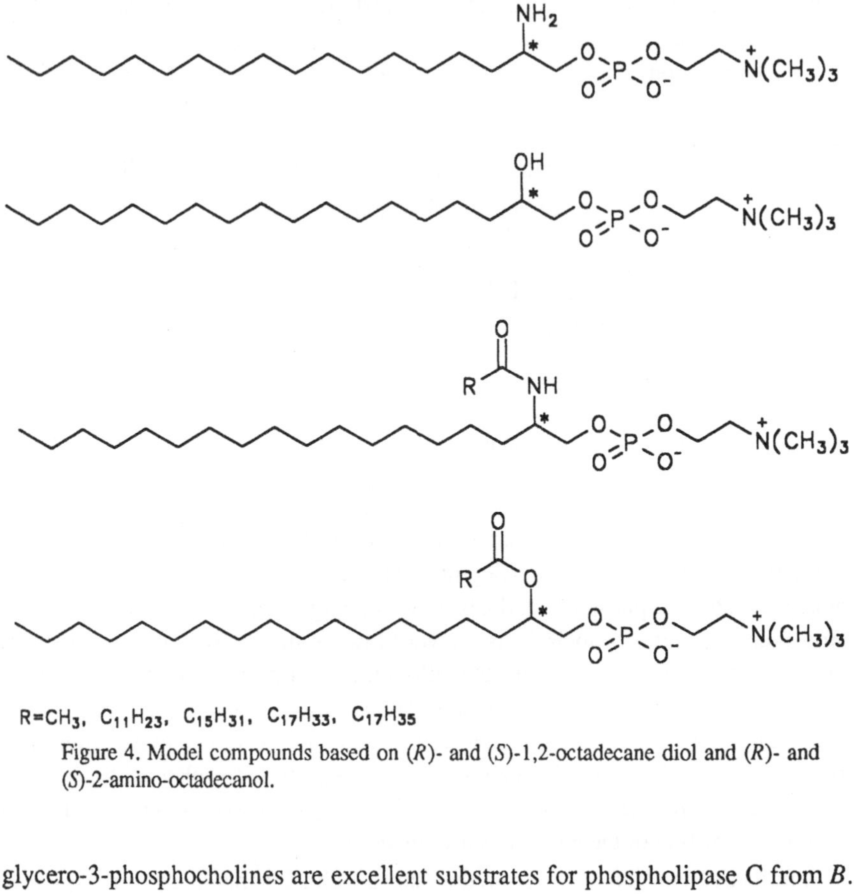

R=CH₃, C₁₁H₂₃, C₁₅H₃₁, C₁₇H₃₃, C₁₇H₃₅

Figure 4. Model compounds based on (R)- and (S)-1,2-octadecane diol and (R)- and (S)-2-amino-octadecanol.

glycero-3-phosphocholines are excellent substrates for phospholipase C from *B. cereus*. Yet simple alkylphosphocholines (compounds 3d, 3e, 3f) are poor substrates, since they lack the structural prerequisites for enzyme hydrolysis; in particular, they lack an ester bond at carbon 2.

According to the discussion up to now, we have learned that the minimal structural elements for phospholipase C hydrolysis are (i) a vicinal diol with phosphocholine attached to the primary hydroxy group and (ii) an acyl ester at the adjacent hydroxy group. This is confirmed neatly by 1-palmitoyl-2-phospho-choline-ethyleneglycol (compound 3c), which is a good substrate for phospholipase C, even though it is not based on glycerol. This compound has no chiral centre, but it still fulfils the criteria formulated above. However, as will be discussed later, the ester or ether bond in the *sn*-1 position in phosphatidylcholines is also of importance for the rate of hydrolysis by phospholipase C. To obtain further proof of this, and also to establish a structural link with sphingomyelins and sphingomyelinase, we prepared model compounds based on

Table 4. Rates of hydrolysis by phospholipase C of enantiomerically pure ester model compounds, 1-O-phosphocholine-2-O-acyl-octadecanes (Figure 4).

Abbreviation	Compound	Activity
2-O-acyl groups (*R*-configuration)		
4a	acetyl	11
4b	lauroyl	1.5
4c	palmitoyl	24
4d	oleyl	1.7
4e	stearoyl	33
2-O-acyl groups (*S*-configuration)		
4f	acetyl	< 0.001
4g	lauroyl	< 0.001
4h	palmitoyl	0.06
4i	oleyl	< 0.001
4j	stearoyl	0.09

Activity in μmol min^{-1} per mg protein.

1,2-octadecane diol (in both enantiomeric forms) along with the corresponding compounds based on 2-amino-octadecanol (Figure 4).

As expected, and as demonstrated in Table 4, the 1,2-octadecane diol compounds with the natural configuration (*R*-enantiomers) and an acyl group at position 2 are substrates for phospholipase C. Neither the free alcohol (no ester at position 2) nor compounds with the (unnatural) *S*-configuration are substrates for this enzyme, irrespective of whether an acyl group is present at the 2-position or not. These results are completely consistent with, and could indeed have been predicted from, the information obtained from Tables 1–3 for phospholipids based on glycerol.

One simple explanation for the difference in substrate recognition between phospholipase C and sphingomyelinase could be that phospholipase C recognises compounds with acyl esters at position 2 while sphingomyelinase recognises compounds with acyl amides at this position. But this is not the case, as is shown strikingly by the activities given in Table 5. For instance, in the case of the stearoyl amide in the natural configuration (*R*-; compound 5e), we observe about the same rate of hydrolysis as for the corresponding ester (compound 4e). Thus, phospholipase C cannot distinguish between an acyl ester and an acyl amide bond at the *sn*-2 position in the ester or amide model compounds.

Table 5. Rates of hydrolysis by phospholipase C of enantiomerically pure amido model compounds, 1-O-phosphocholine-2-N-acyl-octadecanes (Figure 4).

Abbreviation	Compound	Activity
	2-N-acyl groups (*R*-configuration)	
5a	acetyl	0.5
5b	lauroyl	7.4
5c	palmitoyl	16
5d	oleyl	0.56
5e	stearoyl	27
	2-N-acyl groups (*S*-configuration)	
5f	acetyl	< 0.001
5g	lauroyl	< 0.001
5h	palmitoyl	0.09
5i	oleyl	< 0.001
5j	stearoyl	0.13

Activity in μmol$\cdot$min^{-1} per mg protein.

2. STRUCTURAL REQUIREMENTS FOR SPHINGOMYELINASE HYDROLYSIS

The naturally occurring sphingomyelin, of course, is an excellent substrate for sphingomyelinase. Phosphocholine is formed during the enzyme reaction. The rate of hydrolysis is about 250 U/mg protein for sphingomyelin isolated from egg yolk (mainly N-palmitoyl sphingomyelin); this is comparable to the specific activity observed for the system dipalmitoyl-*sn*-glycero-3-phosphocholine and phospholipase C, with values of 375 U/mg protein.

In a search for essential structural elements in substrates for sphingomyelinase, we have introduced structural alterations and studied their influence on the rate of enzymic hydrolysis. Since natural sphingomyelins contain the sphingosine and also the dihydrosphingosine bases, it might be expected that complete hydrogenation of the double bonds in naturally occurring sphingomyelins should leave the substrate properties for sphingomyelinase unaltered. In fact, the rates of hydrolysis lie in the same order of magnitude as for the non-hydrogenated substrates. A comparison of these rates is shown in Table 6. From these data, it is obvious that the *trans* double bond in the sphingosine base is not essential for

Table 6. Hydrolysis rates of sphingomyelins (Figure 1) and of enantiomerically pure synthetic model compounds (Figure 4) that lack the essential hydroxy group at position 3.

Abbreviation	Compounds	Activity
I. Natural compounds		
6a	sphingomyelin	250
6b	dihydrosphingomyelin	270
II. Amido model compounds (1-O-phosphocholine-2-N-acyl-octadecanes)		
2-N-acyl-group (*R*-configuration)		
6c	acetyl	< 0.001
6d	lauroyl	0.02
6e	palmitoyl	0.16
6f	oleyl	0.03
6g	stearoyl	0.19
2-N-acyl-group (*S*-configuration)		
6h	acetyl	< 0.001
6i	lauroyl	< 0.001
6j	palmitoyl	< 0.001
6k	oleyl	< 0.001
6l	stearoyl	< 0.001
III. Ester model compounds (1-O-phosphocholine-2-O-acyl-octadecanes)		
2-O-acyl-group (*R*-configuration)		
6m	acetyl	0.2
6n	lauroyl	< 0.001
6o	palmitoyl	0.08
6p	oleyl	0.004
6q	stearoyl	0.07
2-O-acyl-group (*S*-configuration)		
6r	acetyl	< 0.001
6s	lauroyl	< 0.001
6t	palmitoyl	< 0.001
6u	oleyl	< 0.001
6v	stearoyl	< 0.001

Activity in $\mu mol \cdot min^{-1}$ per mg protein.

hydrolysis. This fact simplifies the design of synthetic model compounds, and further discussion will therefore be concentrated on the substrate properties of dihydrosphingomyelins.

The starting-point for the synthesis of sphingomyelin-like model compounds was 2-amino-octadecanol, which has been prepared in both enantiomeric forms. Together with the naturally occurring sphingomyelins, these models reveal the principal structural elements necessary for hydrolysis by sphingomyelinase. The important difference between the model compounds and the natural sphingomyelins is the models' lack of the hydroxy group at position 3. As demonstrated by the data in Table 6, this structural feature is essential, and substrates for sphingomyelinase depend strictly on the presence of the hydroxy group at position 3 of the sphingosine base. None of the synthetic model compounds lacking the hydroxy group at position 3 show more than marginal hydrolysis by sphingomyelinase from *B. cereus*. The hydrolysis of the compounds 6e (N-palmitoyl), 6g (N-stearoyl) and 6m (O-acetyl), however, is somewhat faster. This very interesting observation will be the object of further experiments.

3. SUBSTRATE REQUIREMENTS OF PHOSPHOLIPASE C AND SPHINGOMYELINASE

A comparison of the hydrolysis rates using substrates that illustrate the substrate requirements of these two enzymes is shown in Table 7. The real rates of sphingomyelin hydrolysis by phospholipase C and of phosphatidylcholine hydrolysis by sphingomyelinase are probably even lower than in the Table, because commercial phospholipase C is always contaminated with sphingomyelinase and *vice versa*. In general, good substrates for phospholipase C are not substrates for sphingomyelinase and, conversely, good substrates for sphingomyelinase are not substrates for phospholipase C.

It is striking that phospholipase C will hydrolyse all substrates with the *R*-configuration, irrespective of whether they possess an ester or an amide bond at position 2 (compounds 7c–7f). The hydrolysis rates are high for molecules with an additional ester or ether bond in the direct neighbourhood of position 2 (*sn*-1 position of glycerol; compounds 7c, 7d). Such an arrangement, obviously, supports substrate properties for phospholipase C. This is demonstrated by the low rates observed for the sphingomyelins 7a and 7b. Instead of an ester or ether bond in the direct neighbourhood of the amido group at position 2, they contain a free hydroxy group, with the result that their ability to be hydrolysed by phospholipase C is close to zero.

As already indicated in sections 1 and 2, the model compounds 7e and 7f are to a certain extent structurally related to phosphatidylcholines and sphingomyelins. It

Table 7. Substrates that illustrate the connection between substrate structure and enzyme hydrolysis by phospholipase C and sphingomyelinase.

Abbreviation	Compound	Activity	
		Phospholipase C	Sphingomyelinase
7a	Sphingomyelin	1.3	275
7b	Dihydro-sphingomyelin	1.0	260
7c	1-O-octadecyl-2-acetyl-*sn*-glycero-3-phosphocholine (PAF)	1312	3.0
7d	1,2-dipalmitoyl-*sn*-glycero-3-phosphocholine (DPPC)	371	< 0.1
7e	(*R*)-1-O-phosphocholine-2-O-palmitoyl-octadecane	24	0.08
7f	(*R*)-1-O-phosphocholine-2-N-palmitoyl-octadecane	16	0.16

Activity in $\mu mol\ min^{-1}$ per mg protein.

is therefore remarkable that phospholipase C cannot differentiate between an ester and an amido group at the 2-position. Substrates 7e and 7f show about the same rates of hydrolysis. In these compounds, the supporting effect of the neighbouring ester or ether linkage is missing. This explains why their hydrolysis rates are lower than those of compounds 7c and 7d.

Sphingomyelinase also requires that its sphingomyelin substrates have the *R*-configuration, as shown in Table 6. The best substrates for sphingomyelinase are naturally occurring sphingomyelins (compounds 7a and 7b). These contain a free hydroxy group at the 3-position of the sphingosine base and, in addition, an amido group at the 2-position. In contrast, compounds 7c and 7d are poor substrates for this enzyme. They contain an ester group at the 2-position and also an ester or ether group at the *sn*-1 position, which compares with the 3-position of sphingomyelins.

The model compounds 7e and 7f are also poor substrates for sphingomyelinase, while the rates for the ester and for the amido compounds are almost the same. The important structural difference between the model compounds and the sphingomyelins is the lack of the free hydroxy group at position 2 of the model compounds, which results in an almost complete loss of enzyme activity.

These experiments with models suggest strongly that certain structural elements are essential in substrates for phospholipase C and sphingomyelinase. These

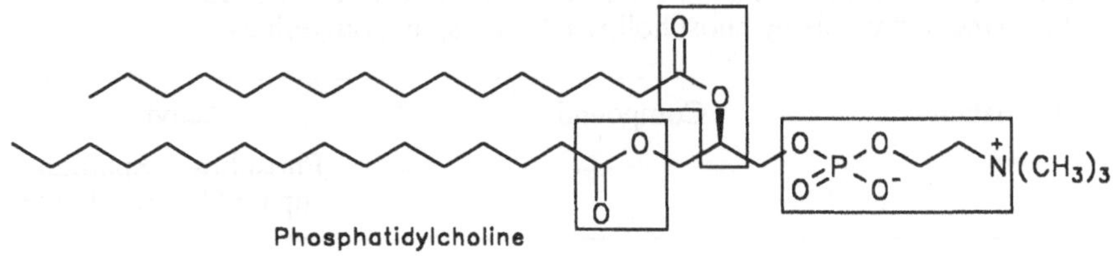

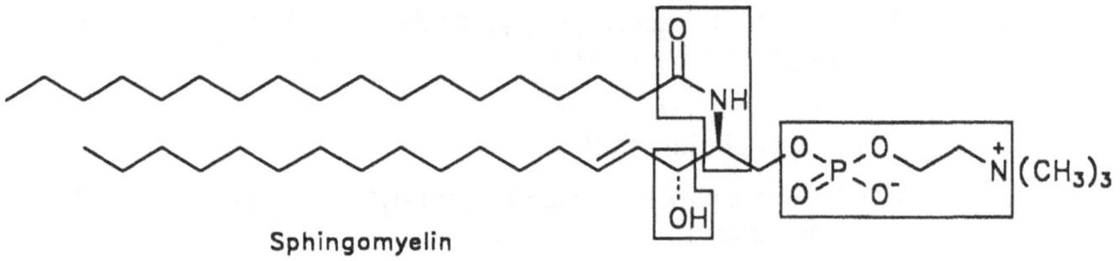

Figure 5. The essential structural elements in substrates for phospholipase C and sphingomyelinase.

essential elements are indicated in Figure 5. A good substrate for phospholipase C must have both an ester at position *sn*-2 in the *R*-configuration and also an ester or ether at the *sn*-1 position of glycerol, which corresponds to the 3-position of sphingomyelins. A good substrate for sphingomyelinase must have both an amide group in the *R*-configuration at position 2 and also a free hydroxy group at position 3. In addition, the phosphocholine ester is by far the best polar group for high activity of phospholipase C.

In spite of the fact that the work discussed here was not carried out with completely pure enzymes, it can still be concluded clearly that phospholipase C and sphingomyelinase from *B. cereus* impose extremely specific requirements on the structure of molecules if these are to qualify as substrates for these enzymes. The data collated in Tables 1–7 allow us to suggest a mechanistic model for the hydrolysis by these two enzymes.

4. MECHANISTIC MODELS FOR ENZYME HYDROLYSIS CATALYSED BY PHOSPHOLIPASE C AND SPHINGO-MYELINASE

Phospholipase C and sphingomyelinase from *B. cereus* are closely related enzymes. Both form phosphocholine from their respective substrates, and this may be a consequence of very similar mechanisms of enzymic catalysis. The basic

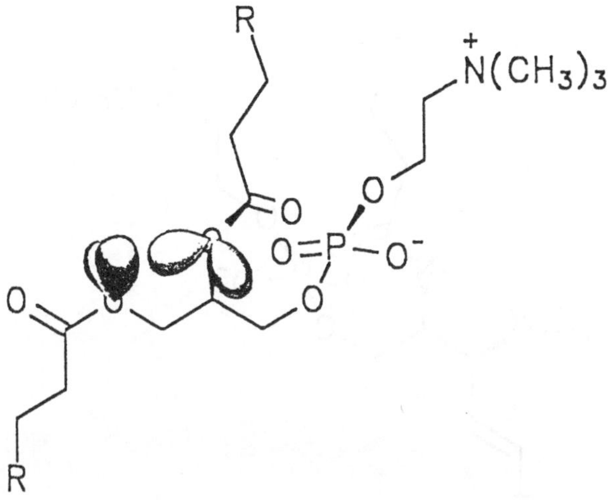

Figure 6. Mechanistic model for phosphatidylcholine which serve as substrate for phospholipase C. The supporting effect on phosphocholine hydrolysis of the ester group at position *sn*-2 in combination with an ester group at position *sn*-1.

hypothesis for the following suggestion is that the substrate, for each enzyme, assumes a specific conformation that facilitates the release of phosphocholine.

In the case of phospholipase C, a stabilised conformation can be obtained by the interaction of the ester group at position 2 with the adjacent functional group (ester or ether) at position 1 of the glycerol molecule. This is depicted in Figure 6, which shows this conformation for a good substrate of phospholipase C, such as 1,2-dipalmitoyl-*sn*-glycero-3-phosphocholine. The acyl group at the *sn*-2 position of phosphatidylcholine reaches a fixed conformation close to phosphate by repulsion between the free electron pairs of the oxygen atoms that belong to the ester group at position 2 and to the ester or ether group at position 1.

This proposal for phosphatidylcholine hydrolysis catalysed by phospholipase C can be tested by detailed consideration of the enzyme's interaction with different substrates. For instance, the hydrolysis rate of the 1-O-phosphocholine-2-O-palmitoyl-octadecane (compound 7e) in the *R*-configuration is decreased by a factor of about 10 in comparison with compound 7d. In this case, the optimal conformation cannot be supported, because of the lack of an ester or ether bond at position 3 (*sn*-1 position in phosphatidylcholines). Consequently, the rate of conversion of this ester model compound is strongly decreased.

For sphingomyelins as substrates for sphingomyelinase, the specific conformation and stabilisation of the 2-amido group in the close vicinity of the phosphate

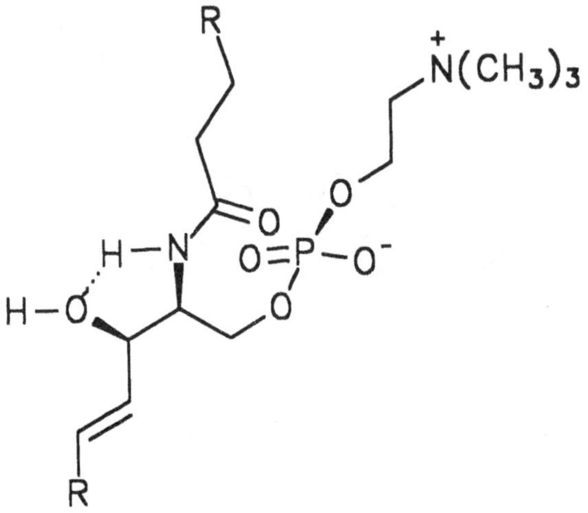

Figure 7. Mechanistic model for sphingomyelin as a substrate for sphingomyelinase. Phosphocholine ester hydrolysis is supported by the amide group in the *R*-configuration at position 2, acting in combination with the free hydroxy group at position 3.

group is achieved by hydrogen bonding between the 2-amido group and the free hydroxy group at position 3 (Figure 7). This arrangement emphasizes the importance of the 3-hydroxy group in combination with the amide group at position 2 for the enzymic release of phosphocholine. As shown earlier, the presence of a free 3-hydroxy group is an absolute condition for sphingomyelinase activity.

As in the case of hydrolysis by phospholipase C, this proposal can also be tested by detailed consideration of the enzyme's interaction with different substrates. An example is the comparison of the hydrolysis of the compounds 7b and 7f by sphingomyelinase. The rate of hydrolysis of compound 7f is lower than that of compound 7b by a factor of about 1600. The optimal configuration cannot be supported by comopound 7f because of the absence of the 3-hydroxy group.

To conclude, the synthesis of model compounds in the lipid field is a powerful tool in discovering the connection between chemical structure and biological activity, and in understanding the effects of structural modification. This is demonstrated in this chapter by the analysis of the properties of synthetic phospholipid molecules as substrates for phospholipase C and sphingomyelinase from *B. cereus*.

REFERENCES

1. Zwaal, R.F.A., Roelofsen, B., Comfurius, P. & van Deenen, L.L.M. (1971). Complete purification and some properties of phospholipase C from *Bacillus cereus*. *Biochim. Biophys. Acta*, 233, 474–479.
2. Little, C. (1981). Phospholipase from *Bacillus cereus*. *Methods in Enzymol.*, 71, 725–730.
3. Ikezawa, H., Mori, M., Ohyabu, T. & Taguchi, R. (1978). Studies on sphingomyelinase of *Bacillus cereus*. I. Purification and properties. *Biochim. Biophys. Acta*, 528, 247–256.
4. Otnæss, A.B., Prydz, H., Bjørklid, E. & Berre, A. (1972). Phospholipase C from *Bacillus cereus* and its use in studies of tissue thromboplastin. *Eur. J. Biochem.*, 27, 238–243.
5. Little, C. & Otnæss, A.B. (1975). The metal ion dependence of phospholipase C from *Bacillus cereus*. *Biochim. Biophys. Acta*, 391, 326–333.
6. Hauser, H., Pascher, I., Pearson, R.H. & Sundell, S. (1981). Preferred conformation and molecular packing of phosphatidylethanolamine and phosphatidylcholine. *Biochim. Biophys. Acta*, 650, 21–51.
7. Hirschmann, H. (1960). The nature of substrate asymmetry in stereoselective reactions. *J. Biol. Chem.*, 235, 2762–2767.
8. Dennis, E.A., Rhee, S.G., Billah, M.M. & Hannun, Y.A. (1991). Role of phospholipases in generating lipid second messengers in signal transduction. *Fed. Am. Soc. Exp. Biol. (FASEB) J.*, 5, 2068–2077.
9. Bell, R.M. (1986). Protein kinase C activation by diacylglycerol second messengers. *Cell*, 45, 631–632.
10. Nishizuka, Y. (1984). Turnover of inositol phospholipids and signal transduction. *Science*, 255, 1365–1370.
11. Gilman, A.G. (1984). G proteins and dual control of adenylate cyclase. *Cell*, 36, 577–579.
12. Okazaki, T., Bielawska, A., Bell, R.M. & Hannun, Y.A. (1990). Role of ceramide as a lipid mediator of 1a,25-dihydroxyvitamin D3-induced HL-60 cell differentiation. *J. Biol. Chem.*, 265, 15823–15831.
13. Kötting, J., Marschner, N.W., Unger, C. & Eibl, H. (1992). Determination of alkylphosphocholines and of alkyl-glycero-phosphocholines in biological fluids and tissues. In *Acetylphosphocholines: New Drugs in Cancer Therapy*, ed. Eibl, H., Hilgard, P. & Unger, C. (in the series *Progress in Experimental Tumour Research*, vol. 34), pp. 6–11. Basel: Karger Verlag.
14. Ries, U., Fleer, E.A.M., Unger, C. & Eibl, H. (1992). Synthetic phospholipids as substrates for phospholipase C from *Bacillus cereus*. *Biochim. Biophys. Acta*, 1125, 166–170.
15. Eibl, H. & Lands, W.E.M. (1969). A new sensitive determination of phosphate. *Anal. Biochem.*, 30, 51–57.
16. Eibl, H. & Woolley, P. (1986). Synthesis of enantiomerically pure glyceryl esters and ethers. I. Methods employing the precursor 1,2-isopropylidene-*sn*-glycerol. *Chem. Phys. Lipids*, 41, 53–63.
17. Eibl, H. & Woolley, P. (1986). Synthesis of enantiomerically pure glyceryl esters and ethers. II. Methods employing the precursor 3,4-isopropylidene-D-mannitol. *Chem. Phys. Lipids*, 47, 47–53.
18. Woolley, P. & Eibl, H. (1988). Synthesis of enantiomerically pure phospholipids including phosphatidylserine and phosphatidylglycerol. *Chem. Phys. Lipids*, 47, 55–62.
19. Eibl, H. & Woolley, P. (1988). A general synthetic method for enantiomerically pure ester and ether lysophospholipids. *Chem. Phys. Lipids*, 47, 63–68.
20. Massing, U. (1992). Synthesen enantiomerenreiner Sphingomyelinanaloga und ihre Verwendung zum Studium der Substraterkennung von Sphingomyelinase und Phospholipase C. Ph.D. Thesis, Technische Universität Braunschweig, Germany.

21. IUPAC-IUB Commission on Biochemical Nomenclature (1977). *The Nomenclature of Lipids (Recommendations 1976). Biochem. J.,* 79, 11–21.
22. Eibl, H. & Lands, W.E.M. (1970). Phosphorylation of 1-alkenyl-2-acylglycerol and the preparation of 2-acylphosphoglycerides. *Biochemistry,* 9, 423–428.

12.

Isolation and purification of lipases

M. RAQUEL AIRES-BARROS, M. ÂNGELA TAIPA
and JOAQUIM M.S. CABRAL

Lipases have been purified from mammalian, bacterial, fungal and plant sources by different methods (1–3). Most of the purification procedures reported involve series of non-specific techniques, such as ammonium sulphate precipitation, gel filtration and ion-exchange chromatography. In recent years, affinity-chromatographic techniques have become frequently used, decreasing the number of steps necessary for lipase purification.

More recently, reversed-micellar and aqueous two-phase systems, ultrafiltration membranes and immunopurification have also been applied to purify some lipases, mainly of microbial origin.

In this chapter, we survey the isolation and purification of lipases, mainly from microbial and mammalian sources. Most of the purification schemes are chromatographic processes. The purification protocols are analysed in order to emphasize the most frequently used chromatographic techniques and to reach the best sequence of steps in designing a multi-step purification scheme.

1. MICROBIAL LIPASES

Microbial lipases are very diverse in their enzymic properties and substrate specificities, which makes them attractive for industrial applications. A large number of lipases has been screened for application as food additives (flavour-modifying enzymes), industrial reagents (glyceride-hydrolysing enzymes) and stain removers (detergent additives), as well as for medical applications (digestive drugs; diagnostic enzymes); for more on industrial uses, see Chapter 13.

Most of the microbial lipases are extracellular, being excreted through the external membrane into the culture medium. Optimisation of fermentation conditions for microbial lipases is of great importance, since culture conditions influence the properties of the enzyme producer as well as the ratio of extracellular to intracellular lipases. The amount of lipase produced is dependent on several environmental factors, such as cultivation temperature, pH, nitrogen composition, carbon and lipid sources, concentration of inorganic salts and the availability of oxygen. Lipase production is stimulated by lipids such as butter, lard, olive oil and fatty acids (4, 5).

The fermentation process is usually followed by the removal of the cells from the culture broth, either by centrifugation or by filtration. The cell-free culture broth is then concentrated by ultrafiltration, ammonium sulphate precipitation or extraction with organic solvents. Most of the purification schemes (80%) used a precipitation step, with 60% of these using ammonium sulphate and 35% using ethanol, acetone or an acid (usually hydrochloric). This concentration step is normally the first isolation step, the crude enzyme preparation being concentrated and at the same time partially purified with purification factors around 3. When microbial lipases are cultivated in a lipid-abundant medium, the first isolation step is normally an extraction or precipitation with organic solvents, such as ethanol, acetone or butanol. Such treatment to dissociate bound lipids is very important for the overall purification and recovery yield. The extracellular microbial lipases, after being concentrated and partially isolated, are then brought to high purity by a combination of several chromatographic methods.

The high degree of purification obtained for some microbial lipases has allowed the successful determination of primary and tertiary structures of some lipolytic enzymes. The three-dimensional structure of *Mucor miehei* lipase was recently reported (6). It contains a trypsin-like Ser...His...Asp catalytic triad with an active serine, similar to the catalytic centre of the human pancreatic lipase (7).

The three-dimensional structure of *Geotrichum candidum* lipase was also described (8). Unlike human pancreatic lipase (7), *Mucor miehei* lipase (6) and serine proteases, the catalytic triad of *G. candidum* lipase is Ser...His...Glu, with glutamic acid replacing the usual aspartate. Although the sequence similarity with the other two lipases is limited to the region near the active-site serine, there is some similarity in their three-dimensional structure. The *G. candidum* lipase is also an α/β protein with a central mixed β sheet whose topology is similar to that of the N-terminal domain of human pancreatic lipase. As in the other two lipases, the catalytic site is buried under surface loops (8).

2. MAMMALIAN LIPASES

Three groups of lipolytic enzymes may be distinguished in mammals: the lipases discharged into the digestive tract by the specialised organs, the tissue lipases and the milk lipases (1). Digestive lipases include *inter alia* lingual, pharyngeal, gastric and pancreatic lipases (9).

Several tissues and organs of mammals such as heart, brain, muscle, arteries, kidney, spleen, lung, liver, adipose tissue and serum contain lipases. Adipose tissue is known to contain several lipases, among which are lipoprotein lipase and hormone-sensitive lipase.

Tissues with the highest levels of lipoprotein lipase (LPL) expression have been shown to be skeletal muscle, heart muscle and adipose tissue (10). Heparin induces the liberation of the enzyme attached to tissues and was also observed to stabilise LPL *in vitro*. Heparin also induces the release into the blood of a hepatic lipolytic enzyme, hepatic triglyceride lipase (HTGL). The purification of human LPL and HGTL from post-heparin plasma has been a difficult task, because of the low concentration of these two enzymes in the starting material and the presence of plasma proteins, *e.g.*, anti-thrombin III, with similar chemical and physical properties (11).

Previous attempts to purify hormone-sensitive lipase have been only partly successful, both because the enzyme interacts with lipids, resulting in high-molecular-weight lipid–protein aggregates, and because it is difficult to extract the lipids with solvents or detergents (12).

A first de-lipidation step by treatment with organic solvents (chloroform, *n*-butanol, acetone, ether) was found to be crucial for reproducible purification of many mammalian lipases, since lipolytic enzymes often seem to exist as large lipid–protein aggregates in organs and tissues.

An efficient purification of porcine pancreatic lipase from fresh pancreas was achieved only after an almost complete de-lipidation of pancreas homogenates by successive extraction with mixtures of chloroform and *n*-butanol in different ratios, followed by an acetone-ether treatment (13). These extractions removed almost all lipid contaminants, with only a small amount of an acidic phosphatide remaining. This contaminant was then removed from the de-lipidated pancreas powder by extraction of the lipase into an aqueous phase, fractionation with ammonium sulphate (0.32 and 0.52 saturation) and partitioning between *n*-butanol and ammonium sulphate solution.

Colipase, a protein that activates lipase *in vivo*, is a frequent contaminant of fresh lipolytic preparations from pancreas. Lack of reproducibility is apparently a common feature of several methods reported for lipase–colipase separation, indicating that lipase–colipase binding is not a simple protein–protein interaction. The presence of variable amounts of colipase in the lipase samples seems to be related to the efficiency of the critical first de-lipidation step (2).

In the last ten years, complete and reproducible purification of mammalian lipases has allowed the amino-acid sequence to be determined for several important lipases such as pancreatic and gastric lipases (2, 9). Recently, it was reported that human pancreatic lipase is a glycoprotein containing 449 amino-acid residues (7). The primary structure of the human enzyme was established by sequencing complementary DNA clones, and its three-dimensional structure was determined by X-ray crystallography.

3. PLANT LIPASES

Although it is known that lipase activity exists in several plant tissues (wheat, oats, corn and palm fruit), few studies have so far been made on the distribution of lipases in whole plants, except in seeds and fruits. Seeds usually contain a considerable amount of triacylglycerides that are processed by lipases. Large amounts of a triacylglyceride hydrolase were found in dormant castor-bean seeds (1).

In the last ten years, little has been reported on the purification of plant lipases. However, germinating oilseeds have been investigated as sources of lipases. Crude lipase preparations were obtained by acetone or buffer-solution extractions from easily growing seedlings of rape (*Brassica napus*), mustard (*Sinapis alba*) and cotyledons of lupin (*Lupinus alba*) (14).

4. PURIFICATION OF LIPASES BY CHROMATOGRAPHIC METHODS

Chromatography has been the technique most frequently used for the purification of lipases. This section describes a few examples of isolation and purification protocols for microbial and mammalian lipases, showing the different types of chromatographic method used.

4.1 Bacterial lipases

Chromobacterium *lipases*

A potent bacterium for lipase production was isolated from soil and identified as *Chromobacterium viscosum* (15). The crude lipase preparation contained more than two species of lipase, which differed from each other in molecular weight and isoelectric point (16).

The lipase with high molecular weight (lipase A) was purified from the crude extract by chromatography using Amberlite CG-50 and Sephadex G-75. The enzyme was purified about 23-fold, with a recovery of 2.8%, from the crude extract. The purified lipase A was found to be homogeneous by discontinuous electrophoresis (16). The other lipase (B) from *C. viscosum* was purified approximately 1000-fold from the original crude powder to give a yield of 18%, by using Sephadex G-100, CM-cellulose and DEAE-Sephadex. The homogeneity of the preparation was demonstrated by discontinuous electrophoresis (17). The respective molecular weights of lipases A and B were found by gel filtration to be 120 and 27 kDa. Sodium dodecyl sulphate polyacrylamide gel electrophoresis (SDS-PAGE) was also carried out to estimate the molecular weight. It was observed that lipase A dissociated into two subunits with molecular weights of 80 and 50 kDa, while lipase B showed only a single protein band, with a molecular weight of 26 kDa (18).

C. viscosum lipases were also purified by hydrophobic chromatography on palmitoyl cellulose (19). This chromatography was found to be simple and effective, especially for large-scale purification of the lipase. An overall recovery of 71% and an increase of 11-fold in the specific activity from the supernatant fluid of bacterial cultures were achieved. Two isoenzymes were obtained, each homogeneous as judged by SDS-PAGE. One had a molecular weight of 120 kDa and a pI of 3.7, and the other a molecular weight of 30 kDa and a pI of 7.3.

Pseudomonas *lipases*

An alkaline and thermostable lipase was purified from the culture supernatant of *Pseudomonas fragi* by chromatography on DEAE-Toyopearl 650M and DEAE-Sepharose CL-6B as final steps in the purification (20). The lipase was purified 68-fold with 48% recovery, and its molecular weight was estimated to be 33 kDa. This relatively high recovery of the lipase activity was probably due to the enzyme's high stability and good separation on the column, promoted by adding Triton X-100 to the elution buffer in the DEAE-Sepharose CL-6B step. The homogeneity of the purified lipase was confirmed by discontinuous and SDS gel electrophoresis.

A *Pseudomonas* strain producing a lipase suitable for the hydrolysis of castor oil was isolated from soil. DEAE-cellulose and DEAE-Toyopearl 650M were used to purify this lipase. A 400-fold purity and a yield of 13% were obtained. The purified enzyme was electrophoretically homogeneous and its molecular weight was 30 kDa (21).

A purification procedure was designed to exploit the very hydrophobic nature of the lipase *Pseudomonas* species ATCC 21808. The crude lipase preparation was purified by anion-exchange chromatography on Q-Sepharose followed by hydrophobic interaction chromatography on octyl-Sepharose, the enzyme being eluted with isopropanol with a high yield (56%) and purity (160-fold). SDS-PAGE revealed a single protein band with a molecular weight of 35 kDa (22).

Pseudomonas species KWI-S 6 produces an extracellular, thermostable lipase. The enzyme was purified 14-fold with 2.9% recovery by gel filtration on HPLC. The purified enzyme, which showed a single band on SDS-PAGE, had a molecular weight of 33 kDa (23).

A triacylglycerol lipase of *Pseudomonas fluorescens* was purified from the crude enzyme by chromatographic steps on DEAE-cellulose and octyl-Sepharose CL-4B. This lipase was purified about 3400-fold and the yield of activity was 21% of the original. The molecular weight of the lipase was estimated by SDS-PAGE to be about 45 kDa (24).

Staphylococcus *lipases*

Several research workers have studied *Staphylococcus* lipases, describing different purification methods and properties of the purified preparations (25–28).

A *Staphylococcus* strain isolated from a case of pyodermitis and named *S. aureus* 226 produces a lipase which was purified by successive chromatographic steps on hydroxyapatite and Sephadex G-200 and G-150. This method gave a 380-fold purification of the enzyme from the culture medium with a yield of 25%. The purified enzyme appeared homogeneous, as shown by polyacrylamide gel electrophoresis and isoelectric focussing. The molecular weight of the enzyme was determined as 34 kDa by gel filtration, and its pI was 9.7 (25).

An *S. aureus* lipase (L-l) has also been purified, by a multi-step procedure involving hydrophobic chromatography on phenyl-Sepharose CL-4B, gel filtration through Sepharose CL-4B and re-chromatography on Sepharose CL-4B (26). A pure enzyme was obtained, which appeared to be homogeneous by molecular sieving, polyacrylamide gel electrophoresis and sucrose-gradient centrifugation. Values for the molecular weight obtained by molecular sieving and electrophoresis in the presence of SDS were 300 and 45 kDa, respectively. Staphylococcal lipase seems to exhibit a complex structure, which may explain this discrepancy. Aggregation of units of lipolytic enzymes, or a subunit structure of these enzymes, is known to occur not only in bacterial lipases (18) but also in pancreatic lipase (2).

Other bacterial lipases

A lipase produced by *Propionibacterium acnes* (hydrolysing native sebum triacylglycerols to free fatty acids) was purified 4800-fold with a recovery of 9.5% from crude culture supernatant. Gel filtration on Sephadex G-100 and ion exchange in CM-Sephadex C-50 columns were used. The molecular weight of the lipase was 46.7 kDa by gel filtration. SDS-PAGE revealed a major protein component (molecular weight 41.2 kDa) together with two minor protein components of higher molecular weight (67 and 125.9 kDa; reference 29).

A thermostable lipase from *Bacillus* species has been purified to homogeneity as judged by discontinuous PAGE, SDS-PAGE, and isoelectric focussing. The enzyme was purified 7800-fold with 9% recovery from the culture supernatant. The purification included sequential column-chromatographic steps on DEAE-Sephadex A-50, Toyopearl HW-SSF, and butyl-Toyopearl 650M. The purified enzyme was found to be a monomeric protein with a molecular weight of 22 kDa and a pI of 5.1 (30).

4.2 Yeast lipases

Candida *lipases*

The lipase from *Candida paralipolytica* was purified by using sequential chromatographic steps on CM-Sephadex C-50 and on DEAE-Sephadex C-50. The purification degree was about 130-fold, with a 32% recovery from the acetone precipitate of the cultivated broth (31).

C. rugosa produces two distinct lipases (I and II), which were identified and separated with a high-resolution anion-exchange column (Mono Q). Lipase I was purified sixfold with 18% recovery and lipase II twofold with 25% recovery. Both proteins have an apparent molecular weight of 58 kDa as judged by SDS-PAGE. Their respective isoelectric points are 5.6 and 5.8 (32).

Three distinct forms of lipolytic enzyme were identified in a commercial *Candida* lipase preparation. Two of these lipases (lipases A and C) were purified on a DEAE-Sepharose CL-6B column. Lipases A and C correspond to lipase I and II described above. Lipase B was not detected by these workers, probably because it had been converted into lipase I by the ethanol treatment of crude lipase powder in their purification procedure. The apparent molecular weights of lipases A, B and C are 362, 200 and 143 kDa, respectively. As results from SDS-PAGE indicated that lipases A and C are composed of subunits with the same molecular weight (62 kDa), it is possible that lipases A, B and C are respectively hexamers, trimers and dimers of this subunit (33).

Other yeast lipases

Two kinds of cell-bound lipases (lipase I and lipase II) from *Saccharomycopsis lipolytica* were purified by chromatographic techniques. Lipase I was adsorbed to CM-Sepharose CL-6B at pH 5.0. The method of purification involved chromatography on CM-Sepharose CL-6B, DEAE-Sepharose CL-6B and Sephadex G-100. Total recoveries were respectively 3.7% and 4.6%, with purification factors of 69 and 58. SDS-PAGE of the purified lipases gave a single band. The molecular weights of lipases I and II were estimated to be 39 and 44 kDa, respectively, and both lipases were monomeric (34).

4.3 Fungal lipases

Aspergillus *lipases*

A lipase secreted by *Aspergillus* was purified from the culture filtrate by successive cation exchange on CM-cellulose CM-52, anion exchange on DEAE-cellulose and hydrophobic chromatography on octyl-Sepharose (or affinity chromatography on ConA-Sepharose). A recovery of 25% was achieved. The *Aspergillus* lipase was resolved as a single band in SDS-PAGE with a molecular weight of 47.8 kDa. Isoelectric focussing showed the preparation to be a mixture of approximately equal amounts of two closely related components of pI *ca.* 4.5 (35).

Another preparation of *Aspergillus niger* lipase was purified by hydrophobic interaction (butyl-Toyopearl 650M), gel filtration (Sephadex G-75), anion-exchange chromatography (DEAE-Sepharose CL-6B) and adsorption on hydroxyapatite. By this procedure, the lipase was purified about 600-fold over the crude extract, with a yield of 34%. The final enzyme preparation gave a single band in SDS-PAGE. The molecular weight of the enzyme was estimated to be 35 kDa from SDS-PAGE and Sephadex G-100 chromatography, and the pI was 4.1. This lipase differs in molecular weight from the lipase described by Tombs *et al.* (36).

Recently, an acid-resistant lipase from *A. niger* was purified from a crude commercial preparation by a procedure including size exclusion on Bio-Gel P-100 and ion exchange on Mono-Q (37). The overall purification was 40-fold with a yield of 36%. The most effective step in the purification procedure is Mono-Q ion-exchange chromatography, which gives a 5.6-fold increase in specific activity.

Geotrichum *lipases*

Geotrichum candidum extracellular lipases have been extensively purified and characterised (38–42). *G. candidum* Link lipase was purified by means of chromatographic methods such as DEAE-Sephadex column chromatography and gel filtration on Sephadex G-100 and G-200, and was finally crystallised from concentrated aqueous solution. The purification resulted in a 40-fold increase in specific activity, with 20% recovery of the original activity. It was confirmed that the crystallised preparation was electrophoretically and ultracentrifugally homogeneous, with a molecular weight of 54 kDa and a pI of 4.3 (38). The lipase cDNA was subsequently cloned and sequenced as an initial step for protein engineering of the lipase. Genomic Southern-blot analysis, with the cloned lipase cDNA as a probe, indicated two genes of the *G. candidum* chromosome that hybridised with the cDNA. This implies that there may be two genes whose expression gives proteins with lipase activity. Hydrophobic-interaction chromatography of the enzyme preparation clearly separated the protein into two fractions with lipase activity, named lipase I (64 kDa) and lipase II (66 kDa) (39). The extracellular lipase from *G. candidum* is highly heterogeneous, as demonstrated by immunoelectrophoretic techniques (43).

Lipase from *G. candidum* has also been purified by chromatography on Sephacryl-200 HR, high-resolution anion exchange resin (Mono Q) and Polybuffer exchanger 94 (40). With this procedure, two forms of lipase from *G. candidum* were obtained. Lipases I (the main component) and II were purified respectively 35-fold, with 62% recovery of activity, and 94-fold, with 18% recovery. Their molecular weights have been estimated, by SDS-PAGE and by molecular sieving under native conditions, to be 56 kDa, with isoelectric points 4.56 (lipase I) and 4.46 (lipase II).

In order to obtain crystals of lipase from *G. candidum* suitable for X-ray analysis, this lipase was purified to homogeneity (41). The purification procedure included ion exchange on Q-Sepharose and hydrophobic interaction on phenyl-Sepharose. The enzyme proved to be a glycoprotein with a molecular weight of 61.6 kDa.

The purification of a *G. candidum* lipase from the commercial GC-4 enzyme powder (Amano Co.) was achieved by a simple two-step method: ion exchange on Q-Sepharose, followed by hydrophobic-interaction chromatography on phenyl-Sepharose CL-4B (44, 45). After this step, the overall purification factor was 13.2. The GC-4 lipase was homogeneous in SDS-PAGE. This purified enzyme focussed in four bands at pH 4.40, 4.48, 4.55 and 4.64. When treated with endoglycosidase H prior to ioselectric focussing, the lipase showed a single band at pH 4.57 and a molecular weight of 57 kDa by SDS-PAGE. This procedure afforded one pure lipase, lipase I, whereas two lipases have been found by other workers (39, 43). It may be that one lipase was lost during the hydrophobic-interaction chromatography, as 50% of the lipase activity disappeared in this step (44).

Mucor *lipases*

A lipase produced by *Mucor lipolyticus* Aac-0102 was separated into three fractions named F-1, F-2 and F-3. Most of the olive-oil-hydrolysing activity of *Mucor* lipase was recovered in F-3. This fraction was separated into two sub-fractions F-3A (molecular weight 25.4 kDa) and F3-B (29 kDa) by CM-Sephadex column chromatography. Both fractions were homogeneous, and F-3A was crystallised. The F-3A crystallised fraction was obtained with a purification factor of 3 and a recovery of 21%, and the F-3B fraction with a purification factor of 3 and 19% recovery (46).

A lipase from *M. javanicus* was purified about 180-fold with 2.4% recovery. Gel filtration on Sephadex G-200 (at low ionic strength) and Sephadex G-75 (at high ionic strength) were used. Discontinuous PAGE showed a major protein peak at 21 kDa (47).

An extracellular lipase, a glycoprotein, produced by a selected strain of *Mucor miehei* has been partially purified in two forms, A and B. Lipase A was purified 15-fold from the crude powder with 57% recovery, by chromatography on DE-cellulose 52 and on a ConA-Sepharose affinity column. Lipase B was purified from the crude powder on DEAE-Sepharose and phenyl-Sepharose columns, with a purification of eightfold and a recovery in activity of 32% (48).

Penicillium *lipases*

A lipolytic enzyme that specifically hydrolyses *p*-nitrophenyl laureate was found in the culture broth of *Penicillium cyclopium* M1. The enzyme was purified with an overall yield of 27% and 1400-fold purification. The purification procedure included chromatographic steps on DEAE-cellulose, DEAE-Sepharose, hexyl-Sepharose and hydroxyapatite, and gel filtration on Cellulofine GC-700. The molecular weight of the enzyme was about 110 kDa as found by gel filtration. SDS-PAGE analysis revealed two subunits with identical molecular weights, 54 kDa (49).

An extracellular lipase from *P. simplicissimum* was purified by hydrophobic-interaction chromatography on phenyl-Sepharose. A recovery of 26% and a purification of about 3000-fold were obtained. The enzyme showed a molecular weight of 56 kDa in SDS-PAGE (50).

P. camembertii U-150 also produces an extracellular lipase, which was purified into four active fractions by a procedure involving aminooctyl-Sepharose, hydroxyapatite and ConA-Sepharose column chromatography. Enzyme 1 was not adsorbed on ConA-Sepharose, but enzymes 2–4 were adsorbed, and were separated into three active fractions. By SDS-PAGE, the molecular weights were found to be approximately 37 kDa for enzyme 1 and about 39 kDa for the others (24). However, after enzymic removal of carbohydrates from enzymes 2–4 by an endoglycosidase, these also exhibit the same molecular weight as enzyme 1 (51).

Rhizopus *lipases*

The purification and properties of three lipases, A, B and C, produced by *Rhizopus delemar* were studied. The purification included chromatography on Ultrogel and Sephadex G-150 columns. Lipase A was purified 110-fold with 11% recovery; lipase B 71-fold with 3% recovery and lipase C 47-fold with 4% recovery. The enzymes were found to be homogeneous by electrophoresis, with respective molecular weights of 76, 60 and 45 kDa (52).

A lipoprotein lipase (LPL) and a lipase could be found in the culture medium of *R. japonicus* KY 521 (53). The two lipases were separated into two peaks by gel filtration on Sephadex G-100. LPL was further purified by chromatography on hydroxyapatite, gel filtration on Sephadex G-150 and isoelectric focussing. The enzyme was purified about 48-fold from the crude enzyme powder, with a recovery of 4.7% of the LPL activity. The ratio of LPL to lipase activities in the purified enzyme preparation was about 1.0–1.5 (53). The highly active lipase was also purified from the culture supernatant; the purification procedure included chromatography on Column-lite, affinity chromatography on heparin-Sepharose 4B and separation into two lipases, I and II, by isoelectric focussing. The lipases were purified about 60-fold, with a recovery of 2%, from the crude enzyme powder. Lipases I and II were found to be homogeneous by discontinuous electrophoresis, and showed pI values of 7.4 and 7.9, respectively. Both lipases had a molecular weight of 42 kDa (54).

A lipase from *R. japonicus* NR 400 was purified to homogeneity by chromatography on hydroxyapatite, octyl-Sepharose and Sephacryl S-200. An overall purification of 93-fold was obtained with 31% yield. The purified lipase preparation showed a single band in SDS-PAGE, with a molecular weight of 30 kDa (55).

Thermomyces *lipases*

Thermomyces lanuginosus (formerly *Humicola lanuginosa*) produces an extra-cellular lipase (5), which was purified by means of successive chromatographic steps on Sephadex G-75, DEAE-Sepharose CL-6B and hydroxyapatite columns. The enzyme was purified about 150-fold with a yield of 15%. The purified enzyme showed a single protein band on polyacrylamide gel electrophoresis. Its molecular weight was estimated to be 39 kDa from both SDS-PAGE and gel filtration on Sephadex G-100, suggesting that the enzyme is a monomer. Its pI value was 6.6 (56).

Other fungal lipases

The purification of a triacylglycerol lipase from the conidia of *Neurospora crassa* was achieved by Sephadex G-100 column chromatography. Isoelectric focussing revealed a single band corresponding to a pI of 6.4. The enzyme has a molecular weight of 54 kDa as determined by gel filtration. SDS-PAGE gave a single band of molecular weight 27 kDa, suggesting the presence of two identical subunits (57).

A crude lipase preparation from *Ustilago maydis* ATCC 14826 was purified by chromatographic methods leading to two lipolytic enzymes. The culture supernatant was purified by ion-exchange chromatography on an S-Sepharose column, giving two lipases, I and II. Lipase I was then purified on a butyl-Sepharose column with an overall yield of 12%. Lipase II was purified on DEAE-Sepharose and butyl-Sepharose columns, with a yield of 14% (58).

4.4 Mammalian lipases

Human pancreatic lipase

A single triacylglycerol lipase from human pancreatic juice was purified with an overall yield of 30%. The method involved column chromatography on Sephadex G-25, DEAE-cellulose, CM-cellulose and Sephadex G-75 (2). Determination of the amino-acid composition showed that the enzyme contained approximately 420 amino-acid residues and had a molecular weight of 46 kDa.

Other studies on the purification of human pancreatic lipase have reported the existence of a protein of 48 kDa that corresponds to two isoenzymes with isoelectric points of 5.80 and 5.85. Both isolipases seemed to be glycoproteins (2).

Porcine pancreatic lipase

After complete de-lipidation of fresh porcine pancreas homogenates, DEAE-cellulose, Sephadex G-100 and CM-cellulose chromatography were used to obtain a pure preparation of porcine pancreatic lipase. Two molecular forms of lipase activity (LA and LB) with similar molecular weight (around 50 kDa) were obtained after separation at low pH on a cation-exchanger CM-cellulose column. Lipase LA is the more acidic protein, and it emerges first from the CM-cellulose column, whereas it migrates behind LB on the anion exchanger DEAE cellulose. With this purification procedure, a 30-fold enrichment of lipase activity was obtained from de-lipidated pancreas powder, and an overall yield of 15% was achieved (13). This method, without the first de-lipidation step, can also be applied to porcine pancreatic juice as a lipase source.

Most porcine preparations obtained by the above procedure were contaminated by variable amounts of colipase (2). A slightly modified method for the purification of porcine pancreatic lipase consisted in using a batch procedure instead of a column for lipase adsorption on DEAE cellulose. The adsorption was performed directly on the pancreatic powder extracts (omitting the butanol-ammonium sulphate de-lipidation step), and it was reported that no colipase impurity could be detected either in the unfractionated isoenzymes after Sephadex G-100 filtration or in the separated lipases (2).

Human gastric lipase

Gel filtration on Sephadex G-75 was used to purify an acid lipase from human gastric juice (59). The enzyme was purified 15-fold to 25-fold, depending on the initial protein concentration in the gastric juice, with an overall recovery yield of 40–60%. On SDS-PAGE the enzyme appeared as a single band, with an apparent molecular weight of 44 kDa. By gel filtration of the purified enzyme on Sephadex G-200, molecular weights of 90 kDa and 45 kDa were obtained at pH 6.0 and pH 3.0, respectively, revealing a pH-dependent dimerisation. However, the enzyme in the original gastric juice at pH 6.0 beahved like a 50-kDa monomer, as determined by Sephadex G-200 gel filtration. These results suggest that factors other than the pH value may influence the aggregation of the lipase.

Rabbit gastric lipase

Rabbit gastric lipase was purified from an acetone powder from rabbit stomach by using gel filtration on Sephadex G-100 and cation exchange chromatography on a Mono-S FPLC system (60). The enzyme was purified 640-fold with a recovery of 48%. SDS-PAGE showed only one band, with a molecular weight of around 48 kDa. However, isoelectric focussing in an ampholine buffer of pH 3.0–9.0 revealed at least 12 bands, with pI values ranging from 5.7 to 7.0, a pattern typical of a glycoprotein.

Lipoprotein lipase and liver lipases

Lipoprotein lipase (LPL) and hepatic triglyceride lipase (HTGL) from postheparin plasma were separated by batch adsorption on heparin-Sepharose-4B. LPL was purified further by heparin-Sepharose column chromatography, hydroxyapatite chromatography and gel filtration on Sephadex G-150 (11). This lipase was purified 350 000-fold with an overall recovery of 6.7%. HTGL was further purified by octyl-Sepharose hydrophobic chromatography and heparin-Sepharose chromatography. The enzyme was purified 9000-fold with a recovery yield of 7.4%.

An alternative scheme proposed for HTGL purification was DEAE-Sephacel and Ultrogel AcA34 gel permeation chromatography. In this case HTGL was purified 100 000-fold in 19% yield (61). In both cases, contamination by antithrombin III was found to be absent in the purified samples.

Hepatic lipase was purified from rat-liver homogenates by octyl-Sepharose 4B, adsorption on heparin-Sepharose, hydroxyapatite chromatography and gel filtration on Ultrogel Ac34. Heparin-Sepharose chromatography was also used as a concentration step between the hydroxyapatite chromatography and the gel filtration. Non-ionic detergents were used in several steps in order to stabilise the enzyme. A 60 000-fold purification of hepatic lipase was obtained with a yield of about 10%. The hepatic enzyme was shown to have a native molecular weight of 200 kDa by gel filtration. On SDS-PAGE, a single band appeared at a molecular weight of 53 kDa, suggesting that the active enzyme is composed of four subunits (62).

Hepatic lipase was also purified from heparin perfusates of rat liver (63). Heparin-Sepharose, hydroxyapatite and dextransulphate-Sepharose column chromatography were used. The enzyme was purified 640-fold with 31% recovery. A molecular weight of 52.7 kDa was obtained for the pure lipase by SDS-PAGE.

Adipose tissue lipases

Human adipose tissue triacylglycerol lipase was purified by affinity chromatography based on substrate–enzyme interactions (presumably hydrophobic bonds). The purification procedure consisted of one-step chromatography on agarose-linked dioleoylglycerol. This chromatography resulted in a tenfold enrichment of lipase-specific activity with a recovery yield of 70% from crude human adipose tissue extract (64).

Purification of hormone-sensitive lipase was carried out, after solubilisation in a non-ionic detergent and complete removal of non-solubilised lipids by flotation in a thin layer of water, by column chromatography on QAE-Sephadex and high-performance ion-exchange chromatography on Mono-Q and Mono-S columns (65). A 2500-fold enrichment in the specific lipolytic activity was obtained, with an overall yield of 11%. A protein purity of 70–75% was achieved.

The purification of a triacylglycerol lipase isolated from adipose tissue of the teleost fish *Salmo gairdneri* has also been described (66). The purification included heparin-Sepharose chromatography. This chromatographic step allowed the isolation of the triacylglycerol lipase from lipoprotein lipase. The triacylglycerol lipase was purified 71-fold, with an overall recovery of 37%.

Milk lipases

The presence of lipolytic and esterolytic activities in cow milk has attracted much attention because of the known effect of endogenous fat hydrolysis on milk flavour. Bovine lipoprotein lipase from raw skimmed milk was purified by batch adsorption on heparin-Sepharose CL-4B followed by dextran-sulphate Sepharose CL-4B column chromatography (67). A very pure enzyme was obtained, as revealed by SDS-PAGE. One major band at an apparent molecular weight of 57 kDa was found.

A rapid and simple procedure was recently described for the purification of lipoprotein lipase from human milk (10). This involved heparin-Sepharose adsorption chromatography followed by phenyl-Sepharose hydrophobic chromatography. The LPL was purified around 30000-fold, with 24% recovery. Immunodiffusion using specific antisera against alpha-lactoferrin and human antithrombin III failed to reveal these contaminants in the pure lipase preparation. The apparent molecular weight of the enzyme (determined by SDS-PAGE) was 60 kDa.

4.5 Comparative analysis of chromatographic processes for lipase purification

From the analysis of 70 articles on the isolation and purification of microbial lipases, published between 1970 and 1991, some conclusions can be drawn about the major chromatographic methods for lipase purification and the sequence in which these methods are generally used in multi-step purification schemes (68).

Ion-exchange chromatography is the most common chromatographic method, used in 67% of the purification schemes analysed. In 29% of these procedures, ion exchange is used more than once. The weak ionogenic groups are the most commonly used ion-exchangers; the most frequently employed are the diethylaminoethyl (DEAE) group in anion exchange (58%) and the carboxymethyl (CM) in cation exchange (20%). Strong ion-exchangers, especially Mono Q (40) and Q-Sepharose (41), are becoming more popular in lipase purification. Both are based on triethylaminoethyl groups.

Gel filtration (apart from in the desalting step) is the second most frequently employed purification method, used in 60% of the purification schemes and more than once in 22% of them.

Affinity chromatography has been used as a purification step in 27% of the purification schemes. Hydrophobic chromatography is regarded as an affinity method in this analysis (used as a purification step in 18% of the purification schemes). The most popular hydrophobic adsorbents are based on agarose with

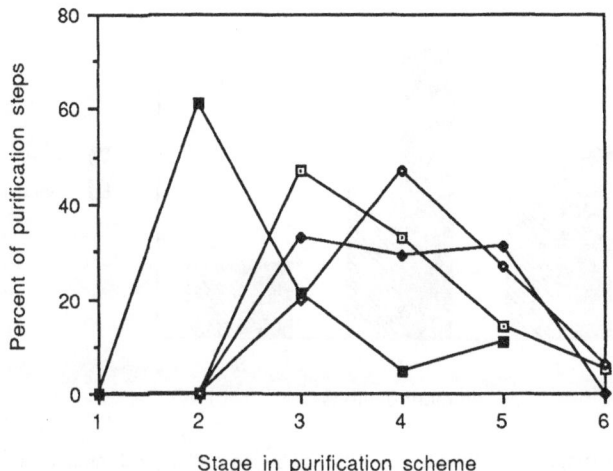

Figure 1. Analysis of the number of times each method is used in purification of microbial lipase (expressed as a percentage of the total number of steps) as a function of the number of steps: (■) precipitation, (□) ion-exchange chromatography, (♦) gel filtration and (◇) affinity chromatography.

octyl or phenyl functional groups. Concavalin-A (ConA) and heparin are employed for purification of fungal and mammalian lipases (35, 48), on account of the glycolytic character of these lipases.

Adsorption chromatography is applied in 16% of the purification schemes, and the adsorbent hydroxyapatite is used in most cases.

The purification protocols did not show adherence to any particular sequence of steps or methods. Methods chosen to purify microbial lipases depend on the initial lipase preparation, and even for the same crude lipase preparation different purification schemes have been used. Figure 1 shows the number of times each method has been used as a function of the number of steps in a purification scheme.

Precipitation is usually used as a fairly crude separation step, often during the early stages of a purification procedure, and is followed by chromatographic separation. Large quantities of material can be handled, and this step is less affected by interfering non-protein materials than chromatographic methods are.

Affinity methods can be applied at an early stage, but the materials are expensive, and the less costly ion exchange and gel filtration are preferred after the precipitation step. Although gel filtration has the lower capacity for loaded protein,

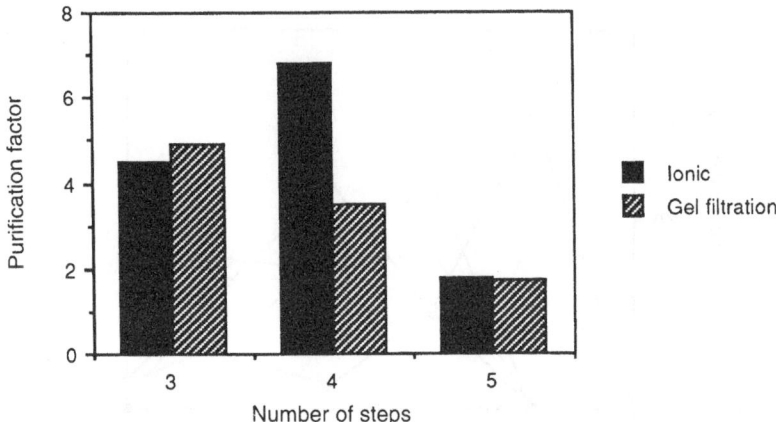

Figure 2. Average purification factors achieved by ion-exchange chromatography (■) and gel filtration (▨), for microbial lipases.

it can be used at an early stage in the purification (after the precipitation step) instead of ion exchange, or as one of the last steps. Figure 2 shows the purification achieved per step by ionic exchange and gel-filtration chromatography.

A large variation was observed in the relative purification factors attained by these two methods, being the purification dependent on the stage in the purification scheme (Figure 2). In a purification scheme with five steps the purification factor obtained in the last step is lower (by 50%) for both these methods compared with factors attained when the same methods are used earlier in the purification scheme. The purification factors for ion-exchange chromatography and gel filtration are approximately 5 when these methods are used as the third step in the purification scheme; these values change to 7 for ion exchange and 3.5 for gel filtration, when they are used as a fourth step. The value obtained for the precipitation method (used as second step in the purification scheme) is 3. With affinity techniques (including hydrophobic interaction chromatography), the purification factor achieved by an affinity step varies from 2 to 10. Precipitation methods had the highest average yield (87%), while the other techniques gave lower yields in comparison (60–70%).

To purify the protein to homogeneity, with an overall yield of 30% and purification factors of 320, four or five purification steps are usually required. These values are averages of the overall yields and purification factors obtained for each purification scheme.

The purification of mammalian lipases is usually achieved in three or four steps, with the exception of pancreatic isoenzymes from fresh pancreas, for which seven steps are required for total separation. One-step affinity-chromatographic purification was also described for a few lipases (2).

The analysis of 28 protocols describing the purification of mammalian lipases and the comparison with the typical purification procedures used for microbial lipases purification shows that the main difference between them lies in the choice of affinity chromatographic techniques. In fact, half of the purification schemes for mammalian lipases that were analysed involved affinity-chromatographic steps (including hydrophobic chromatography), some of the processes being based exclusively on affinity separation (10–12, 61, 62). This is explained by the lower concentration of mammalian proteins in plasma or in tissue extracts when compared with the enzyme concentration in microbial fermentation media.

Gel filtration is used in 54% of the cases, in 36% as a final purification step. In 43% of the purification strategies, ion-exchange chromatography is used as the first chromatographic step, either directly from the enzyme extract or after a protein-precipitation step using acid, ammonium sulphate, ethanol, acetone or polyethyleneglycol.

5. NOVEL PURIFICATION TECHNOLOGY APPLIED TO LIPASES

The usual procedures of lipase purification are sometimes troublesome and time-consuming, and usually result in low final yields. Alternative, novel purification technologies, such as reversed-micellar and aqueous two-phase systems, ultrafiltration and immunopurification techniques, have recently been applied to the purification of lipases. Some of the applications of these techniques are described here.

5.1 Reversed micellar systems

Liquid–liquid extraction of biomolecules using reversed micelles is a promising method when the traditional techniques with organic solvents are limited by protein denaturation and solubilisation (69, 70). The selective separation of a synthetic protein mixture (71), the concentration of α-amylase by continuous extraction using two mixer-settler units (72), the recovery of an extracellular alkaline protease from a whole fermentation broth (73) and the selective separation and purification of a lipolytic preparation from *Chromobacterium viscosum* (3, 74, 75) have been some of the successful applications of reversed micellar systems to the isolation and purification of proteins.

This method involves a very simple procedure and only requires two steps. The first step is based on the ability of reversed micelles to solubilise proteins from an aqueous phase, into the water pool of the surfactant aggregates. In the second step, the solubilised proteins are back-extracted into a new aqueous phase by changing the interactions between the protein and the reversed micellar system.

Selective solubilisation of a mixture of proteins can be achieved by manipulating the parameters of the system both in the micellar and aqueous phases, the most important parameters being the pH and ionic strength of the aqueous phase. The pH value influences electrostatic interactions between the polar head groups of the surfactant and the charged protein. Hydrophobic interactions may also act on the transfer of proteins, especially those proteins, such as lipases, that bear hydrophobic regions on their surface. The ionic strength of the system also allows a selective transfer, based on size exclusion effects.

5.2 Selective separation of *Chromobacterium viscosum* lipases

Selective separation and purification of a crude lipolytic preparation from *C. viscosum* was carried out by liquid–liquid extraction using reversed micellar systems (3, 74, 75). This lipolytic preparation is a mixture of two lipases (19): lipase A with a molecular weight of 120 kDa and pI of 3.7; and lipase B which has a molecular weight of 30 kDa and pI of 7.3.

The effect of pH on the transfer of lipases from the aqueous phase to a micellar solution of sodium di-isoctyl sulphosuccinate (Aerosol OT®; AOT) in *iso*-octane was studied (3, 75). At pH 6.0 and low ionic strength (50 mM KCl), lipase B was completely solubilised in the micellar solution, while lipase A remained in the aqueous solution. At this pH lipase B is positively charged and is easily solubilised in the AOT reversed micelles, while lipase A is negatively charged and is not solubilised in reversed micelles of anionic surfactants.

Protein solubilisation seems to require an attractive electrostatic interaction between the micelle and the protein, and therefore pH values lower than the pI of a given protein are required to solubilise it in a micellar solution of an anionic surfactant. However, approximately, 50% of lipase B is solubilised at pH values higher than its pI, indicating that hydrophobic interactions between the lipase and the surfactant and/or organic solvent are also important. Lipase A is a relatively large protein, and the size exclusion effect must be more important than the protein–surfactant interactions.

The extraction of lipase B to the micellar phase was also studied at pH 6.0, at various values of w_0 ([H$_2$0]/[surfactant]) by changing the KCl concentration (from 50 to 350 mM). Lipase B was completely solubilised in the micellar phase at low ionic strengths, 50 and 100 mM KCl, corresponding to w_0 values of 20 and 15.8, respectively. A sharp decrease in solubilisation was observed at 250 mM KCl ($w_0 = 10.8$), while at $w_0 < 9.6$, which corresponds to a KCl concentration greater than 350 mM, no protein solubilisation occurred.

Table 1. Purification of lipases from *C. viscosum* using reversed micelles (74).

Purification stage	Total protein (mg/ml)	Total activity (μmol/ml·min)	Specific activity (μmol/mg·min)	Purification factor	Recovery yield (%)
Crude extract (Lipases A+B)	1.41	21.7	15.4	1.0	100
Aqueous phase after extraction (Lipase A)	0.299	19.8	66.2	4.3	91
Aqueous phase after back-extraction (Lipase B)	0.290	16.6	57.2	3.7	76

It can be predicted that increasing the ionic strength of the protein feed solution should diminish the protein–surfactant interactions, leading to the formation of smaller reversed micelles (71). A reduction in w_0 from 20 to 9.6 can result in a decrease in the solubilisation capacity of lipase B through size exclusion (75).

5.3 Recovery of lipase B from the micellar solution

Back-extraction of lipase B from the micellar phase was tested at different values of the aqueous-phase pH and ionic strength, but no re-extraction of protein was achieved. The enzyme was only recovered from the organic phase by introducing a water-miscible organic solvent (75). The best results were obtained by adding 2.5% (v/v of the total volume of the phases) ethanol to the micellar phase at pH 9.0, using the same ionic strength as for extraction. Ethanol was used to minimise the hydrophobic interactions between the lipase and the surfactant and/or organic solvent, allowing the lipase to be recovered into the aqueous phase. A yield of 77% was obtained for lipase B, as shown by analysing the protein concentration in the organic phases that revealed lipolytic activity. An FPLC gel-filtration chromatogram of the aqueous phase after back-extraction confirmed that only lipase B was in the aqueous phase (74). The results of the purification procedure are summarised in Table 1. Lipase A was purified 4.3-fold and 91% of lipase activity was recovered, while Lipase B was purified 3.7-fold with a recovery of 76% from the crude extract.

Purification of lipases A and B of *Chromobacterium viscosum* using reversed micelles (74) can be compared with the results obtained from purification in successive chromatographic steps on various materials (16, 17; Table 2). The purification factor obtained for lipase A by using reversed micelles was similar to that obtained by chromatographic methods; however, for lipase B the highest

Table 2. Comparison of the purification results obtained with reversed micelles and other methods for *Chromobacterium viscosum* lipases (74).

Lipase	Method	Number of Steps	Purification Factor	Recovery yield (%)
A	Sucessive chromatographic processes (16)	1	5.0	21
A	Reversed micellar system	1	4.3	91
B	Sucessive chromatographic processes (17)	2	104	51
B	Reversed micellar system	2	3.7	76

purification factor was achieved using chromatography. Best results for the recovery of lipolytic activity were achieved by reversed micellar methods for both lipases. These results demonstrate the possibility of using reversed micellar systems as a promising pre-chromatographic step, as an alternative to traditional methods.

5.4 Aqueous two-phase systems

Enzymes and other proteins have been purified by using aqueous two-phase systems (76). Most studies have been carried out using systems containing polyethyleneglycol (PEG) and dextran, while large-scale isolation has predominantly been carried out with PEG-salt systems. The later was found useful in separation of enzymes from cell debris (77).

Liquid–liquid extraction with aqueous two-phase systems was applied to the purification of a phthalate-ester-hydrolysing lipase produced by *Nocardia erythropolis*; this technique was combined with chromatographic methods. The lipase was purified using a mixture of 30% (w/w) Dextran 500 and 30% (w/w) PEG 6000. The lower phase, containing the enzyme, was loaded onto a DEAE-Sephadex A-50 column. The last step of the purification procedure was gel filtration on a Sephadex G-100 column, the purified lipase being obtained with a recovery of 2.6% and a purification of 192-fold. The purified enzyme appeared homogeneous on polyacrylamide gel discontinuous-electrophoresis, and its molecular weight was 15 kDa (78). These aqueous-two phase systems have recently been applied to the purification of other lipases (79, 80).

A procedure was developed to purify the lipase from *Mucor miehei* using PEG and phosphate (79). Three factors were varied, the molecular weight of PEG and the total concentrations of phosphate and PEG, resulting in different lengths of tie-lines and different pH values. Low-molecular-weight PEG and longer tie-lines favour the partition of lipase, as well as the total protein, from *M. miehei* into the upper phase. In systems with longer tie-lines at pH 3.5, corresponding to higher concentrations of phosphate and PEG 6000, a different partition of lipase and protein was obtained. In systems with shorter tie-lines at pH 4.5, the partitioning of lipase from *M. miehei* could be reversed from the upper to the lower phase. Crude lipase from *M. miehei* was purified in two steps: first, it was extracted into the top phase of the system (3.3% (w/w) PEG 6000, 15% Na-phosphate, 1.9% NaCl, pH 3.4); then it was extracted into a bottom phase by addition of 7% (w/w) Na-phosphate, pH 6.3, 1% NaCl, 42% H_2O, to the primary top phase. This two-step procedure enables up to 69% purification of lipase at a yield of > 80% within less than one hour (79).

The culture supernatant of *Pseudomonas cepacia* contained a lipase, which was concentrated by cross-flow filtration, with a recovery of 80% of the lipase activity (80). A purification of this concentrated lipase was not possible by chromatography methods such as ion exchange or hydrophobic interaction. With liquid–liquid extraction using two immiscible aqueous phases of polymers (10%-PEG 6000 and 10% Dextran 500), a purification factor of 7.7 was obtained. The PEG phase was then chromatographed on an ion-exchange column (Q-Sepharose), leading to a yield of 30% and a purification factor of 55. The enriched lipase solution showed a main protein band at pH 7.1 (80).

Liquid-liquid extraction using two-aqueous phase systems, PEG and potassium phosphate, was also used (81) to partially purify a crude lipolytic preparation of *Chromobacterium viscosum* (lipase A with molecular weight of 120 kDa and pI 3.7; lipase B with molecular weight 30 kDa and pI 7.3). Both lipases partitioned preferentially into the PEG phase (the low-density phase), apparently owing to hydrophobic interactions with the ethylene groups of the polymer. The effects of molecular weight of PEG, pH, ionic strength, and the type of salts added to the system as well as the amount of protein upon the yield of extraction were investigated. Reducing the molecular weight of the polyethylene glycol, from 6000 to 400, led to a increase of the recovery of protein and activity and higher purification factors.

The influence of pH on lipase extraction showed an increase of the purification factor from 1.9 to 2.4 when the pH value was increased from 6.3 to 8.5. At pH 8.5, both lipases are negatively charged, which increases their affinity for the PEG phase.

The best conditions for high yields of lipase were: low molecular weight of polymer (PEG 400), a high pH value (above the isoelectric point of both lipases,

Table 3. Percentage of total (TP) and active protein (AP) in permeate (P) and concentrate (C) after ultrafiltration.

Medium	Sample		Membrane	
			PAN	PS
Lipase solution	P	TP	10.2	12.7
		AP	40.2	10.0
	C	TP	74.0	65.8
		AP	30.6	49.6

PAN, polyacetonitrile; PS, polysulphone.

pH 8.5) and addition of 0.5 M NaCl. A purification of 2.8-fold relative to the crude extract was obtained, with a recovery of 50% based on protein (81).

5.5 Membrane processes

Cross-flow membrane filtration has been used in the downstream processing of lipases, namely for microbial cell removal and concentration of the supernatant of fermentation media containing lipases.

The possibility of purifying *Pseudomonas fluorescens* lipase by ultrafiltration (UF) capillary membrane was studied (82). Two types of capillary membranes were tested: polyacrylonitrile (PAN), and polysulphone (PS) with a 10 kDa molecular-weight cut-off and a lumen diameter of 1.1 mm.

The permeate fluxes through both membranes were reduced greatly in the presence of the lipase. This effect was particularly strong in hydrophilic polyacrylonitryle (PAN) membranes, where the presence of lipase solution lowered the flux to 1/15 of its value without lipase. With polysulphone (PS) membranes this decrease was only by a factor of 3. Table 3 shows the protein and activity in permeate and concentrate, after ultrafiltration using PAN and PS membranes.

After the ultrafiltration, the large amount of active protein in PAN membrane permeate was assayed for activity. In the case of PS membrane, this effect was observed only in the concentrate. PS membranes were suitable for concentration of enzyme, while PAN membranes were better for preliminary fractionation. About 15% of total protein, with 30% total activity, was adsorbed on PAN membrane while 30% total protein (40% total activity) was retained on the PS membrane.

Table 4. Purification factors for PAN and PS membranes in permeate and concentrate.

Membranes	Purification factor	
	Permeate	Concentrate
PS	1.04	1.0
PAN	2.69	0.28

The purification factors obtained with the two ultrafiltration membranes, were relatively low especially for the PS membrane, as seen in Table 4. These results may be explained by the more hydrophobic character of the PS membrane.

5.6 Immunopurification

Immunopurification is one of the most selective and powerful affinity techniques for protein purification, leading to purifications of 1000-fold to 10 000-fold in a single step (83). Highly specific antibodies can distinguish between very similar antigens and thus can overcome difficulties in separation that other methods cannot.

Immunopurification has been perceived as one of the most expensive affinity methods, particularly when monoclonal antibodies are used. However, the expansion of the technology of monoclonal antibodies, including methods for economic, large-scale production has already reduced significantly the cost of immunopurification, allowing the development of production-scale applications (84).

Most immunoaffinity purifications are carried out with monoclonal antibodies or affinity-purified polyclonal antibodies. The choice depends on the availability of a monoclonal antibody against the target protein and the composition of the crude preparation, *i.e.*, the type and the concentration of contaminants.

Very few examples of immunopurification of lipases are found in the literature. A hepatic lipase preparation from rat liver homogenates, partly purified by column chromatography on heparin Sepharose, was purified by immunoabsorbent column chromatography (62). Highly purified specific immunoglobulin bound to Sepharose 4B was used in this affinity step, in which 95% of the applied protein containing less than 10% of lipase activity was eluted in the break-through fraction.

6. CONCLUDING REMARKS

Most purification schemes for lipases are based on multi-step sequences of chromatographic methods. However, in recent years, new techniques have been developed that may be used as a first step in purification procedures leading to high recovery. Similar yields and purification factors may thus be achieved with fewer steps, when these novel methods are combined with chromatographic processes. Now that the tertiary structures of several lipases are known, one can also envisage that the design of novel, highly selective purification procedures, based upon molecular recognition, will be developed, allowing even easier and more efficient separation and recovery.

REFERENCES

1. Desnuelle, P. (1972). The lipases. In *The Enzymes*, vol. III, ed. Boyer, P.D., pp. 575–616. New York: Academic Press.
2. Verger, R. (1984). Pancreatic lipases. In *Lipases*, ed. Borgström, B. & Brockman, H.L., pp. 83–150. Amsterdam: Elsevier Science Publishers.
3. Aires-Barros, M.R. & Cabral, J.M.S. (1991). Purification and kinetic studies of lipases using reversed micellar systems. In *Lipases: Structure, Mechanism and Genetic Engineering*, GBF monographs, vol. 16, ed. Alberghina, L., Schmid, R.D. & Verger, R., pp. 407–416. Weinheim: VCH.
4. Suzuki, T., Mushiga, Y., Yamane, T. & Shimizu, S. (1988). Mass production of lipase by fed-batch culture of *Pseudomonas fluorescens*. *Appl. Microbiol.*, 27, 417–422.
5. Omar, I.C., Nishio, N. & Nagai, S. (1987). Production of a thermostable lipase by *Humicola lanuginosa* grown on sorbitol-corn steep liquor medium. *Agric. Biol. Chem.*, 51, 2145–2151.
6. Brady, L., Brzozowski, A.M., Derewenda, Z.S., Dodson, E., Dodson, G. & Tolley, S. (1990). A serine protease triad (Ser.His.Asp) forms the catalytic centre of a triacylglycerol lipase. *Nature*, 343, 767–770.
7. Winkler, F.K., D'Arcy, A. & Hunziker, W. (1990). Structure of human pancreatic lipase. *Nature*, 343, 771–774
8. Schrag, J.D., Li, Y., Wu, S. & Cygler, M. (1991). Ser-His-Glu triad forms the catalytic site of the lipase from *Geotrichum candidum*. *Nature*, 351, 761–764.
9. Gargouri, Y., Moreau, H. & Verger, R. (1989). Gastric lipases: biochemical and physiological studies. *Biochim. Biophys. Acta*, 1006, 255–271
10. Zechner, R. (1990). Rapid and simple isolation procedure for lipoprotein lipase from human milk. *Biochim. Biophys. Acta*, 1044, 20–25.
11. Ikeda, Y., Takagi, A. & Yamamoto, A. (1989). Purification and characterization of lipoprotein lipase and hepatic triglyceride lipase from human postheparin plasma: production of monospecific antibody to the individual lipase. *Biochim. Biophys. Acta*, 1003, 254–269.
12. Fredrikson, G., Stralfors, P., Nilsson, N.O. & Belfrage, P. (1981). Hormone-sensitive lipase of rat adipose tissue: purification and some properties. *J. Biol. Chem.*, 256, 6311–6320.
13. Verger, R., De Haas, G.H., Sarda, L. & Desnuelle, P. (1969). Purification from porcine pancreas of two molecular species with lipase activity. *Biochim. Biophys. Acta*, 188, 272–282.

14. Antonian, E. (1988). Recent advances in the purification, characterization and structure determination of lipases. *Lipids*, 23, 1101–1106
15. Yamaguchi, T., Muroya, N., Isobe, M. & Sugiura, M. (1973). Production and properties of lipase from a newly isolated *Chromobacterium*. *Agric. Biol. Chem.*, 37, 999–1005.
16. Sugiura, M., Isobe, M., Muroya, N. & Yamaguchi, T. (1974). Purification and properties of a *Chromobacterium viscosum* lipase with a high molecular weight. *Agric. Biol. Chem.*, 38, 947–952.
17. Sugiura, M. & Isobe, M. (1974). Studies on the lipase of *Chromobacterium viscosum*. III. Purification of a low molecular weight lipase and its enzymatic properties. *Biochim. Biophys. Acta*, 341, 195–200.
18. Isobe, M. & Sugiura, M. (1977). Studies on the lipase of *Chromobacterium viscosum*. V. Physical and chemical properties of the lipase. *Chem. Pharm. Bull.*, 25, 1980–1986.
19. Horiuti, Y. & Imamura, S. (1977). Purification of lipase from *Chromobacterium viscosum* by chromatography on palmitoyl cellulose. *J. Biochem.*, 81, 1639–1649.
20. Nishio, T., Chikano, T. & Kamimura, M. (1987). Purification and some properties of lipase produced by *Pseudomonas fragi* 22.39B. *Agric. Biol. Chem.*, 51, 181–187.
21. Yamamoto, K. & Fujiwara, N. (1988). Purification and some properties of a castor-oil-hydrolysing lipase from *Pseudomonas* sp. *Agric. Biol. Chem.*, 52, 3015–3021.
22. Kordel, M. & Schmid, R.D. (1990). Purification and characterization of the lipase from *Pseudomonas* spec. ATCC 21808. *Abstracts of the 5th. European Congress in Biotechnology* (July 8–13, Copenhagen, Denmark), p. 269.
23. Iizumi, T., Nakamura, K. & Fukase, T. (1990). Purification and characterization of a thermostable lipase from newly isolated *Pseudomonas* sp. KWI-56. *Agric. Biol. Chem.*, 54, 1253–1258.
24. Sztajer, H., Borkowski, J. & Sobiech, K. (1991). Purification and some properties of *Pseudomonas fluorescens* lipase. *Biotechnol. Appl. Biochem.*, 13, 65–71.
25. Muraoka, T., Ando, T. & Okuda, H. (1982). Purification and properties of a novel lipase from *Staphylococcus aureus* 226. *J. Biochem.*, 92, 1933–1939.
26. Tyski, S., Hryniewicz, W. & Jeljaszewicz, J. (1983). Purification and some properties of the staphylococcal extracellular lipase. *Biochim. Biophys. Acta*, 749, 312–317.
27. Van Oort, M.G., Deveer, A.M.Th.J., Dijkman, R., Tjeenk, M.L., Verheij, H.M., de Haas, G.H., Wenzig, E. & Götz, F. (1989). Purification and substrate specificity of *Staphylococcus hyicus* lipase. *Biochemistry*, 28, 9278–9285.
28. Götz, F. (1991). Staphylococcal lipases and phospholipases. In *Lipases: Structure, Mechanism and Genetic Engineering*, GBF monographs, vol. 16, ed. Alberghina, L., Schmid, R.D. & Verger, R., pp. 285–292. Weinheim: VCH.
29. Ingham, E., Holland, K.T., Gowland, G. & Cunliffe, W.J. (1981). Partial purification and characterization of lipase (EC 3.1.1.3). from *Propionibacterium acnes*. *J. Gen. Microbiol.*, 124, 393–401.
30. Sugihara, A., Tani, T. & Tominaga, Y. (1991). Purification and characterization of a novel thermostable lipase from *Bacillus* sp. *J. Biochem.*, 109, 211–216.
31. Ota, Y., Nakamiya, T. & Yamada, K. (1970). Lipase from *Candida paralipolytica*. Part IV. Purification, some properties and modification of the purified enzyme with the concentrated solution of sodium chloride. *Agric. Biol. Chem.*, 34, 1368–1374.
32. Veeraragavan, K. & Gibbs, B.F. (1989). Detection and partial purification of two lipases from *Candida rugosa*. *Biotechnol. Letters*, 11, 345–348.
33. Shaw, J.F., Chang, C.H. & Wang, Y.J. (1989). Characterization of three distinct forms of lipolytic enzyme in a commercial *Candida* lipase preparation, *Biotechnol. Letters*, 11, 779–784.
34. Ota, Y., Gomi, K., Kato, S. Sugiura, T. & Minoda, Y. (1982). Purification and some properties of cell-bound lipase from *Saccharomvcopsis lipolytica*. *Agric. Biol. Chem.*, 46, 2885–2893.

35. Tombs, M.P. & Blake, G.G. (1982). Stability and inhibition of *Aspergillus* and *Rhizopus* lipases. *Biochim. Biophys. Acta*, 700, 81–89.
36. Sugihara, A., Shimada, Y. & Tominaga, Y. (1988). Purification and characterization of *Aspergillus niger* lipase. *Agric. Biol. Chem.*, 52, 1591–1592.
37. Torossian, K. & Bell, A.W. (1991). Purification and characterization of an acid-resistant triacylglycerol lipase from *Aspergillus niger*, *Biotechnol. Appl. Biochem.*, 13, 205–211.
38. Tsujisaka, Y., Iwai, M. & Tominaga, Y. (1973). Purification, crystallization and some properties of lipase from *Geotrichum candidum* Link. *Agric. Biol. Chem.*, 37, 1457–1464.
39. Sugihara, A., Shimada, Y. & Tominaga, Y. (1990). Separation and characterization of two molecular forms of *Geotrichum candidum* lipase. *J. Biochem.*, 107, 426–430.
40. Veeraragavan, K., Colpitts, T. & Gibbs, B.F. (1990). Purification and characterization of two distinct lipases from *Geotrichum candidum*. *Biochim. Biophys. Acta*, 1044, 26–23.
41. Menge, U., Hecht, H.-J., Schomburg, D. & Schmid, R.D. (1990). Crystallization and preliminary X-ray studies of lipase from *Geotrichum candidum*. In *Lipases: Structure, Mechanism and Genetic Engineering*, GBF monographs, vol. 16, ed. Alberghina, L., Schmid, R.D. & Verger, R., pp. 59–62. Weinheim: VCH.
42. Charton, E., Davies, C., Sidebottom, C.M., Sutton, J.L., Dunn, P.P.J., Slabas, A.R. & Macrae, A.R. (1991). Purification and substrate specificities of lipase from *Geotrichum candidum*. In *Lipases: Structure, Mechanism and Genetic Engineering*, GBF monographs, vol. 16, ed. Alberghina, L., Schmid, R.D. & Verger, R., pp. 335–338. Weinheim: VCH.
43. Jacobsen, T., Olsen, J. & Allermann, K. (1989). Production, partial purification, and immunochemical characterization of multiple forms of lipase from *Geotrichum candidum*. *Enzyme Microb. Technol.*, 11, 90–95.
44. Spener, F., Hedrich, H.C., Menge, U. & Schmid, R.D. (1991). Intrinsic activity and catalytic residues of the lipase from *Geotrichum candidum*. In *Lipases: Structure, Mechanism and Genetic Engineering*, GBF monographs, vol. 16, ed. Alberghina, L., Schmid, R.D. & Verger, R., pp. 325–334. Weinheim: VCH.
45. Hedrich, H.C., Spener, F., Menge, U., Hecht, H.-J. & Schmid, R.D. (1991). Large-scale purification, enzyme characterization, and crystallization of the lipase from *Geotrichum candidum*, *Enzyme Microb. Technol.*, 13, 840–847.
46. Nagaoka, K. & Yamada, Y. (1973). Purification of *Mucor* lipases and their properties. *Agric. Biol. Chem.*, 37, 2791–2796.
47. Ishihara, H., Okuyama, H., Ikezawa, H. & Tejima, S. (1975). Studies on lipase from *Mucor iavanicus*. I. Purification and properties. *Biochim. Biophys. Acta*, 388, 413–422.
48. Huge-Jensen, B., Galluzzo, D.R. & Jensen, R.G. (1987). Partial purification and characterization of free and immobilized lipases from *Mucor miehei*. *Lipids*, 22, 559–565.
49. Isobe, K., Akiba, T. & Yamaguchi, S. (1988). Crystallization and characterization of lipase from *Penicillium cyclopium*. *Agric. Biol. Chem.*, 52, 41–47.
50. Sztajer, H., Erdmann, H., Isobe, K., Morelle, G. & Schmid, R. (1991). Production and some properties of partially purified lipase from *Penicillium simplicissimum*. In *Lipases: Structure, Mechanism and Genetic Engineering*, GBF monographs, vol. 16, ed. Alberghina, L., Schmid, R.D. & Verger, R., pp. 339–344. Weinheim: VCH.
51. Isobe, K. & Nokihara, K. (1991). Physiochemical properties of mono-and diacylglycerol lipase from *Penicillium camembertii*. In *Lipases: Structure, Mechanism and Genetic Engineering*, GBF monographs, vol. 16, ed. Alberghina, L., Schmid, R.D. & Verger, R., pp. 345–348. Weinheim: VCH.
52. Tahoun, M.K. & Ali, H.A. (1986). Specificity and glyceride synthesis by micelial lipases of *Rhizopus delemar*. Enzyme Microb. Technol., 8, 429–432.
53. Aisaka, K. & Terada, O. (1980). Purification and properties of lipoprotein lipase from *Rhizopus japonicus*. *Agric. Biol. Chem.*, 44, 799–805.
54. Aisaka, K. & Terada, O. (1981). Purification and properties of lipase from *Rhizopus japonicus*. *J. Biochem.*, 89, 817–822.

55. Suzuki, M., Yamamoto, H. & Mizugaki, M. (1986). Purification and general properties of a metal-insensitive lipase from *Rhizopus japonicus* NR 400. *J. Biochem.*, 100, 1207–1213.
56. Omar, I.C., Hayashi, M. & Nagai, S. (1987). Purification and some properties of thermostable lipase from *Humicola lanuginosa* No 3. *Agric. Biol. Chem.*, 51, 37–45.
57. Kundu, M., Basu, J., Guchhait, M. & Chakrabarti, P. (1987). Isolation and characterization of an extracellular lipase from the conidia of *Neurospora crassa*. *J. Gen. Microbiol.*, 133, 149–153.
58. Lang, S., Katsiwela, E., Kleppe, F. & Wagner, F. (1991). *Ustilago maydis* lipolytic enzymes: characterization and partial purification. In *Lipases: Structure, Mechanism and Genetic Engineering*, GBF monographs, vol. 16, ed. Alberghina, L., Schmid, R.D. & Verger, R., pp. 361 364. Weinheim: VCH.
59. Tiruppathi, C. & Balasubramanian, K.A.(1982). Purification and properties of an acid lipase from human gastric juice. *Biochim. Biophys. Acta*, 712, 692–697.
60. Moreau, H., Gargouri, Y., Lecat, D., Junien, J.-L. & Verger, R. (1988). Purification, characterization and kinetic properties of the rabbit gastric lipase. *Biochim. Biophys. Acta*, 960, 286–293.
61. Cheng, C.-F., Bensadoun, A., Bersot, T., Hsu, J.S.T. & Melford, K.H. (1985). Purification and characterization of human lipoprotein lipase and hepatic triglyceride lipase. *J. Biol. Chem.*, 260, 10720–10726.
62. Twu, J.-S., Garfinkel, A.S. & Schotz, M.C. (1984). Hepatic lipase: purification and characterization. *Biochim. Biophys. Acta*, 792, 330–337.
63. Ben-Zeev, O., Ben-Avram, C.M., Wong, H., Nikazy, J., Shively, J.E. & Schotz, M.C. (1987). Hepatic lipase: a member of a family of structurally related lipases. *Biochim. Biophys. Acta*, 919, 13–20.
64. Verine, A., Giudicelli, H. & Boyer, J. (1974). Affinity chromatography of human adipose tissue triacylglycerol lipase on agarose-linked dioleoylglycerol. *Biochim. Biophys. Acta*, 369, 125–128.
65. Nilsson, S. & Belfrage, P. (1986). Purification of hormone-sensitive lipase by high-performance ion exchange chromatography. *Analyt. Biochem.*, 158, 399–407.
66. Sheridan, M.A. & Allen, W.V. (1984). Partial purification of a triacylglycerol lipase isolated from steelhead trout (*Salmo gairdneri*) adipose tissue. *Lipids*, 19, 347–352.
67. Goers, J.W.F., Pedersen, M.E., Kern, P.A., Ong, J. & Schotz, M.C. (1987). An enzyme-linked immunoassay for lipoprotein lipase. *Analyt. Biochem.*, 166, 27–35.
68. Bonnerjea, J., Oh, S., Hoare, M. & Dunnill, P. (1986). Protein purification: the right step at the right time. *Biotechnology*, 4, 954–958.
69. Hatton, T.A. (1989). Reversed micellar extraction of proteins. In: *Surfactant-Based Separations*, vol. 3, ed. Scamehorn, J.F. & Harwell, J.H., pp. 55–90. New York, USA: Marcel Dekker.
70. Castro, M.J.M. & Cabral, J.M.S. (1988). Reversed micelles in biotechnological processes. *Biotech. Advan.*, 6, 151–167.
71. Goklen, K.E. & Hatton, T.A. (1987). Liquid-liquid extraction of low molecular-weight proteins by selective solubilization in reversed micelles. *Sep. Sci. Technol.*, 22, 831–841.
72. Dekker, M., Van't Riet, K., Weifers, S.R., Baltussen, J.W.A., Laane, C. & Bijsterbosch, B.H. (1986). Enzyme recovery by liquid-liquid extraction using reversed micelles. *Chem. Engin. J.*, 33, B27–B33.
73. Rahaman, R.S., Chee, J.Y., Cabral, J.M.S. & Hatton, T.A. (1988). Recovery of an extracellular alkaline protease from whole fermentation broth using reversed micelles. *Biotechnol. Progr.*, 4, 218–224.
74. Vicente, M.L.C., Aires-Barros, M.R. & Cabral, J.M.S. (1990). Purification of *Chromobacterium viscosum* lipases using reversed micelles. *Biotechnol. Techniques*, 4, 137–142.

75. Aires-Barros, M.R. & Cabral, J.M.S. (1991). Selective separation and purification of two lipases from *Chromobacterium viscosum* using AOT reversed micelles. *Biotechnol. Bioengin.*, 38, 1302–1307.

76. Albertsson, P.-A., Johansson, G. & Tjerneld, F. (1990). Aqueous two-phase separations. In *Separation Processes in Biotechnology*, vol. 10, ed. J.A. Asenjo, pp. 287–317. New York, USA: Marcel Dekker.

77. Hustedt, H., Kroner, K.H. & Kula, M.-R. (1985). Applications of phase partitioning in biotechnology. In *Partitioning in Aqueous Two-Phase Systems: Theory, Methods, Uses and Applications to Biotechnology*, ed. Walter, H., Brooks, D.E. & Fisher, D., pp. 529–587. Orlando FA, USA: Academic Press.

78. Kurane, R., Suzuki, T. & Fukuoka, S. (1984). Purification and some properties of a phthalate ester hydrolyzing enzyme from *Nocardia erythropolis*. *Appl. Microbiol. Biotechnol.*, 20, 378–383.

79. Menge, U. & Schmid, R.D. (1989). Extraction and crystallization of lipase from *Mucor miehei*. In *Proceedings of 15th. Scandinavian Symposium on Lipids*, (June 11–15, Skørping, Denmark), pp. 305–316.

80. Dünhaupt, A., Lang, S. & Wagner, F. (1991). Properties and partial purification of *Pseudomonas cepacia* lipase. In *Lipases: Structure, Mechanism and Genetic Engineering*, GBF monographs, vol. 16, ed. Alberghina, L., Schmid, R.D. & Verger, R., pp. 389–392. Weinheim: VCH.

81. Queiroz, J.A.S.R. (1991). Isolamento e purificação de lipases, Tese de Mestrado, Universidade Técnica de Lisboa, Instituto Superior Técnico, Lisboa, Portugal.

82. Sztajer, H. & Bryjak, M. (1989). Capillar membranes for purification of *Pseudomonas fluorescens* lipase. *Bioprocess Engin.*, 4, 257–259.

83. Harlow, E. & Lane, D. (1988). *Antibodies*. Cold Spring Harbor Publications, Cold Spring Harbor NY, USA.

84. Hill, C.R., Thompson, L.G. & Kenney, A.C. (1989). Immunopurification. In *Protein Purification Methods*, ed. Harris, E.L.V. & Angal, S., pp. 282–292. New York, USA: IRL Press.

13.

Industrial applications of lipases

EVGENY N. VULFSON

Lipases (triacylglycerol acylhydrolase EC 3.1.1.3) are often perceived by research scientists as being one of the most important class of industrial enzymes. In terms of their versatility in use this perception is probably justified, but it is worth bearing in mind that annual sales of lipases account at present for only 20 million US dollars, corresponding to less than 4% of the world-wide enzyme market, which was recently estimated at 600 million US dollars (1). Although this figure excludes several newly developed processes, it still looks unimpressive when compared with the market for other hydrolytic enzymes (*e.g.*, proteases and carbohydrases), which is currently at least ten times larger. There are probably two main reasons for this apparent misconception of the economic significance of lipases. First, lipases have been investigated extensively as a route to novel biotransformations, and hundreds of elegant bio-organic syntheses using them have been described in recent years. Secondly, the diversity of the current and proposed industrial applications of lipases exceeds by far that of proteases or carbohydrases. Thus, the significance of lipases rests on their potential rather than their current level of use.

The major obstacle to practical realisation of the potential of lipases remains the cost of these enzymes. However, according to the enzyme manufacturers, 'progress in genetics and in process technology enables the enzyme industry to offer products with improved properties and often at reduced costs' (2). Indeed, application of genetic engineering techniques has simplified down-stream processing and increased the productivity of micro-organisms manyfold, while the availability of the three-dimensional structure of several lipases (3) should lead to the design and introduction of mutants with greatly superior properties, tailored for specific uses. It is also worth mentioning that the major manufacturers of enzymes, such as Novo Nordisk, have put a substantial effort into developing novel lipase-based technologies for their potential customers (for an excellent recent review on the future impact of industrial lipases, see reference 4).

There is a consensus of opinion in industry and academia that the scope for commercial exploitation of lipases is likely to widen significantly in the coming years. One of the objectives of this chapter is to identify some promising areas of future lipase application. Its primary purpose is to survey the current industrial uses of lipases, without, however, going into details of particular manufacturing processes. Instead, reference will be made to comprehensive reviews or original patents and papers.

1. LIPASES IN THE DAIRY INDUSTRY

Lipases are used extensively in the dairy industry for the hydrolysis of milk fat. Current applications include the flavour enhancement of cheeses, the acceleration of cheese ripening, the manufacture of cheese-like products, and the lipolysis of butterfat and cream (5). The free fatty acids that are generated by the action of lipases on milk fat endow many dairy products, particularly soft cheeses, with their specific flavour characteristics. Thus, the addition of lipases that primarily release short-chain (mainly C4 and C6) fatty acids lead to the development of a sharp, tangy flavour, while the release of medium-chain (C12, C14) fatty acids tend to impart a soapy taste to the product. In addition, the free fatty acids take part in simple chemical reactions, as well as being converted by the microbial population of the cheese. This initiates the synthesis of other flavour ingredients such as acetoacetate, β-keto acids, methyl ketones, flavour esters and lactones (6).

The intensive use of lipases in cheese-making started in the USA after the Second World War. It was engendered by the Food and Drug Administration's ban on the import of rennet paste from Europe because of the impurity and unsatisfactory microbiology of the product. This paste was used by manufacturers of traditional Italian cheeses (Provolone, Romano, Mozzarella, Parmesan) and the first lipase cocktails were introduced as a substitute to create the typical lipolytic flavour profiles of these varieties.

The traditional sources of lipases for cheese-flavour enhancement are animal tissues, especially pancreatic glands (bovine and porcine) and the pre-gastric tissues of young ruminants (kid, lamb, calf). The latter are most commonly used in cheese-making. The commercial pre-gastric lipases are available in the form of liquid extracts, pastes and vacuum- or freeze-dried powders. Each type of pre-gastric lipase gives rise to its own characteristic flavour profile: a buttery and slightly peppery flavour (calf); a sharp 'piccante' flavour (kid); a strong 'pecorino', also described as a 'dirty sock' flavour (lamb) (5).

More recently, a whole range of microbial lipase preparations has been developed for the cheese-manufacturing industry: *Mucor miehei* (Piccantase, Gist-Brocades; Palatase M, Novo Nordisk), *Aspergillus niger* and *A. oryzae* (Palatase A, Novo Nordisk; Lipase AP, Amano; Flavour AGE, Chr. Hansen) and several others. These microbial lipases are used not only for flavour enhancement and the acceleration of the ripening of specific cheeses such as blue, but in some cases

Table 1. Selected examples of the use of lipases in cheese-making and accelerated cheese-ripening.

Cheese type	Lipase source
Romano Domiati Feta	Kid/lamb pre-gastric *Mucor miehei*
Mozarella Parmesan Provolone	Calf/kid pre-gastric
Fontina Ras Romi	*Mucor miehei*
Cheddar Manchego Blue	*Aspergillus oryzae/niger*

they have also successfully replaced pre-gastric lipases (Table 1). A range of cheeses of good quality was produced by using individual microbial lipases or mixtures of several preparations (5).

Apart from substitution for rennet paste and flavour enhancement, lipases are widely used for imitation of cheeses made from ewe's or goat's milk. Addition of lipases to cow's milk generates a flavour rather similar to that of ewe/goat milk. This is used for producing cheeses like Feta, Manchego and Romano from cow's milk. When added to certain blue cheeses, lipases imitate the taste of Roquefort, which is normally produced from sheep's milk. Similarly, the addition of lipases to pasteurised milk leads to the development of the normal flavour profile of Ras or Kopanisti, which traditionally are produced from raw milk.

Lipases also play a crucial rôle in the preparation of so-called enzyme-modified cheeses (EMC). EMC is a cheese that has been incubated in the presence of enzymes at elevated temperature in order to produce a concentrated flavour for use as an ingredient in other products (dips, sauces, dressings, soups, snacks *etc.*) Typically, the concentration of free fatty acids is ten times higher in EMC than in that of the corresponding young cheese. EMC technology is widely used in the USA and is expected to be approved in E.C. countries.

2. LIPASES IN HOUSEHOLD DETERGENTS

Enzyme sales for use in washing powders still remain the single biggest market for industrial enzymes (1, 2). At present, lipases have not played a significant rôle in household detergents, mainly because of the lack of enzymes that are sufficiently stable and active under alkaline conditions. However the world-wide trend towards lower laundering temperatures has led to a much higher demand for such preparations. Understandably, fat stains are more difficult to remove in cold water that their hydrolysis products, which are more easily removed under alkaline conditions.

Recent intensive screening programmes, followed by genetic manipulation, have resulted in the introduction of several suitable preparations, for instance Novo Nordisk's 'Lipolase' (*Humicola* lipase expressed in *Aspergillus oryzae*). This market is expected to grow, and undoubtedly we shall see the introduction of lipases from novel sources as well as genetically engineered enzymes with improved characteristics. The commercial worth of the latter approach for enhanced stability of detergent proteases has already been proved by the research teams of Genex Corporation and Novo Nordisk.

3. LIPASES IN THE OLEOCHEMICAL INDUSTRY

The scope for the application of lipases in the oleochemical industry is enormous. Fats and oils are produced world-wide at a level of approximately 60 million tonnes per annum and a very substantial part of this (more than 2 million tonnes p.a.) is utilised in high-energy-consuming processes such as hydrolysis, glycerolysis and alcoholysis (Figure 1; references 1–3). The conditions for steam fat-splitting and conventional glycerolysis of oils involve temperatures of 240–260°C and high pressures (methanolysis is currently performed under slightly milder conditions). The resulting products are often unusable as obtained and require re-distillation to remove impurities and products of degradation. In addition to this, highly unsaturated heat-sensitive oils cannot not be used in this process without prior hydrogenation.

The saving of energy and minimisation of thermal degradation are probably the major attractions in replacing the current chemical technologies with biological ones. However, in spite of their apparent superiority, enzymic methods have not as yet attained a level of commercial exploitation commensurate with their potential. There have been several communications about relatively small-scale enzymic fat-splitting processes for the production of some high-value polyunsaturated fatty acids (section 4) and the manufacture of soap. For instance, Miyoshi Oil & Fat Co., Japan, reported the commercial use of *Candida cylindracea* lipase in the production of soaps (7). The company claimed that the enzymic method yielded a superior product and was cheaper overall than the conventional Colgate–Emery process.

There are probably several reasons for the generally disappointing level of commercial exploitation of lipases in this sector at present. First of all, the

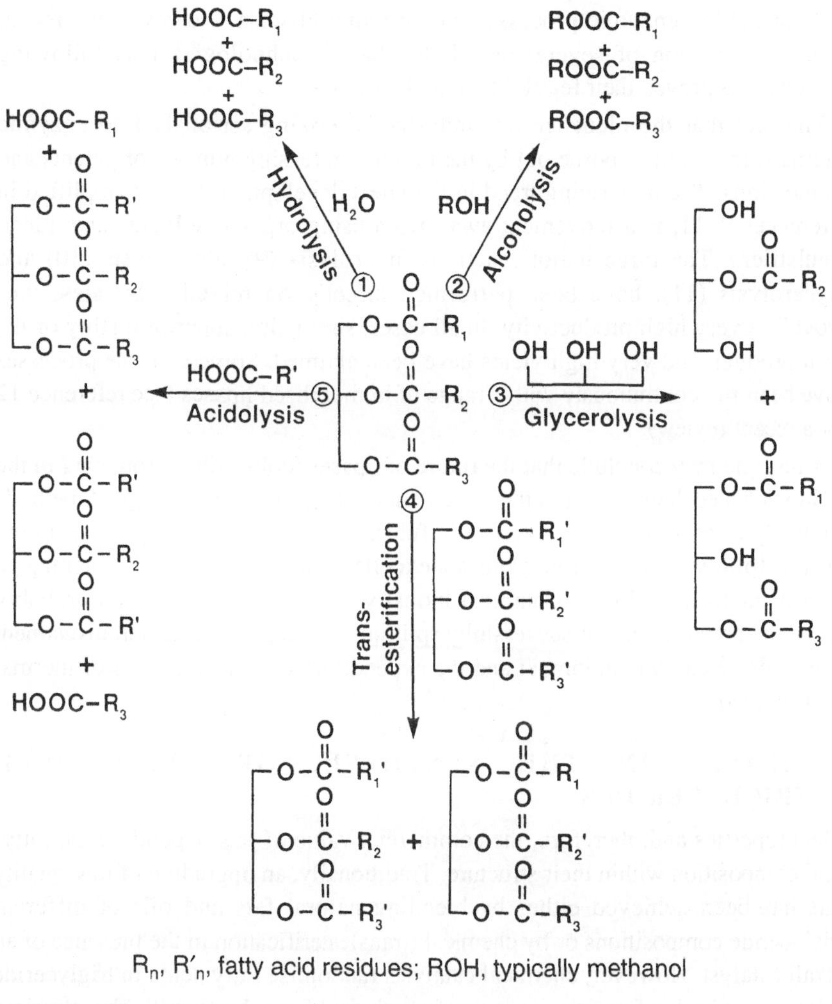

HOOC–R$_1$
+
HOOC–R$_2$
+
HOOC–R$_1$ HOOC–R$_3$

ROOC–R$_1$
+
ROOC–R$_2$
+
ROOC–R$_3$

Hydrolysis   H$_2$O   ROH   *Alcoholysis*

HOOC–R' — Acidolysis ⑤   Glycerolysis ③ — OH OH OH

Trans-esterification ④

R$_n$, R'$_n$, fatty acid residues; ROH, typically methanol

Figure 1. Biotransformations of lipids (fats and oils).

oleochemical industry is very conservative, owing to the huge capital investments involved. Therefore, one cannot expect rapid changes in this area. Secondly, until recently the high cost of lipases remained prohibitive for the manufacturing of bulk products. The introduction of the new generation of cheap and very thermostable enzymes should change the economic balance in favour of lipase use. Indeed, it has already become quite normal to see lipase-based bioreactors in continuous operation for 400–600 hours (see, for example, reference 8). Thirdly, some concern has been expressed by chemical engineers with regard to running and controlling enzymic processes on the required scale. However, the recent commercialisation of several new lipase-based technologies (see following sections) has proved their feasibility unambiguously.

The fact that the oleochemical industry is looking seriously into enzymic alternatives is further borne out by the rapidly increasing number of patents and publications. The underlying trend in the latest developments, as exemplified in references 9–11, is a movement away from using organic solvents and added emulsifiers. The three major reactions, hydrolysis (9), alcoholysis (10) and glycerolysis (11), have been performed directly on mixed substrates, thus providing very high productivity. In all cases, simplicity, superior quality of the final product, and very high yields have been claimed. Some of these processes have been run continuously with a range of immobilised lipases (see reference 12 for a recent review).

Thus, one may conclude that the future of lipases looks rather promising in the context of oleochemistry, and there are indications that biotechnological methods are likely to become more important in future years. This statement is supported by the following quotation from an excellent paper by N.O.V. Sonntag, a consultant for the US oleochemical industry: 'Enzymatic hydrolysis probably offers the greatest hope of successfully splitting fat without substantial investment in expensive capital equipment and the expenditure of large amounts of thermal energy' (13).

4. LIPASES IN THE SYNTHESIS OF STRUCTURED TRIGLYCERIDES

The properties and, therefore, the commercial value of fats depend on the fatty-acid composition within their structure. Traditionally, an upgrading of low-quality fats has been achieved either by blending natural fats and oils of different triglyceride compositions or by chemical (trans)esterification in the presence of an alkali catalyst. However, chemical catalysts randomise fatty acids in triglyceride mixtures, which often prevents the formation of products with the required physico-chemical characteristics.

A typical example of a high-value asymmetric triglyceride mixture is cocoa butter. It consists predominantly of 1,3-disaturated-2-oleoyl-glycerol, where palmitic, stearic and oleic acids account for more than 95% of the total fatty acids. The most important features of cocoa butter, vital for its use and responsible for its unique sensory characteristics, are its crystal structure and a very sharp melting profile between 25 and 35 °C. Thus, it is a brittle solid at room temperature, but melts completely just below body temperature leaving no 'greasy' sensation in the mouth.

The potential of 1,3-specific lipases for the manufacture of cocoa-butter substitutes was clearly recognised more than a decade ago when Unilever (14) and Fuji Oil (15) filed their patent applications. In both cases, the process relied on lipase-catalysed transesterification (Figure 1.4) or acidolysis (Figure 1.5) of cheap oil such as palm mid-fraction, which contains a significant amount of 1,3-dipalmitoyl-2-oleoyl-glycerol with tristearin or stearic acid respectively. The reactions can be conducted continuously or batch-wise with or without organic solvents. Pilot-scale trials have been performed successfully on tonne scale, but Unilever has had to postpone the commercialisation because of a sharp drop in the price of cocoa butter on the world market. The opposite decision has been taken by Fuji Oil, and the company is currently manufacturing enzymically produced cocoa-butter substitute. Comprehensive reviews on this technology, including the analysis of the product composition, are available (8, 16).

In principle, the same approach is applicable to the synthesis of many other structured triglycerides possessing valuable dietetic or nutritional properties. For example, milk-fat substitutes have been prepared as a partial replacement for the milk fat in baby foods. In contrast to plant oils, triglycerides of human milk contain palmitic acid almost exclusively in the 2-position, with 1,3-dioleoyl-2-palmitoyl-glycerol being the major individual component. This triglyceride and functionally similar fats are readily obtainable by acidolysis of palm-oil top fraction, rich in 2-palmitoyl glycerides with unsaturated fatty acid(s) (17).

Acidolysis, catalysed by 1,3-specific lipases, is likely to become the key to the manufacture of yet another group of nutritionally important products: medium-chain triglycerides of octanoic and decanoic acids. It is common medical practice to administer these as a concentrated form of calories to patients with pancreatic insufficiency and other forms of malabsorbtion. This is due to the ease of hydrolysis of these substrates, as compared with long-chain fats, by pancreatic esterases, resulting in practically complete absorbtion. However, medium-chain fats do not provide essential fatty acids, and this dietary intake may lead to the development of deficiency in the patients. Recently it has been shown that this problem can be overcome by using structured triglycerides containing octanoic acid in the 1- and 3-positions and an essential fatty acid in the 2-position (18).

Lipases are also being investigated intensively with regard to the modification of oils rich in high-value polyunsaturated fatty acids such as arachidonic, eicosapentaenoic, docosahexaenoic acids and some others. Many of the n–3 polyunsaturated fatty acids have been reported to have beneficial therapeutic and nutritional effects (see references 19, 20 and others cited therein). The commercial interest centres mainly around mild methods of their recovery in a form suitable for incorporation into specific products (dietary formulations, skin creams *etc.*). Substantial enrichment in the polyunsaturated fatty-acid content of the monoglyceride fraction has been achieved by lipase-catalysed alcoholysis (21, 22) or hydrolysis (22). This approach has already found a small-scale commercial application.

5. LIPASES IN THE SYNTHESIS OF SURFACTANTS

This is another area where lipase-based technology is waiting to be commercialised. The overriding issue in the current use of industrial and household detergents is their environmental safety. Similarly, more stringent regulations are being introduced in the food industry to minimise or eliminate the adverse allergic effects of artificial food ingredients, particularly food emulsifiers. In short, there is much increased consumer awareness, and a trend towards 'green' and 'natural' products.

(Poly)glycerol and carbohydrate fatty-acid esters are widely used as industrial detergents and as emulsifiers in a great variety of food formulations (low-fat spreads, sauces, ice-creams, mayonnaises *etc.*). The chemical manufacturing methods in current use typically suffer from the same drawbacks that apply to the processing of oils: high consumption of energy and the formation of undesirable by-products. Additionally, a whole range of structures is usually obtained, owing to the presence of multiple hydroxyl groups of similar reactivity in carbohydrate substrates. For example, food-grade sorbitan monoesters, obtained from several manufacturers, have recently been re-examined in our laboratory (23). It has been shown that a typical preparation consists of up to 65 individual compounds, many of which were identified by gas chromatography and mass spectroscopy as various isomers of sorbitan, isosorbide and their mono-, di- and tri-esters. Structures of some possible isomers are shown in Figure 2.

Enzymic synthesis of functionally similar surfactants has been carried out at moderate temperatures (60–80 °C) and with excellent regioselectivity (Figure 3(*a*)). Adelhorst *et. al.* (24) have performed the solvent-free esterification of simple alkyl-glycosides using molten fatty acid and immobilised *Candida antarctica* lipase. A range of 6-O-acylglucopyranosides has been prepared in up to 90% yield from the equimolar mixture of substrates, and Novo Nordisk has taken this process to pilot-scale trials (4). The synthesis is claimed to be economical and the products are claimed to be non-toxic and rapidly biodegradable.

A similar process has been developed in the author's laboratory (Figure 3(*b*)). Mono- and di-esters of monosaccharides have been prepared in high yields using sugar acetals as starting materials (25). As above, prior 'hydrophobisation' of sugars was required to carry out the reaction in the absence of added solvents, but, in contrast to the Novo Nordisk approach, this kind of method can be readily extended to the synthesis of disaccharide esters (D.B. Sarney & E.N. Vulfson, unpublished results). Although acetalisation and subsequent hydrolysis do not seem to present any technological difficulties, the overall production may be complicated. It remains to be seen whether these additional steps can be justified by energy savings, improved quality and the superior emulsifying properties of the final product.

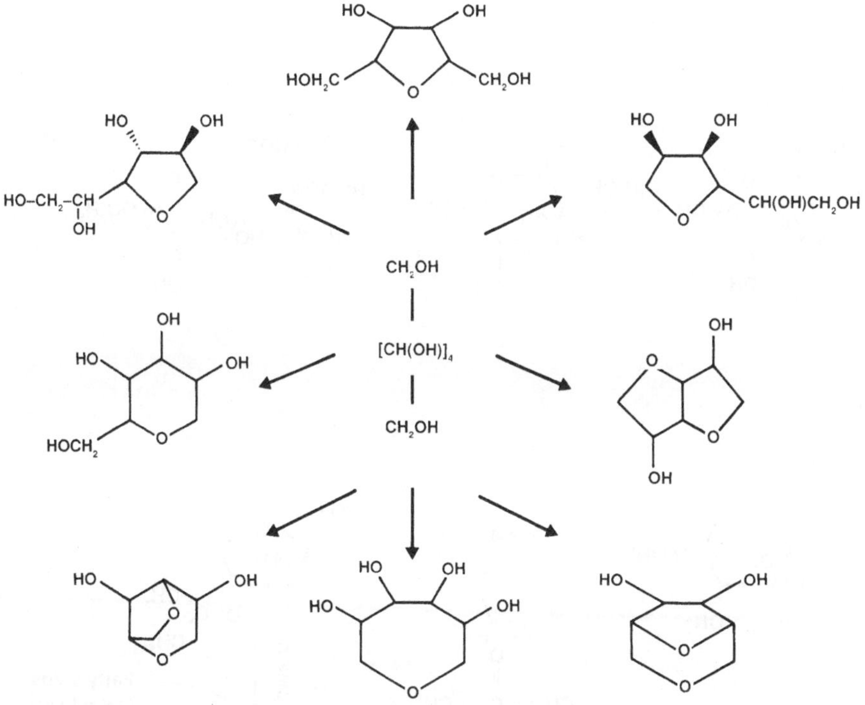

Figure 2. Possible dehydration products of sorbitol in food grade sorbitan esters.

(Poly)glycerol-based surfactants can also be produced enzymically at ambient temperatures and, ultimately, in solvent-free processes. Apart from glycerolysis of oils (section 3), mono- and di-glycerides were prepared by the direct coupling of glycerol and free fatty acids in batch mode (26) and in membrane bioreactors (27). All these methods are also applicable to the production of polyglycerol–fatty-acid esters (C.J. Kirby & E.N. Vulfson, unpublished results).

Lipases may also replace phospholipases in the production of lysophospholipids. The current manufacturing method is rather complex: phospholipase-catalysed hydrolysis is performed in an emulsion containing up to 30% (w/w) of phospholipid and a high concentration of $CaCl_2$. As a result, the enzyme cannot be re-used and has to be inactivated by subsequent heat or protease treatment. Alternatively, *Mucor miehei* lipase has been used for the transesterification of phospholipids in a range of primary and secondary alcohols (28). This approach

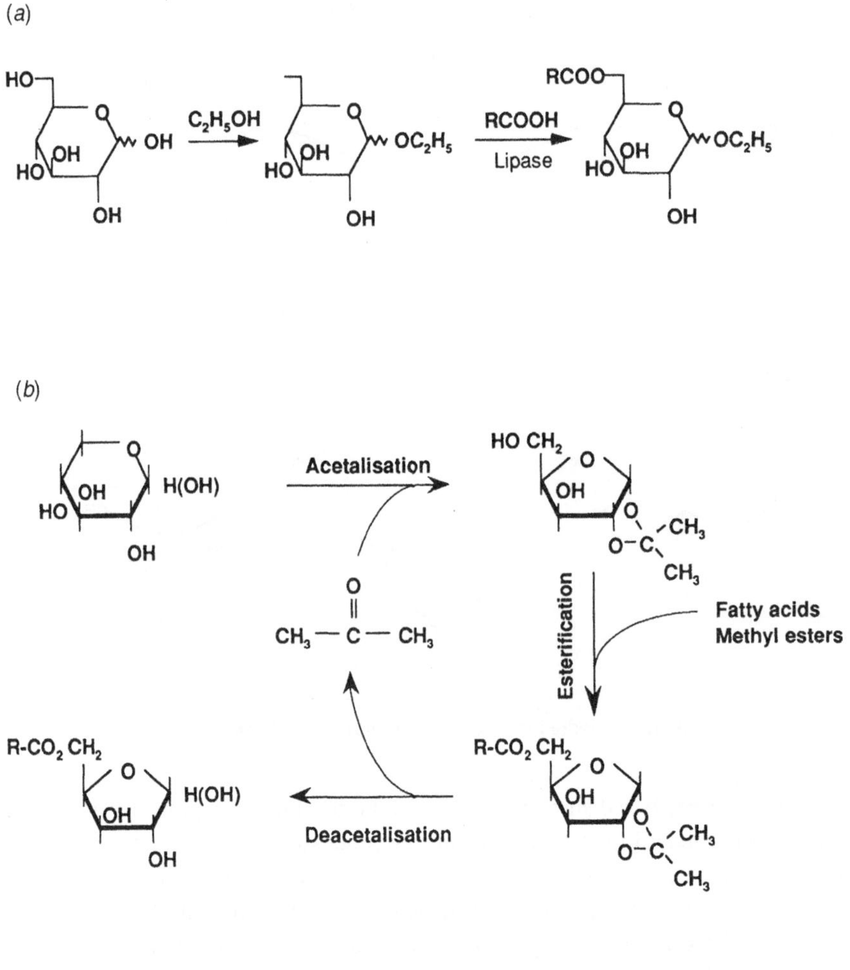

(a)

(b)

Figure 3. Lipase-catalysed synthesis of sugar-based surfactants.

facilitates product recovery and does not require co-factors, and the reaction can be run continuously in a packed-column bioreactor.

Lipases may also be useful in the synthesis of a whole range of amphoteric bio-degradable surfactants, for instance, amino-acid-based esters and amides (29, 30).

6. LIPASES IN THE SYNTHESIS OF INGREDIENTS FOR PERSONAL-CARE PRODUCTS

Although the cost of lipase-catalysed esterification typically remains too high for the manufacturing of many bulk products, the synthesis of several speciality esters has found its way into the market-place. Unichem International has recently launched the production of isopropyl myristate, isopropyl palmitate and 2-ethylhexyl palmitate (31) for use as an emollient in personal-care products such as skin and sun-tan creams, bath oils *etc.* Immobilised *Mucor miehei* lipase was used as a biocatalyst in the solvent-free esterification, which was driven to completion by vacuum distillation of the water produced during the reaction. The company claims that the use of the enzyme in place of the conventional acid catalyst gives products of much higher quality, requiring minimum downstream refining. Batches of several tonnes have been successfully produced at Unichem's factory in Spain.

Wax esters (esters of fatty acids and fatty alcohols) have similar application in personal care products and are also being manufactured enzymically (Croda Universal Ltd.). The company uses *Candida cylindracea* lipase in a batch bioreactor, but no specific details have yet been released. According to the manufacturer, the overall production cost is slightly higher than that of the conventional method, but the cost is justified by the improved quality of the final product.

7. LIPASES IN THE SYNTHESIS OF PHARMACEUTICALS AND AGROCHEMICALS

Over the last decade we have witnessed a noticeable trend, especially within the pharmaceutical industry, towards the manufacture of optically pure products in preference to racemic mixtures. Nevertheless, the vast majority of synthetic pharmaceuticals and agrochemicals, containing one or more chiral centres, are still sold as racemates. This is despite the fact that the desired biological activity usually resides in one particular enantiomer. Although it is no longer disputed that a single isomer is preferable to a racemic mixture, there are severe technical and/or economic problems associated with the production of single isomers.

The usefulness of lipases in the preparation of chiral synthons is well recognised and documented. There has been an explosion in the number of patents and publications over the last few years, and it is simply impossible to survey the area within the scope of this chapter. Therefore, further consideration will be restricted to several processes that have recently been commercialised, while comprehensive reviews of the subject (32–34) can be found elsewhere.

The resolution of 2-halopropionic acids, starting materials for the synthesis of phenoxypropionate herbicides, is being carried out on a 100-kg scale by Chemie

Linz Co. (Austria) under a license from the Massachusetts Institute of Technology (35). The process is based on the selective esterification of (S)-isomers with butanol catalysed by porcine pancreatic lipase in anhydrous hexane (Figure 4(a)). Typically, >99% enantiomeric excess (e.e.) is obtained at 75% of the theoretical yield, and the resolution is complete in several hours.

Generally, the lipase-mediated resolution of 2-substituted propionic acids, and especially 2-aryl derivatives, has been the subject of intensive investigation. A substantial body of literature exists on the production of both (R)- and (S)-isomers of α-phenoxypropionic acids, which are useful synthons for the preparation of enantiomerically pure herbicides and non-steroidal anti-inflammatory drugs (ibuprofen, naproxen) respectively (32). The required optically pure derivative can be obtained directly via lipase-catalysed (trans)esterification or hydrolysis of the corresponding ester. These resolutions have been performed on a multi-kilogramme scale by several companies world-wide (see references 36, 37 as typical examples).

Another impressive instance of commercial application of lipases to the resolution of racemic mixtures is the hydrolysis of epoxy alcohol esters. The highly enantioselective hydrolysis of (R,S)-glycidyl butyrate (Figure 4(b)) has been developed by DSM-Andeno (the Netherlands) and is currently being carried out on a multi-tonne scale (38). The reaction products, (R)-glycidyl esters and (R)-glycidol, are readily converted to (R)- and (S)-glycidyltosylates, which are very attractive intermediates for the preparation of optically active β-blockers and a wide range of other products (38).

A similar technology has been commercialised by Sepracor Inc. (USA). This company has successfully operated a multi-kilogramme-scale membrane bioreactor to produce 2(R),3(S) methyl methoxyphenyl glycidate, the key intermediate in the manufacture of the optically pure cardiovascular drug diltiazem (37). It is interesting to note that 2(S),3(R)-methoxyphenyl glycidic acid, the product of enzymic hydrolysis, was found to be unstable under the conditions of the reaction, and the resultant aldehyde both inhibited the lipase activity and reduced the lifetime of the enzyme. Both problems were overcome by the introduction of a multi-phase membrane reactor where the aldehyde by-product reacted in situ with bisulphite to form a non-inhibitory, water-soluble adduct, extracted into the aqueous phase.

Lipases have also been found useful as industrial catalysts for the resolution of racemic alcohols. Enantiomerically pure endo-2-norborn-2-ol is an important chiral intermediate in the preparation of some prostaglandins, steroids and carbocyclic nucleoside analogues. Bend Research Inc. (USA) have recently developed a two-step resolution process depicted in Figure 4(c). The process involved acylation of the (R)-alcohol with butyric anhydride, mediated by Candida cylindracea lipase, followed by the hydrolysis of the (R)-ester catalysed by the

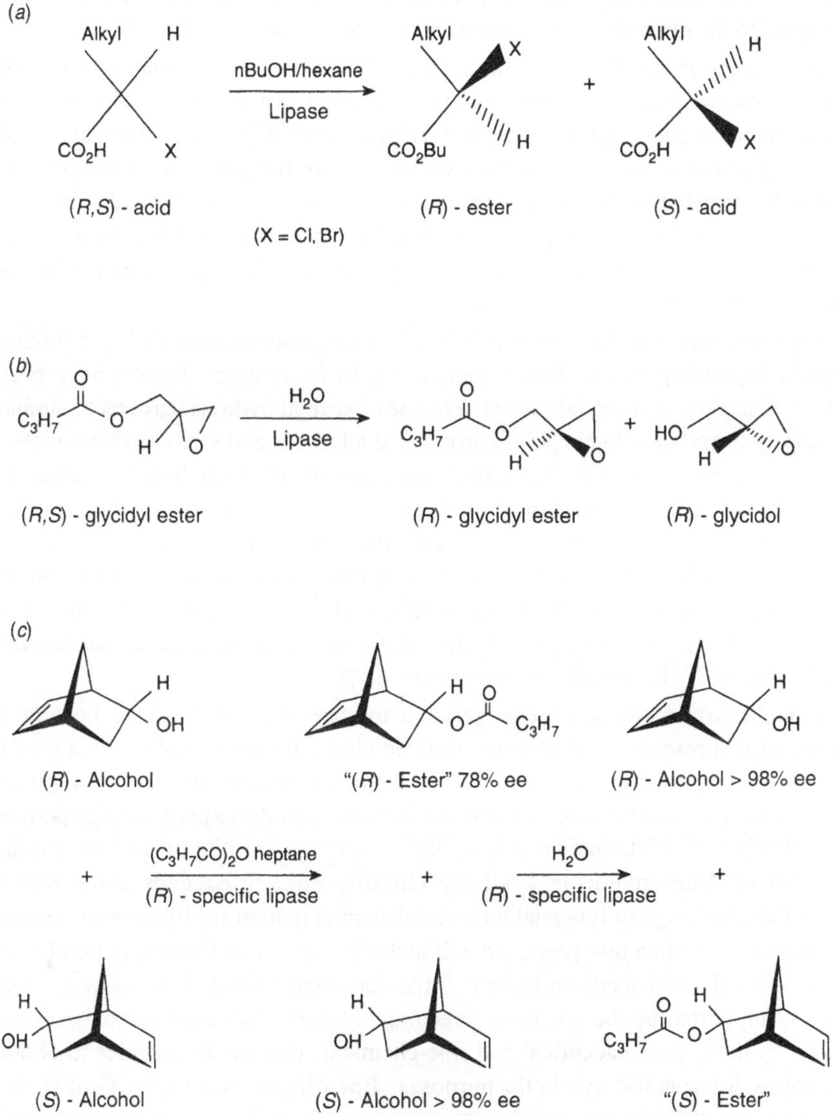

Figure 4. Lipase-mediated resolution of racemic mixtures. BuOH, butanol.

same enzyme (39). The first resolution resulted in enantiomerically pure (S)-alcohol (e.e. > 98%) and (R)-ester (e.e. = 78%) which was further enriched by the back conversion to (R)-alcohol (e.e. > 98%). The resolution was performed on a multi-kilogramme scale in a permselective membrane bioreactor specially designed to facilitate product recovery and to minimise product inhibition.

It should be stressed once again that these lipase-mediated resolutions are just selected examples of successful commercialisation and are not by any means a comprehensive survey of this rapidly developing area. Lipases are currently used by many pharmaceutical companies world-wide for the preparation of optically active intermediates on a kilogramme scale. Moreover, a number of relatively small biotechnological companies, such as Enzymatix in the UK, specialise in biotransformations and offer a whole variety of intermediates prepared *via* lipase-mediated resolution.

Regioselective modification of polyfunctional organic compounds is yet another rapidly expanding area of lipase application. In many cases, lipases have been shown to acylate or deacylate selectively one of several hydroxyl groups of similar reactivity in carbohydrates, polyhydroxylated alkaloids and steroids (for reviews, see references 33 and 40). Apart from the synthesis of sugar-based surfactants, described briefly in section 5, lipases were successfully applied in the regioselective modification of castanospermine, a promising drug for the treatment of AIDS (40). It is also worth mentioning in this context the attempt by Tate & Lyle to prepare 4,1',6'-trihydroxy-pentaacetyl-sucrose, potentially the key intermediate in the synthesis of the artificial sweetener sucralose, by the regioselective hydrolysis of octaacetylsucrose (41).

Thus, it would not be an exaggeration to state that lipases have become a conventional research tool in many organic chemistry laboratories. As a result, they are readily incorporated into synthetic routes, especially where optical purity of the final product is essential. However, according to the experts, a large pool of generically useful intermediates is unlikely to emerge, in spite of the fact that the number of drugs requiring synthetic chirality already exceeds some recent estimates. Although in financial terms a substantial market for lipase sales cannot be expected within a few years, we will undoubtedly see an increasing number of diverse small- and medium-scale commercial applications. This view is most clearly supported by the intensive screening programmes currently being undertaken by many pharmaceutical and fine-chemicals companies in order to obtain esterases for specific synthetic purposes. Boehringer Mannheim GmbH, for instance, has recently reported the preliminary results of such a programme: more than 300 enzymes from 6000 micro-organisms were studied with regard to their stereo- and regio-selectivity (42).

8. LIPASES IN POLYMER SYNTHESIS

Optically active polymers have found several applications: as asymmetric reagents, as absorbents and in the field of liquid crystals. Polymers of this type have generally been prepared by chemical means, but a few recent reports have demonstrated clearly the scope of lipase-catalysed reactions in polymer science. Excellent stereoselectivity has been demonstrated in the lipase-mediated poly-transesterification of racemic diesters and a diol (43, 44), but in all cases only oligomeric products were formed. An alternative strategy for the production of chiral polymers involves the use of prochiral monomers that can be polymerised in a single enantiomeric form (a so-called enantioconvergent synthesis; reference 45) in the presence of a number of enzymes including lipases.

High-molecular-mass polymers are more easily prepared by conventional means, *e.g.*, by free-radical polymerisation from vinyl monomers. However, suitable monomers can be prepared by lipase-catalysed transesterification of alcohols (46), which with racemic alcohols may be accompanied by resolution (47). The use of chiral glycidyl tosylates for the preparation of ferro-electric liquid crystals (38) and acylation of hydroxyl groups present in the side-chain of comb-like polymethacrylates (48) have also been reported.

It is important to stress that the economic feasibility of enzymic pilot-scale synthesis of speciality polymers has already been proved by the MEAD Corporation (USA). The company uses polyphenol oxidase in the manufacture of coatings for transparencies and computer paper under license from the Massachusetts Institute of Technology (49). The first review of this emerging area of enzyme application has just been published (50).

9. CONCLUSIONS

It is hoped that this brief survey has given a fair impression of the current and potential applications of lipases in modern industry. The striking feature of lipases, as compared with other classes of enzymes, is the diversity of their commercial use in terms both of scale and of processes. Indeed, lipases are being employed successfully in traditional areas of the food industry (cheese-making) as well as in the high-tech production of fine chemicals and pharmaceuticals.

However, there is no shortage of new potential opportunities. Thus, lipases have been reported to be useful in paper manufacturing: apparently treatment of pulp with lipases leads to a higher-quality product and reduced cleaning requirements (4). Similarly, lipases have been used in conjunction with a microbial cocktail (commercial name 'Combizyme') designed by Biocatalysts Ltd. (UK) for the treatment of fat-rich effluents from an ice-cream plant. The company claims that the addition of lipases helped to solve a waste problem at Windsor Creameries and that the technology should be widely applicable to the food industry in waste-processing (51). Novel lipase-based technologies are also likely to have some impact on traditional chemical synthesis. The most remarkable recent example is the lipase-mediated epoxidation of alkenes with hydrogen peroxide in the presence of catalytic amounts of carboxylic acids (52).

The commercial exploitation of enzymes, and lipases in particular, has traditionally been hampered by their inadequate stability and the limitations imposed by reaction conditions (typically neutral buffers, moderate temperatures and so on). However, the recent developments in genetic engineering and in biocatalysis in unconventional media, especially anhydrous organic solvents (33, 53), have transformed the industrial perception of enzymes às very delicate catalysts, generally unsuitable for large-scale chemical synthesis. One of the objectives of this chapter has been to demonstrate that lipase-based technology is beginning to compete with conventional processes in many industries, and that we shall probably see a substantial growth in the commercial use of lipases in the years to come.

REFERENCES

1. Arbige, M.V. & Pitcher, W.H. (1989). Industrial enzymology: a look towards the future. *Trends Biotech.* 7, 330–335.
2. Falch, E.A. (1991). Industrial enzymes – developments in production and application. *Biotech. Advan.* 9, 643–658.
3. Winkler, F. & Gubernator, K. (1992); Dodson, G., Hubbard, R. & Lawson D. (1992). This book, chapters 4 and 7.
4. Björkling, F., Godtfredsen, S.E. & Kirk, O. (1991). The future impact of industrial lipases. *Trends Biotech.* 9, 360–363.
5. (a) Bech, A.M. (1992). Enzymes for the acceleration of cheese-ripening. In *Fermentation-produced Enzymes and Accelerated Ripening in Cheesemaking* (bulletin of the International Dairy Federation No. 269/1992), pp. 24–28. Brussels: International Dairy Federation. (b) Birschbach. P. (1992). Pregastric lipases. *ibid.*, pp. 36–39.
6. Godfrey, T. & Hawkins, D. Enzymic modification of fats for flavour. (1991). *European Food & Drink Review*, 103–107.
7. Hoq, M.M., Yamane, T., Shimizu, S., Funada, T. & Ishida, S. (1985). Continuous hydrolysis of olive oil by lipase in microporous hydrophobic membrane bioreactor. *J. Am. Oil Chem. Soc.*, 62, 1016–1021.
8. Macrae, A.R. & Hammond, R.C. (1985). Present and future applications of lipases. *Biotech. Genet. Engin. Rev.*, 3, 193–217.
9. Buhler, M. & Wandrey, C. (1987). Continuous use of lipases in fat hydrolysis. *Fat Sci. Technol.*, 89, 598–605.
10. Zaks, A. (1990). Monoglyceride production by enzymatic transesterification. International Patent WO 90/04033 (Enzytech Inc.).
11. McNeill, G.P. & Yamane, T. (1991). Further improvements in the yield of monoglycerides during enzymatic glycerolysis of fats and oils. *J. Am. Oil Chem. Soc.*, 68, 6–10.

The author wishes to thank Professor A.M. Klibanov (Massachusetts Institute of Technology), Drs J. Bosley and P. Quinlan (Unilever Research, Colworth Laboratory) and Drs K. Coupland and N. Langley (Croda Universal Ltd.) for providing information used in this chapter. Professor B.A. Law and Drs M. Whitcombe, J.A. Khan and I. Gill (AFRC Institute of Food Research) are also thanked for helpful discussions and suggestions during the preparation of the manuscript.

12. Malcata, F.X., Reyes, H.R., Garcia, S.G., Hill, C.G. & Amundson, C.H. (1990). Immobilized lipase reactors for modification of fats and oils – a review. *J. Am. Oil Chem. Soc.*, 67, 890–910.
13. Sonntag, N.O.V. (1984). New developments in the fatty acid industry in America. *J. Am. Oil Chem. Soc.* 61, 229–232.
14. Coleman, M.H. & Macrae, A.R. (1980). Fat process and composition. UK Patent No. 1 577 933 (Unilever Limited).
15. Matsuo, T., Sawamura, N., Hashimoto, Y. & Hashida, W. (1981). Method for enzymatic transesterification of lipid and enzyme used therein. European Patent No. 0 035 883 (Fuji Oil Co.).
16. Macrae, A.R. (1983). Lipase-catalysed interesterification of oils and fats. *J. Am. Oil. Chem. Soc.*, 60, 291–294.
17. King, D.M. & Padley, F.B. (1990). Milk fat substitutes. European Patent No 0 209 327 (Unilever plc).
18. Jandacek, R., Whiteside, J.A., Holcombe, B.N., Volpenheim, R.A. & Taulbee, J.D. (1987). The rapid hydrolysis and efficient absorption of triglycerides with octanoic acid in the 1 and 3 positions and long-chain fatty acid in the 2 position. *Am. J. Clin. Nutr.*, 45, 940–945.
19. Kinsella, J.E., Lokesh, B. & Stone, R.A. (1990). Dietary n–3 polyunsaturated fatty acids and amelioration of cardiovascular disease: possible mechanisms. *Am. J. Clin. Nutr.*, 52, 1–28.
20. Young, V. (1990). The usage of fish oils in food. *Lipid Tech.*, 2, 7–10.
21. Mukherjee, K.D. & Kiewitt, I. (1991). Enrichment of γ-linolenic acid from fungal oil by lipase-catalysed reactions. *Appl. Microbiol. Biotechnol.*, 35, 579–584.
22. Zaks, A. & Bross, A.T. (1990). Production of glycerides rich in omega–3 fatty acids by lipase-catalysed transesterification and recrystallisation. International Patent WO 90/13656 (Enzytech Inc.).
23. Fregapane, G., Sarney, D.B., Greenberg, S.G., Knight, D.J. & Vulfson, E.N. (1992). Chemo-enzymatic synthesis of monosaccharide fatty acid esters and their preliminary characterization. In *Proceedings of Biocatalysis in Non-Conventional Media*, ed. Tramper, J., Vermue, M.H., Beeftink, H.H. & von Stockar, U., pp. 563–568. Amsterdam: Elsevier.
24. Adelhorst, K., Björkling, F., Godtfredsen, S.E. & Kirk, O. (1990). Enzyme-catalyzed preparation of 6-O-acylglucophyranosides. *Synthesis*, 112–115.
25. Fregapane, G., Sarney, D.B. & Vulfson, E.N. (1991). Enzymic solvent-free synthesis of sugar acetal fatty acid esters. *Enzyme Microb. Technol.*, 13, 796–800.
26. Yamaguchi, S., Mase, T., Asada, S. (1986). Mono- and di-glyceride production from glycerol by reaction with fatty acid or ester in presence of lipase enzyme. European Patent No. 0 191 217 (Amano Pharmaceutical Co.).
27. Hoq, M.M., Yamane, T., Shimizu, S., Funada, T. & Ishida, S. (1984). Continuous synthesis of glycerides by lipase in a microporous membrane bioreactor. *J. Am. Oil Chem. Soc.*, 61, 776–781.
28. Sarney, D.B., Fregapane, G. & Vulfson, E.N. (1993). Continuous synthesis of lysophospholipids catalysed by immobilized lipase. *Enzyme. Microb. Technol.*, 14, in the press.
29. Nagao, A. & Kito, M. (1989). Synthesis of O-acyl-L-homoserine by lipase. *J. Am. Oil Chem. Soc.*, 66, 710–713.
30. Montet, D., Servat, F., Pina, M., Graille, J., Galzy, P., Arnaud, A., Ledon, H. & Marcou, L. (1990). Enzymatic synthesis of N-acyllysines. *J. Am. Oil Chem. Soc.*, 67, 771–774.
31. Hills, G.A., Macrae, A.R. & Poulina, R.R. (1990). Esters preparation from acids and alcohols with lipase catalyst, and azeotropic distillation of alcohol and obtained water. European Patent No 0 383 405 (Unichema Chem BV). See also: Macrae, A., Roehl, E.L. & Brand, H.M. (1990). Bio-esters in Cosmetic. *Drug Cosmet. Ind.*, 147, 36–39.

32. Sih, C.J. & Wu, S.H. (1989). Resolution of enantiomers *via* biocatalysis. *Topics in Stereochemistry*, 19, 63–125.
33. Klibanov, A.M. (1990). Asymmetric transformations catalysed by enzymes in organic solvents. *Accts Chem. Res.*, 23, 114–120.
34. Davis, H.G., Green, R.H., Kelly, D.R. & Roberts, S.M. (1990). *Biotransformations in Preparative Organic Synthesis*. London: Academic Press.
35. Kirchner, G., Scollar, M.P. & Klibanov, A.M. (1985). Resolution of racemic mixtures *via* lipase catalysis in organic solvents. *J. Am. Chem. Soc.*, 107, 7072–7076.
36. Barton, M.J., Hamman, J.P., Ficher, K.S. & Calton, G.J. (1990). Enzymatic resolution of (*R,S*)-2-(4-hydroxyphenoxy)-propionic acid. *Enzyme Microb. Technol.*, 12, 577–583.
37. Young, J.W. & Bratzler, R.L. (1990). Membrane reactors for the biocatalytic production of chiral compounds. In *Proceedings of 'Chiral-90' Symposium*, pp.23–28, Manchester: Spring Innovation Ltd.
38. Kloosterman, M., Elferink, V.H.M., van Lersel, J., Roskam, J.-H., Meijer, E.M., Hulshof, L.A. & Sheldon, R.A. (1988). Lipases in the preparation of β-blockers. *Trends. Biotech.*, 6, 251–256.
39. van Eikeren, P., Brose, D.J., Muchmore, D.C., West, J.B. & Colton, R.H. (1992). Membrane-assisted synthesis methods for the large-scale production of enantiomerically pure pharmaceutical intermediates. In *Proceedings of 'Chiral-92' Symposium*, pp.63–69, Manchester: Spring Innovation Ltd.
40. Margolin, A.L. (1990). Exploiting enzyme selectivity for the synthesis of biologically active compounds. In *Chemical Aspects of Enzyme Biotechnology*, ed. Baldwin, T.O., Rauschel, F.M. & Scott, A.I., pp. 197–202. New York: Plenum Press.
41. Bornemann, S., Cassells, J.M., Combes, C.L., Dordick, J.S., Hacking, A.J., De-acylation of sucrose esters. (1989). UK Patent No: 2 224 504. (Tate & Lyle plc).
42. Johann, P. & Razor, P. (1992). Biocatalysis in the synthesis of chiral intermediates. In *Proceedings of 'Chiral-92' Symposium*, pp.47–50, Manchester: Spring Innovation Ltd.
43. Wallace, J.S. & Morrow, C.J., (1989). Biocatalytic Synthesis of Polymers. Synthesis of an optically octave epoxy-substituted polyester by lipase-catalyzed polymerisation. *J. Polym. Sci. Polym. Chem.* 27, 2553–2567.
44. Margolin, A.L., Crenne J.-Y. & Klibanov, A.M., (1987). Stereoselective oligomerizations catalyzed by lipases in organic solvents. *Tetrahedron Letters* 28, 1607–1610.
45. Gutman, A.L. & Bravdo, T. (1989). Enzyme-catalyzed enantioconvergent polymerisation of β-hydroxyglutarate in organic solvents. *J. Org. Chem.*, 54, 5645–5646.
46. Pavel, K. & Ritter, H. (1991). Enzymes in polymer chemistry 5. Radical polymerisation of different 11-methacryloylaminoundecanoic acid-esters and oligoesters esterified by lipases. *Makromol. Chem.*, 192, 1941–1949.
47. Margolin, A.L., Fitzpatrick, P.A., Dubin, P.L. & Klibnanov, A.M. (1991). Chemoenzymatic synthesis of optically active (meth)acrylic polymers. *J. Am. Chem. Soc.*, 113, 4693–4694.
48. Pavel, K. & Ritter, H. (1992). Enzymes in polymer chemistry. 6. Lipase-catalyzed acylation of comb-like methacrylic polymers containing OH groups in the side-chains. *Makromol. Chem.*, 193, 323–328.
49. Dordick, J.S., Marletta, M.A. & Klibanov, A.M., (1987). Polymerization of phenols catalysed by peroxidase in nonaqueous media. *Biotech. Bioengin.*, 30, 31–36.
50. Dordick, J.S. (1992). Enzymatic and chemoenzymatic approaches to polymer synthesis. *Trends Biotech.*, 10, 287–293.
51. West, S. Enzymes solve waste problem, (1987). *Food Processing*, 56, 35–39.
52. Björkling, F., Godtfredsen, S.E. & Kirk, O. (1990). *J. Chem. Soc. Chem. Commun.*, 1301–1303.
53. Klibanov, A.M. (1989). Enzymatic catalysis in anhydrous organic solvents. *Trends Biochem. Sci.*, 14, 141–144.

14.

Lipases and phospholipases in organic synthesis

JOCHEM KÖTTING and HANSJÖRG EIBL

The application of lipases and phospholipases as catalysts in organic synthesis has several advantages for synthetic chemists. In general, these enzymes show stereospecificity as well as stereo- and regioselectivity (for definitions of these terms, see reference 1). The conditions for enzyme-catalysed reactions are mild, and yields are usually high. Furthermore, these enzymes work at hydrophilic–lipophilic interfaces and thus tolerate organic solvents in the incubation mixture. Both the activity of the enzyme and the identity of the product depend upon the solvents used, which may vary from aqueous buffer systems through biphasic emulsions and microemulsions, to organic solvents (2–9).

A large variety of lipases and phospholipases are commercially available. Many of them are of microbial origin and supplied in large quantities at relatively low cost. Some are cloned or processed to highest purity. For synthetic purposes, crude enzyme preparations are often convenient. Because the catalyst is always an expensive factor in a chemical reaction, strategies for enzyme recycling are being developed (10, 11). The immobilisation of lipases or phospholipases on suitable materials may lead to procedures for the continuous preparation of lipids in the laboratory or even on an industrial scale.

Lipases are of importance for several types of industrial application, which have been reviewed in several articles (12–14, see also Chapter 13). They are used for the production of detergents, pharmaceuticals, perfumes, flavour enhancers and texturising agents in cosmetic products. Lipases are crucial for the production of a wide variety of foods, especially for products from milk, fat and oil (15, 16).

Apart from molecular biology, much biological and pharmaceutical research in the last decade has been focussed on the immune response, on agonist–receptor interactions and on enzyme–substrate interactions in intra- and intercellular signal chains and cellular growth control. The discovery of single molecular species of lipids and phospholipids that modulate cellular signals was an important stimulus for basic research. Potential new drugs against cancer and AIDS have been conceived, and some projects have already entered the stage of pharmaceutical drug development. Several of the classes of molecules involved in signal transduction, growth regulation and inflammatory and allergic diseases are lipids or lipid analogues. Biological inter- and intracellular signalling is closely connected with single lipid species such as phosphatidylinositol and its breakdown product, 1,2-diacyl-sn-glycerol (17). 1-O-alkyl-2-acetyl-sn-glycero-3-phosphocholine (platelet activating factor, PAF) is a normal intermediate in the bioconversion of phospholipids, catalysed by phospholipases and other phospholipid-metabolising enzymes (18–20), and provides another example of a phospholipid with unique biological properties.

These facts present an enormous challenge for chemists to create biologically active analogues of naturally occurring messenger molecules as antagonists or inhibitors in biological systems. As described in this chapter, synthetic strategies are available for the large-scale synthesis of such biologically active molecules. Lipases and phospholipases can be used in such strategies to prepare molecules of high positional and configurational purity.

The main aim of this chapter, however, is to show how important reactions of chemical synthesis can be combined with enzymic conversions, especially for the preparation of lipids and phospholipids. On the laboratory scale (10–200 g), the products are formed in bioconversions catalysed by lipases and phospholipases, in high yield and with only small amounts of by-products. These procedures have still to be applied on a technical scale. However, they have proved their value for research workers in the lipid field in the initiation of projects with potential medical application. This has been especially important for the development of new drugs based on lipid or phospholipid molecules. In August 1992, the first drug based on a phospholipid molecule was approved for cancer therapy by the German public health authority, the Bundesgesundheitsamt (21–23). In addition, liposomal carrier systems for the tissue-directed application of drugs promise to be an attractive means of reducing side-effects (24, 25).

1. TRIGLYCERIDE LIPASES

The nomenclature of lipolytic enzymes is a common source of confusion, especially in connection with substrate specificity. Different terms for what appear to be the same or very similar enzymes are used in the literature. In addition, commercial substrates are often poorly characterised and may contain impurities, which can lead to contradiction in the classification of lipases (26). The use of defined synthetic substrates may be helpful in this context (27, 28).

Triglyceride lipases can be found as cellular and extracellular proteins in all types of living cells and organisms. Most lipases used as catalysts in organic synthesis are of microbial origin (Table 1). Microbial lipases are easy to obtain, by fermentation processes and a few basic purification steps (29). Most of the lipases listed in Table 1 are commercially available.

Table 1. Specificity of triglyceride lipases of different origin.

Source of lipase	Specificity	Reference
Microbial and fungal		
Candida cylindracea	None	30
Staphylococcus aureus	None	31
Pseudomonas fluorescens		
Rhizopus arrhizus	1,3	32
Aspergillus niger	None	33
Mucor miehei	1,3	34
Geotrichum candidum	None	33
Mammalian		
Porcine pancreatic lipase		35
Human/rat lingual lipase	*sn*-3	36
Posthepatic plasma lipase	*sn*-1	
Lipoprotein lipase (EC 3.1.1.34)	1,3	37

Triglyceride lipases (EC 3.1.1.3) catalyse the hydrolysis of water-immiscible triglycerides at the water–lipid interface (38). Water-soluble short-chain carboxylic acid esters are hydrolysed only slowly (Figure 1). With excess of water, complete hydrolysis is observed; the course of this reaction is usually followed by thin-layer chromatography. Educts such as triglycerides form mixtures of products such as 1,2- or 2,3-diglycerides, 2-monoglycerides, 1- or 3-monoglycerides and finally glycerol. Depending on the experimental conditions, hydrolysis reaches an equilibrium, and the intermediates in the hydrolytic chain can be isolated by crystallisation or column chromatography. Lipase activity can be estimated spectrophotometrically with substrates carrying chromophoric residues, with radiolabelled substrates and by a recently developed HPTLC procedure (39). Lipase activity can also be estimated by pH-stat titration of the liberated fatty acids during hydrolysis (40). The method is described in Appendix 4.1.

Lipases show varying degrees of positional specificity with respect to the primary acyl-ester bond (1,3 specificity, Table 1). The 1,3-specificity of lipases leads to 2,3-diacylglycerides in step 1 of hydrolysis (Figure 1). The intermediate rearranges to give the thermodynamically favoured 1,3-diacylglyceride by acyl

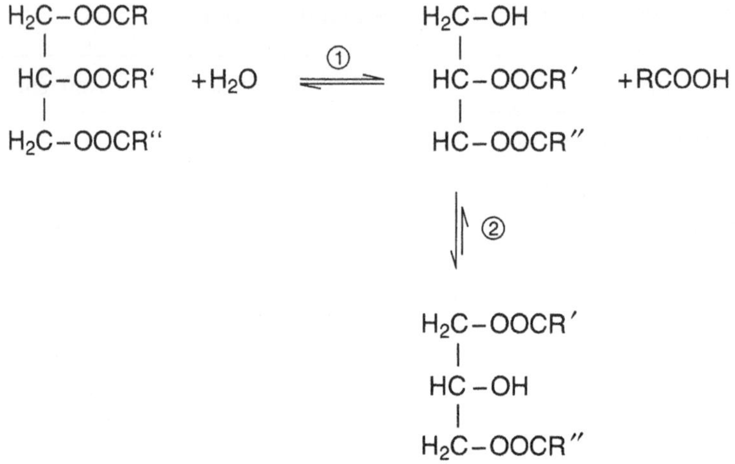

$$
\begin{array}{ccccc}
\text{H}_2\text{C}-\text{OOCR} & & & \text{H}_2\text{C}-\text{OH} & \\
| & & & | & \\
\text{HC}-\text{OOCR}' & +\text{H}_2\text{O} & \xrightleftharpoons{\;①\;} & \text{HC}-\text{OOCR}' & +\text{RCOOH} \\
| & & & | & \\
\text{H}_2\text{C}-\text{OOCR}'' & & & \text{HC}-\text{OOCR}'' &
\end{array}
$$

$$
\Big\Updownarrow ②
$$

$$
\begin{array}{c}
\text{H}_2\text{C}-\text{OOCR}' \\
| \\
\text{HC}-\text{OH} \\
| \\
\text{H}_2\text{C}-\text{OOCR}''
\end{array}
$$

Figure 1. Hydrolysis of triglycerides by lipases. Only esters in the 1- and 3-positions of glycerol are hydrolysed by these enzymes.

and warming (41, 42).

A low water content favours the reverse reaction, the esterification of glycerol. Lipases also tolerate other nucleophilic acceptor molecules than water, for instance alcohols, amines and thioesters. Use of alcohols in place of water results in transesterification (Figure 2).

$$
\text{R}-\text{COOR}' + \text{HO}-\text{R}'' \rightleftharpoons \text{R}-\text{COOR}'' + \text{HO}-\text{R}'
$$

$$
\text{R}'-\text{COOR} + \text{HOOC}-\text{R}'' \rightleftharpoons \text{R}''-\text{COOR} + \text{R}'-\text{COOH}
$$

Figure 2. Transesterification catalysed by lipases.

For good yields, the reaction must be performed under nearly anhydrous conditions, either with the alcohol as solvent or in another appropriate organic medium. To maintain catalytic activity, lipases must contain trace amounts of water. With nearly dry enzyme powder, both porcine pancreas lipase and *Candida cylindracea* lipase were highly active in heptanol–tributyrin at 100 °C (43).

Under given experimental conditions, the amount of water in the reaction mixture will determine the direction of the lipase-catalysed reaction. With water absent, or present in trace quantities only, esterification and transesterification are favoured; with excess water, hydrolysis occurs.

On the laboratory scale, the most useful property of lipases is their regio- and stereospecificity, resulting in products with a better-defined chemical composition than those obtained by chemical hydrolysis. Fatty-acid specificity in the cleavage of triglycerides is observed for some lipases, *e.g.*, that from *Geotrichum candidum*, which is highly specific for *cis*-9-unsaturated fatty-acid esters of oleic and lineic acids. A detailed study, with structurally and configurationally well-defined synthetic substrates such as mixed dialkyl/monoacyl glycerides, was undertaken with *S. aureus* lipases (28, 29). A remarkable stereospecificity was found for the hydrolysis of *sn*-1- and *sn*-3- oleic acid esters of glycerol, for which the respective hydrolysis rates were in the ratio 10:1.

The positional specificity of *Aspergillus niger* and *Rhizopus delemar* lipases also applies in the synthesis of triacylglycerols (44). Whereas *A. niger* lipase catalyses the esterification of glycerol with maleic and succinic acids, *G. candidum* lipases catalyse preferentially the reactions of long-chain fatty acid esters. All lipases tested resulted in the formation of oleic acid esters with primary alcohols. *G. candidum* lipase is also active with secondary alcohols as acceptor molecules. Tertiary alcohols are not substrates for any of these lipases (44). This has several possible reasons; the main one is probably steric hindrance of the formation of the enzyme–substrate complex.

In the absence of water, or in systems with limited amounts of water, properties of lipases such as catalytic rate and selectivity depend strongly on solvents and incubation conditions. Details of the kinetic parameters of lipase-catalysed reactions are reviewed by Chen and Sih (4). Since most research on the physical properties and substrate specificity of lipases has been carried out under physiological conditions (45, 46), results from aqueous systems cannot be applied directly to lipase catalysis in organic solvents. Furthermore, the use of lipases in organic synthesis is limited to the enzymes that are stable in organic solvents. Physico-chemical factors influencing lipase activity have recently been reviewed by John and Abraham (13). Because of the crucial rôle of water in lipase-catalysed reactions, the examples below are presented in sections dealing respectively with lipase catalysis in aqueous media and in organic solvents.

1.1 Bioconversions in aqueous media

A typical lipase-catalysed reaction in aqueous media is ester hydrolysis. This enzymic conversion can be used for the synthesis of triglyceride and phospholipid, as shown for the preparation of platelet-activating factor in section 2 (Figure 6). Another application of the hydrolytic specificity of lipases (14) is the partial hydrolysis of triglycerides to di- and monoglycerides in the food industry. Di- and monoglycerides serve as biocompatible emulsifiers and food additives. Other applications of lipases in industry and research are summarised in Table 2. This subject is discussed in the review by Iwai and Tsujisaka (15).

Table 2. Some industrial applications of lipase-catalysed hydrolysis and transesterifications.

Industry / application	Process / product
Dairy	Hydrolysis of milk fat
Natural oils	Transesterifications of oils for fats (formulated cocoa butter)
Bakery, brewery, food technology in general	Improvement of flavour and quality in beverages, meat, fish products *etc.*
Leather production	Removal of fats from skins
Detergent industry	Removal of oil stains and lipid spots

1.2 Bioconversions in organic media

The synthetic potential of lipases in organic solvents has been widely recognised and is documented in several publications. An important prerequisite for this development was the recognition that lipases work in organic solvents with low water content (5, 48). The main application of lipases in organic chemistry is the resolution of enantiomeric compounds, making use of the enantioselectivity of these enzymes. The theoretical background of kinetic resolution is described elsewhere (49, 50). The use of organic solvents for lipase-catalysed resolutions has four main advantages in comparison with water as solvent.

a) Racemic mixtures of alcohols or acids need not be esterified before resolution into enantiomers.

b) The chosen enzymes are stabler in organic solvents than in water.

c) Lipases need not be immobilised for recovery, owing to their insolubility in organic solvents; they can be collected by filtration in their active state.

d) Furthermore, substrates and products may be unstable in aqueous solution; in this case, reaction in organic solvents is essential for formation and isolation of the product.

Resolution of racemic alcohols

Racemic alcohols in non-aqueous media can be resolved in a biphasic system with lipase from *C. cylindracea* (51). Aqueous lipase solutions, together with tributyrin, were absorbed by the pores of Sepharose or Chromosorb gels and suspended in tributyrin saturated with phosphate buffer at the desired pH value:

Tributyrin + HO–CHRR′ → CH₃CH₂CH₂COO-CHRR′ + dibutyrin

Transesterification was performed with primary (R = H) and secondary (R = CH₃) racemic alcohols such as 1-chloro-2-propanol (R = CH₃, R′ = CH₂Cl) and 2,3-dichloropropanol (R = H, R′ = CH(Cl)CH₂Cl), respectively. In both cases, only the *R*-isomer served as acceptor of butyric acid, leading to the formation of (*R*)-butyric acid esters, which are easily separated from the remaining *S*-alcohols by vacuum distillation. Epoxidation of *R*- and of *S*-alcohols leads to useful chiral intermediates, *e.g.*, for *S*-sulcatol (52) or *R*-(–)- and *S*-(+)- epichlorohydrin for pharmaceuticals (53).

Simpler transesterification reactions were performed in diethyl ether or heptane as solvent (54) with porcine pancreas lipase as catalyst. Porcine pancreas lipase is a very stereoselective enzyme under these conditions. However, the authors report that the enzyme also tolerates paraffins, toluene, CCl₄, acetone, acetonitrile and dioxan as anhydrous solvents. Even small amounts of water suppress the transesterification reaction and cause hydrolysis to predominate. Detailed studies on the influence of the solvents upon lipases from *C. cylindracea*, *M. miehei* and porcine pancreas are summarised in references 4, 48 and 55. These enzymes have a broad range of application and can be used at temperatures between 25 and 60°C.

Transesterification was also used to prepare useful synthons, for instance optically active monoesters of 3-hydroxyglutarate and related compounds (56). Large-scale production of the pure, biologically active *S*-enantiomer of 6-methylhept-5-en-2-ol (sulcatol), which is the aggregation pheromone of an ambrosia beetle, is of crucial importance for the mass trapping of this insect. The transesterification catalysed by a lipase seems to be the most satisfactory approach for the synthesis of enantiomerically pure sulcatol (57). Porcine pancreas lipase was the enzyme of choice, with enol esters as acylating agents (58). The enantioselectivity of the porcine pancreas lipase used was affected strongly by the water content of the enzyme preparation. Dehydration of the enzyme leads to a doubling of the enantiomeric ratio (49). Use of a carefully selected starting ester in the transesterification of sulcatol, such as trifluoroethyl laurate, not only increases the enantiomeric ratio yet further, but also facilitates the separation of the reaction products (+)-(*S*)-sulcatol and (–)-(*R*)-sulcatol laurate (57).

Because the leaving enols in the transesterification tautomerise to aldehydes or ketones, respectively, they shift the equilibrium in the desired direction (59). A practical example of this, the resolution of cyanohydrins with lipoprotein lipase of *Pseudomonas* species (60), is shown in Figure 3.

The optical purities were 98% and 96% for 1a and 2a, respectively. The transesterification was performed in dichloromethane, in which the cyanohydrins

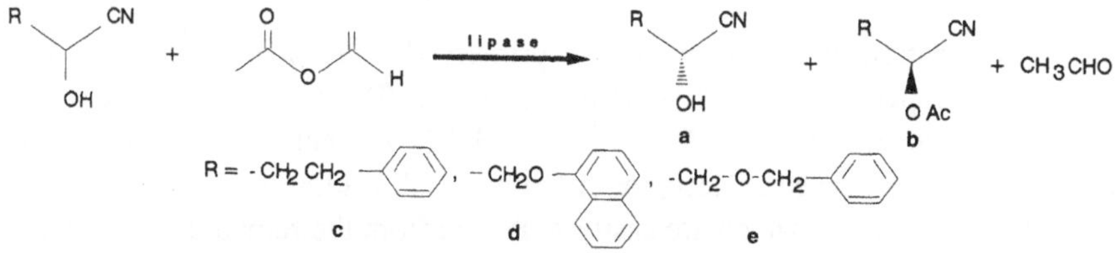

Figure 3. Enantioselective transesterification of cyanohydrin compounds.

show chiral stability. Decomposition of the reactants in water was the rationale for the enzymic transesterification of ferrocenylethanol with *Pseudomonas fluorescens* lipase in benzene (61):

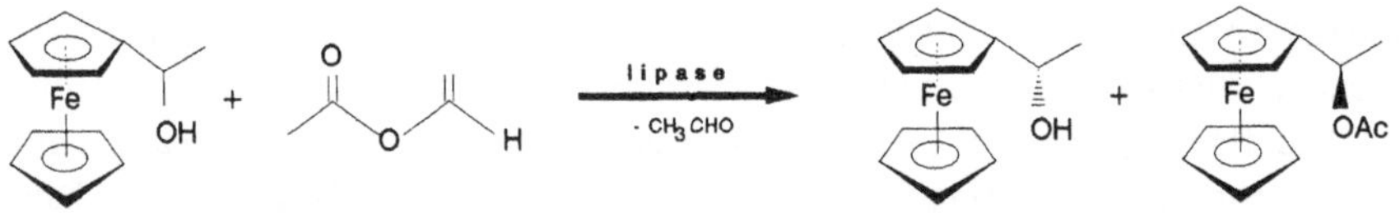

α-Substituted cyclohexanols, including menthol, were resolved with *C. cylindracea* lipase in heptane. The optical purity of the enantiomer varied between 80% and 100% (62, 63). The enzymic hydrolysis in water or transesterification in organic solvents yield complementary stereoisomers, which is shown with cycloalkanediol acetates, chiral intermediates in the synthesis of carbacyclins (64):

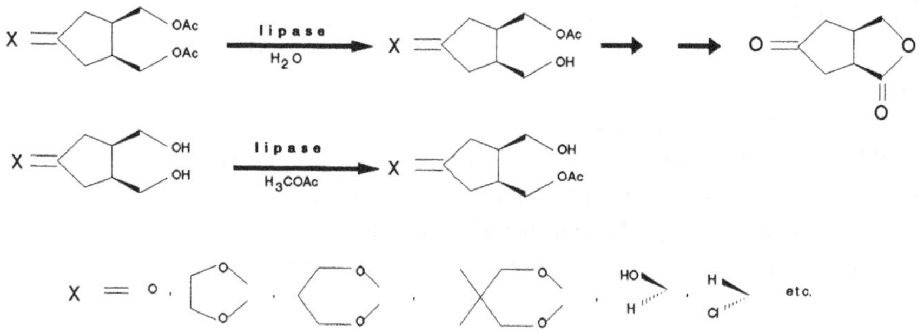

Chiral epoxy alcohols can be made by the hydrolysis of racemic epoxy esters by porcine pancreatic lipase. An enantiomeric excess (e.e.) of 90% was reported (65):

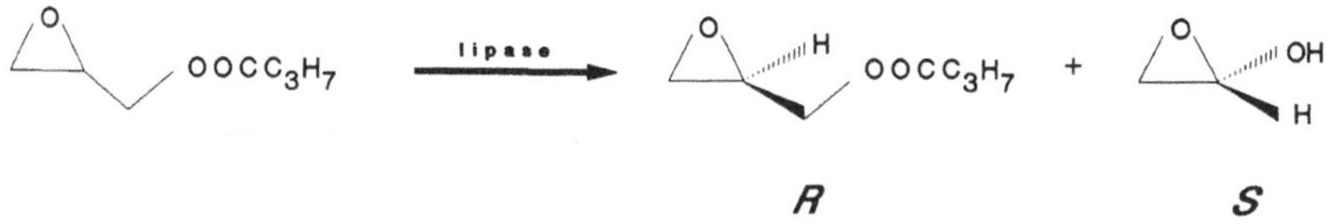

The resolution of alkynyl esters and alcohols has been described for *Bacillus subtilis* lipase (66):

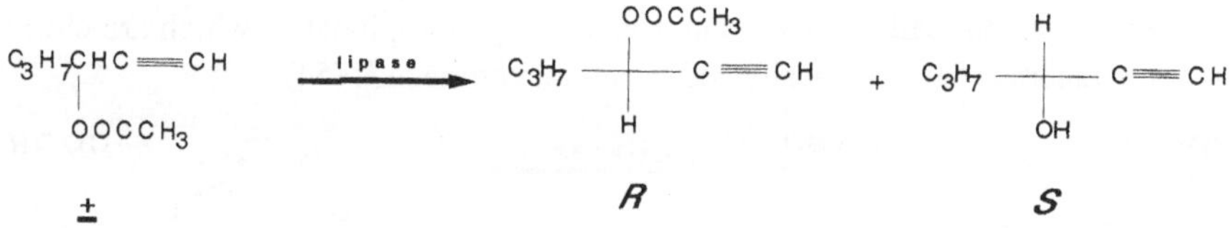

Other applications of lipases for the preparation of chiral intermediates by the resolution of chiral alcohols are described in the literature (3, 67).

Resolution of racemic acids by asymmetric esterification and transesterification

Another application of lipases is the use of their stereoselectivity for the resolution of racemic acid mixtures, as demonstrated by Zaks and Klibanov (51) in immiscible biphasic systems. However, one year later the same group published an improved method in nearly anhydrous organic medium (54). Reactions similar to the equation below, where R varied from methyl to tetradecyl, X was Cl or Br, and R'OH was *n*-butanol, were performed in *n*-hexane and *n*-hexane-diethyl ether mixtures with lipases from porcine pancreas or, better, from *C. cylindracea*. Vigorously shaken suspensions were used because of the insolubility of the enzymes in organic solvents. The reaction resulted in (*R*)- esters with an e.e. of 95–98%. Consequently, (*S*)-α-halogen fatty acid remained unesterified. It is remarkable that the loss of catalytic activity after each application was less than 10%.

$$(rac)\text{-RCHX-COOH} + \text{R'OH} \rightarrow (R)\text{-RCHX-COOR'} + (S)\text{-RCHX-COOR'}$$

Lipase from *C. cylindracea* has also been applied to the resolution of 2-bromo- and 2-chloropropionic acids, which are the starting materials for the synthesis of phenoxypropionic herbicides (54):

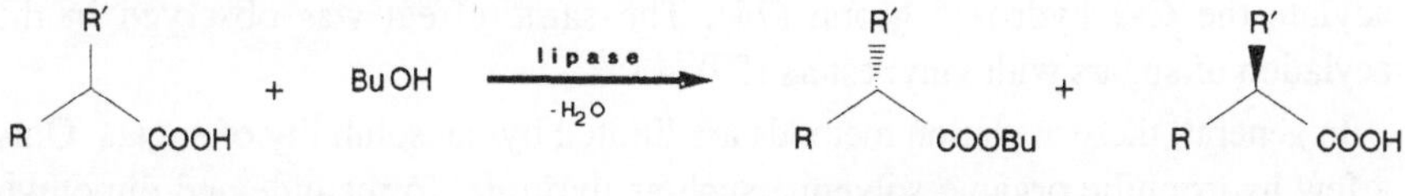

$$R = Cl, Br, p\text{-}ClC_6H_4O, \quad R' = CH_3, CH_3(CH_2)_3, CH_3(CH_2)_{13}, Ph$$

Pseudomonas fluorescens lipase was used for asymmetric ring opening of substituted prochiral glutaric anhydrides of dicarboxylic acids (68):

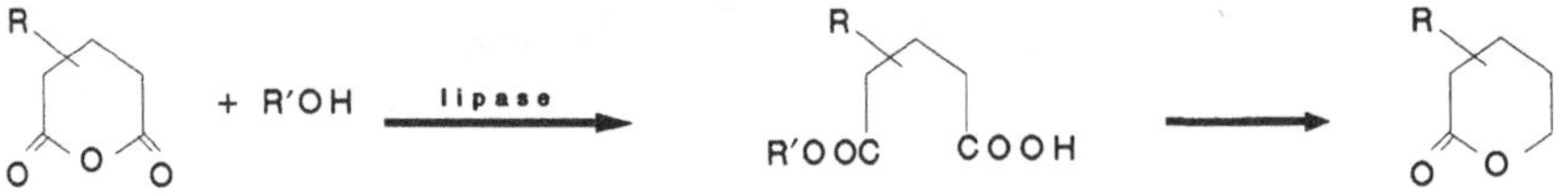

R = 3-methyl, 3-ethyl, 3-propyl, 3-isopropyl

The reaction was performed in toluene or, with better results, in isopropyl ether. With porcine pancreatic lipase (69) demonstrated stereoselective lactonisations and polycondensations from prochiral hydroxy diester, γ-hydroxypimelates were converted into optically active γ-butyrolactone γ-3-propionates, which are chiral starting materials for biologically active natural compounds (54):

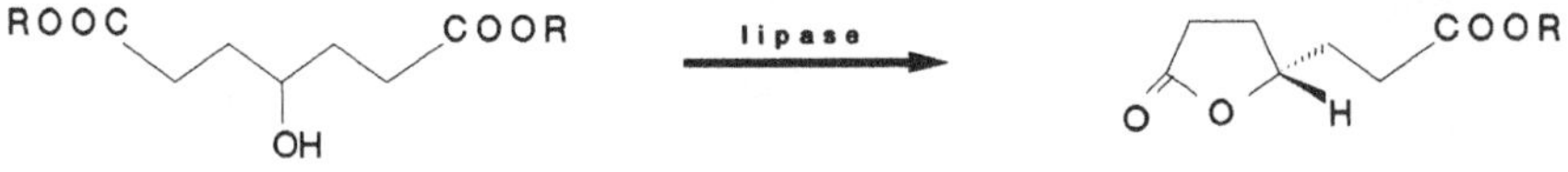

R = H, -C$_2$H$_5$, -benzyl, -isopropyl

Macrocyclic lactones such as (–)-pyrenophorin can be synthesized stereo-selectively by *Pseudomonas* species lipase (70) from (ω–1)-hydroxy alkanoic esters in anhydrous isooctane at 65 °C:

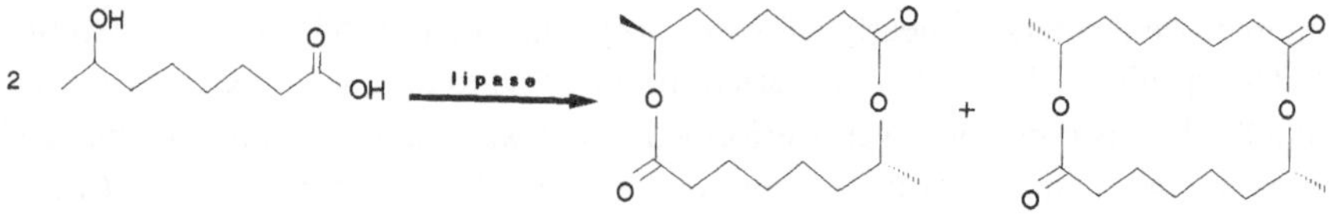

R,S   *R,R*

Although the yield was low for *R,S* isomers (5%) and *R,R/S,S* isomers (12%), the enantioselectivity of the lipase was high, with an *R,R:S,S* ratio of 96:4.

Regioselective acylations

Lipase-catalysed acylations are not limited to simple alcohols. Lipases also acylate with high regioselectivity certain steroids, sugars and sugar derivatives (71–73); for example, monoacylated sugars have been produced in anhydrous pyridine from triethyl carboxylates and various monosaccharides. Some of the lipases acylate exclusively the C-3 hydroxyl group in C-6-protected glucose, while others acylate the C-2 hydroxyl group (74). The same effect was observed in the acylation of sugars with vinyl esters (59).

In general, these acylation methods are limited by the solubility of sugars. Only a few hydrophilic organic solvents, such as pyridine, formamide and dimethylformamide, can be used. Among the commercially available lipases (6, 72), only lipases from porcine pancreas and *Chromobacterium viscosum* are active in pyridine.

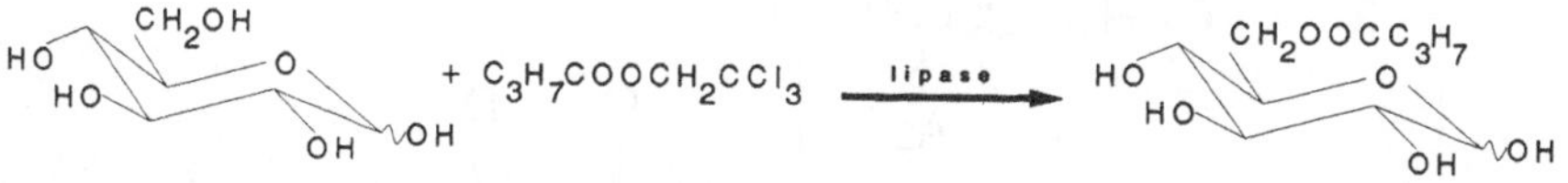

The regioselective acylation of hydroxy groups in glycals, which are versatile chiral intermediates, is catalysed by lipases. The transesterification of selected hydroxy groups was successful with *C. cylindracea* and *P. fluorescens* lipase in vinyl acetate (75) with high yields (>90%):

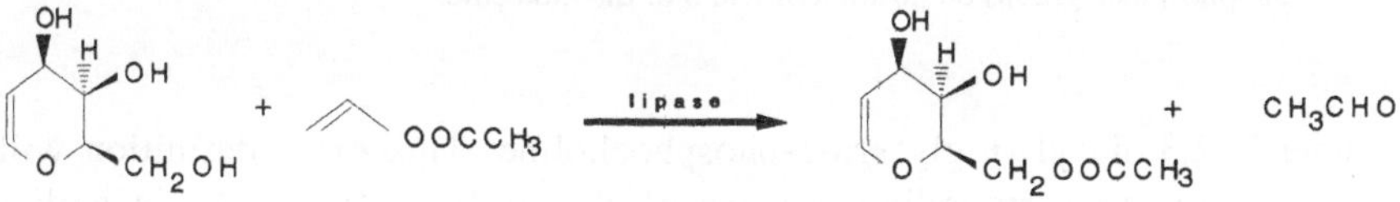

Another example of the high regiospecificity of lipases in acylation reactions in anhydrous organic solvents is shown for hydroxy steroids:

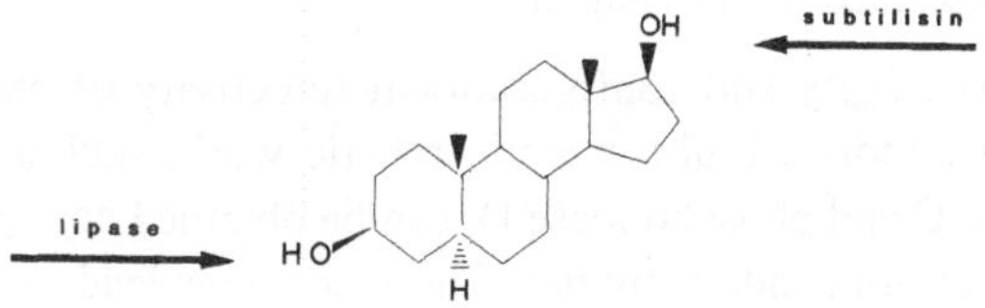

Several lipases are able to discriminate between the secondary hydroxyl groups in the A ring and the D ring of steroids, *e.g.*, regioselective acylation by lipase from *C. viscosum* was used for many steroid structures (76) as protecting groups (77).

2. PHOSPHOLIPASES

Phospholipases are enzymes that hydrolyse phospholipids at specific positions of the glycerol molecule. In this discussion, the standard *sn* nomenclature for lipids and phospholipids introduced by Hirschmann (78) is used. The chemical structure of lipids and phospholipids is based on glycerol. The 1- and the 2- positions of glycerol in naturally occurring phospholipids are usually esterified with fatty acids and phosphate is bound to the 3- position. As shown in Figure 4, if the substituents in the positions 1 and 3 are different, this specific distribution will result in enantiomers because of the asymmetry of carbon atom 2 in glycerols. According to the recommendations for the nomenclature of lipids proposed by the IUPAC-IUB Commission (79), the systematic name for natural phosphatidyl-choline is 1,2-diacyl-*sn*-glycero-3-phosphocholine, and that of the configurational

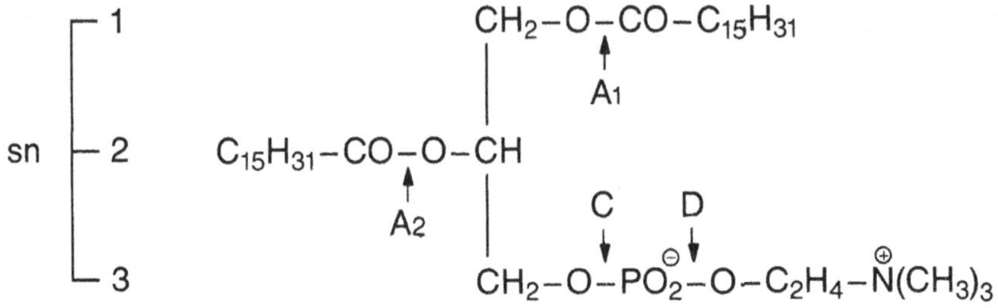

$$sn \begin{cases} 1 \\ 2 \\ 3 \end{cases}$$

$$CH_2-O-CO-C_{15}H_{31}$$
$$\uparrow$$
$$A_1$$

$$C_{15}H_{31}-CO-O-CH$$
$$\uparrow$$
$$A_2$$

$$C \quad D$$
$$\downarrow \quad \downarrow$$
$$CH_2-O-\overset{\ominus}{PO_2}-O-C_2H_4-\overset{\oplus}{N}(CH_3)_3$$

Figure 4. Structure of 1,2-dipalmitoyl-*sn*-glycero-3-phosphocholine. This is a typical substrate for phospholipases, as indicated by A1, A2, C and D. The opposite enantiomer is not found in nature; schematically, it is obtained by exchanging the acyl and phosphate groups on positions 1 and 3 of the molecule.

isomer is 2,3-diacyl-*sn*-glycero-1-phosphocholine. This exact definition will avoid confusion surrounding the use of the prefixes D- and L- in earlier publications. If racemic mixtures are described, the prefix *sn*- before glycerol is replaced by the prefix *rac*. Figure 4 shows the sites of attack of the different phospholipases described in this chapter.

The positional specificity and configurational selectivity of phospholipases is high. The enzymes most useful for semisynthetic work, such as phospholipase A$_2$, phospholipase C and phospholipase D, can be obtained commercially in high purity with no interfering side-activities. These enzymes lead to relatively stable end-products or to intermediates that can be used in subsequent reactions without intramolecular rearrangement, avoiding isomeric product mixtures that cannot be resolved by chromatography.

This is not the case for phospholipase A$_1$. Beside the fact that this enzyme is not available commercially, its application in semisynthetic routes for the preparation of lipids is questionable. For instance, the hydrolysis of 1,2-dipalmitoyl-*sn*-glycero-3-phosphocholine by phospholipase A$_1$ gives 2-palmitoyl-*sn*-glycero-3-phosphocholine. This represents an unstable intermediate, owing to the tendency of fatty-acid esters to migrate from position *sn*-2 to position *sn*-1 of glycerol (41), which occurs because fatty-acid esters of primary alcohols are thermodynamically stabler than those of secondary alcohols. Therefore, we prefer intermediates with fatty acid esters at the *sn*-1 position, which are obtained by hydrolysis with phospholipase A$_2$.

In general, the use of phospholipases for the preparation of lipids or phospholipids is a combination of techniques provided by organic chemists and biochemists. Optimal strategies require continual adjustment to new developments in both fields. Recently, for instance, 1,2-isopropylidene-*sn*-glycerol and also the

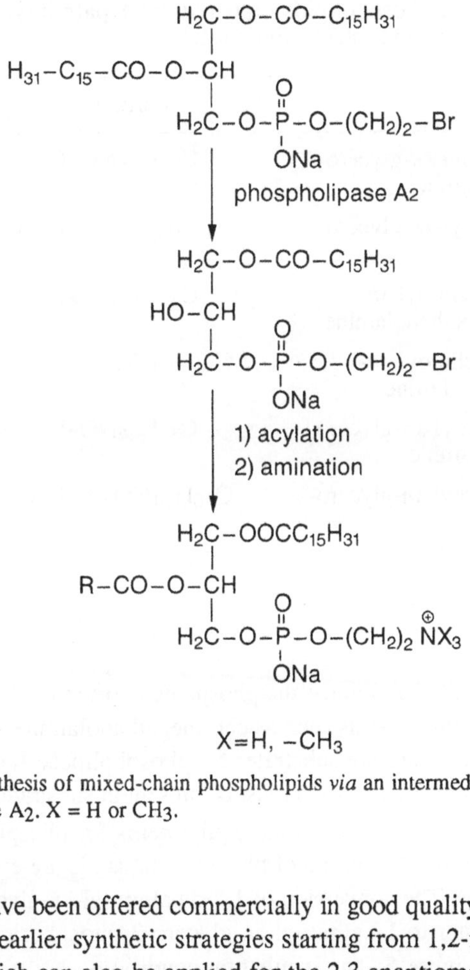

$$H_2C-O-CO-C_{15}H_{31}$$
$$|$$
$$H_{31}-C_{15}-CO-O-CH$$
$$|\qquad O$$
$$\qquad\qquad ||$$
$$H_2C-O-P-O-(CH_2)_2-Br$$
$$|$$
$$ONa$$

phospholipase A2

$$\downarrow$$

$$H_2C-O-CO-C_{15}H_{31}$$
$$|$$
$$HO-CH$$
$$|\qquad O$$
$$\qquad\qquad ||$$
$$H_2C-O-P-O-(CH_2)_2-Br$$
$$|$$
$$ONa$$

1) acylation
2) amination

$$\downarrow$$

$$H_2C-OOCC_{15}H_{31}$$
$$|$$
$$R-CO-O-CH$$
$$|\qquad O$$
$$\qquad\qquad ||$$
$$H_2C-O-P-O-(CH_2)_2\ \overset{\oplus}{N}X_3$$
$$|$$
$$ONa$$

$$X=H,\ -CH_3$$

Figure 5. Synthesis of mixed-chain phospholipids *via* an intermediate prepared with phospholipase A$_2$. X = H or CH$_3$.

2,3 enantiomer have been offered commercially in good quality and at reasonable cost. This makes earlier synthetic strategies starting from 1,2-isopropylidene-*sn*-glycerol (80), which can also be applied for the 2,3 enantiomer, more important for lipid synthesis than those starting from D-mannitol (81).

2.1 Phospholipase A$_2$

The substrate specificity of phospholipase A$_2$ has been studied extensively (82, 83). This experience has already been applied to the stereospecific and position-selective hydrolysis of phosphatidylcholines and of analogues by phospholipase A$_2$, to generate versatile intermediates for the synthesis of new phospholipid molecules of well-defined structure and configuration. According to these pioneering investigations, the structural requirements for phospholipase A$_2$ hydrolysis are phosphate in the *sn*-3 position of glycerol and an ester bond in the *sn*-2 position. Better substrates than phosphate monoesters are diesters. In

Table 3. Mixed-chain phospholipids prepared via 1-palmitoyl-sn-glycero-3-phosphoric acid bromoethyl ester, sodium salt.

Compound	Formula	Mol. wt.
1-palmitoyl-2-oleoyl-sn-glycero-3-phosphoethanolamine	$C_{39}H_{76}NO_8P$	718.00
1-palmitoyl-2-oleoyl-sn-glycero-3-phosphocholine	$C_{42}H_{82}NO_8P \cdot H_2O$	778.10
1-palmitoyl-2-arachinoyl-sn-glycero-3-phosphoethanolamine	$C_{41}H_{82}NO_8P$	748.07
1-palmitoyl-2-arachinoyl-sn-glycero-3-phosphocholine	$C_{44}H_{88}NO_8P \cdot H_2O$	808.17
1-palmitoyl-2-lauroyl-sn-glycero-3-phosphoethanolamine	$C_{33}H_{66}NO_8P$	635.86
1-palmitoyl-2-lauroyl-sn-glycero-3-phosphocholine	$C_{36}H_{74}NO_8P \cdot H_2O$	695.95

naturally occurring phospholipids, the phosphate is often doubly esterified, both with glycerol and with alcohols such as choline, ethanolamine, a second glycerol etc. Phosphate triesters are not substrates for phospholipase A_2, indicating that a negatively charged phosphate group must be present in substrate molecules.

In line with these minimal substrate requirements for phospholipase A_2 is the observation that bromoalkyl esters of phosphatidic acids are excellent substrates for this enzyme (84). The conversion of 1,2-dipalmitoyl-sn-glycero-3-phosphoric acid bromoethyl ester to 1-palmitoyl-sn-glycero-3-phosphoric acid bromoethyl ester is shown in Figure 5. The incubation conditions as a typical example for phospholipase A_2 hydrolysis are given in Appendix 4.2.

The versatility of this synthetic route for the preparation of a large variety of phospholipid molecules is demonstrated in Table 3. This contains some of the conversions that have been realised experimentally in our laboratory. For instance, 1-palmitoyl-2-oleoyl-sn-glycero-3-phosphocholine was obtained from the intermediate 1-palmitoyl-sn-glycero-3-phosphoric acid bromoethyl ester by acylation with oleoyl chloride in the presence of triethylamine/4-dimethylamino-pyridine and subsequent amination with trimethylamine (84).

Other molecular species of phospholipids may be obtained from intermediates that vary in the fatty acid attached to position sn-1 of glycerol, such as the bromoethyl esters of 1-stearoyl-, 1-myristoyl-, 1-oleoyl- or 1-linoleoyl-sn-glycero-3-phosphoric acids. After acylation with different acyl chlorides and subsequent amination with amines such as trimethylamine, dimethylamine, methylamine, ammonia *etc.*, phospholipids can be obtained with natural or unnatural fatty-acid substitution patterns at the *sn*-1 and *sn*-2 positions of glycerol.

In addition, for the synthesis of mixed-chain phosphatidylcholines, relatively cheap and configurationally pure starting materials such as 1,2-dipalmitoyl-*sn*-glycero-3-phosphocholine and 1,2-dipalmitoyl-*sn*-glycero-3-phospho-(*N,N*-dimethyl)-ethanolamine are now commercially available. Both are good substrates for phospholipase A$_2$, and both intermediates can be converted to the respective mixed-chain phosphatidylcholines or phosphatidyl-(*N,N*-dimethyl)-ethanolamines according to the reaction scheme shown in Figure 5 for phosphoric acid bromoethyl esters.

In summary, phospholipase A$_2$ is useful for the preparation of 1-acyl-*sn*-glycero-3-phospho-bromoalkyl esters, -phosphoalkyl esters and -phosphocholines. One general advantage of such a semisynthetic approach, using an enzyme with high positional and configurational specificity, is the fact that configurational isomers are not interconverted, and enzymic hydrolysis will thus lead, even with enantiomerically impure starting materials, to products with a configurational and positional purity above 99%. Since optically pure starting materials are now available, it is of no advantage to start with racemic educts. According to the reaction scheme depicted in Figure 5, the intermediate 1-acyl-compounds can be converted to mixed-chain phospholipids in high yield. The products of these reactions can be modified further by hydrolysis with phospholipase C or phospholipase D. In addition, transesterification with phospholipase D will result in a cascade of new products, as described below.

2.2 Phospholipase C

Since the products of hydrolysis by phospholipase C are important molecules in the signal chain, the substrate properties of phospholipids for phospholipase C are the objects of intensive study. We describe elsewhere the substrate requirements of phospholipase C (Chapter 11). In short, there are three structural conditions for optimal activity: phosphocholine attached to the *sn*-3-position of the glycerol, an ester bond in the *sn*-2 and an ether- or an ester bond in the *sn*-1 positions. The length of the ester or ether chains is of minor importance if at least one of the substituents is longer than 14 carbon atoms. Any deviation from these structural requirements results in a large decrease in the rate of hydrolysis. These essential structural and configurational requirements for optimal activity of the phospholipase C from *Bacillus cereus* are combined perfectly in PAF (platelet-activating factor, 1-O-hexadecyl-2-acetyl-*sn*-glycero-3-phosphocholine). This molecule is one of the best substrates for hydrolysis by bacterial phospholipase C (22, 85).

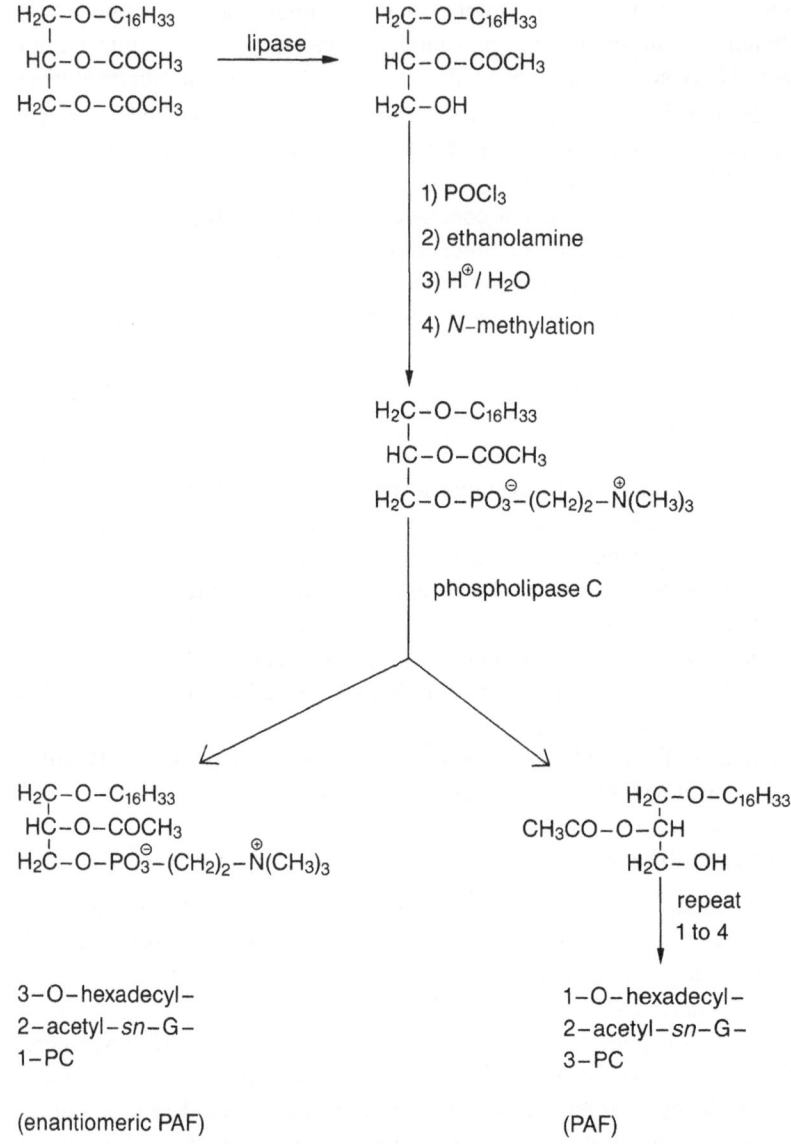

Figure 6. Synthesis of platelet-activating factor from racemic starting materials with lipase and enantiomeric resolution by phospholipase C.

The advantage of phospholipase C is its strong stereospecificity combined with extreme structural selectivity. For instance, platelet-activating factor is a molecule based on glycerol with 3 residues – hexadecyl or octadecyl, acetyl and phosphocholine – attached to the hydroxyl groups of glycerol. Of the six possible isomers, only the biologically active and naturally occurring isomer, 1-O-hexadecyl-2-acetyl-*sn*-glycero-3-phosphocholine, is a substrate for phospholipase C.

If both enantiomers of a compound are desired, for a comparison of the biological activity, then a synthetic approach using phospholipase C can provide both enantiomers from the racemate. In the case of platelet-activating factor, this can be achieved by the sequence of reactions described in Figure 6.

1-O-hexadecyl-*rac*-glycerol is acylated with acetic anhydride in the presence of triethylamine and 4-dimethylaminopyridine to yield 1-O-hexadecyl-*rac*-glycero-2,3-diacetate, which is converted by triglyceride lipase (step 1) to 1-O-hexadecyl-*rac*-glycero-2-acetate. This intermediate is converted to the respective phosphocholine by a sequence of reactions (step 2) including phosphorylation with phosphorus oxychloride, oxazaphospholane ring formation with ethanolamine, ring scission at acidic pH (86) and N-permethylation with dimethyl sulphate (87). In step 3, the racemate is resolved into enantiomeric structures, as 3-O-hexadecyl-2-acetyl-*sn*-glycero-1-phosphocholine is not a substrate for phospholipase C. Thus, only 1-O-hexadecyl-2-acetyl-*sn*-glycero-3-phosphocholine is converted to 1-O-hexadecyl-2-acetyl-*sn*-glycerol, which is easily extracted in the ether phase of the incubation mixture at pH 6.5 (see Appendix 4.3 for details). Then, PAF of natural configuration is obtained from 1-O-hexadecyl-2-acetyl-*sn*-glycerol in the sequence of reactions (Figure 6, 1–4) already described for the conversion of the respective racemic compound, 1-O-hexadecyl-2-acetyl-*rac*-glycerol.

If only one isomer is desired, the sequence of reactions described in Figure 6 can also be used. Instead of the racemate, the synthesis is started with 1-O-hexadecyl-*sn*-glycero-2,3-diacetate for the generation of platelet-activating factor or with 3-O-hexadecyl-*sn*-glycero-1,2-diacetate for the preparation of the enantiomer. In these cases, hydrolysis by phospholipase C is not required. However, phospholipase C may be used for the confirmation of structure and configuration.

In summary, hydrolysis by phospholipase C can be used for the resolution of racemic mixtures of phospholipid molecules. A great variety of phospholipids can be prepared starting from 1-O-alkyl-2-acyl-*rac*-glycerols or 1-acyl-2-acyl-*rac*-glycerols. After phosphorylation and phospholipase C hydrolysis, the phospholipids with unnatural configuration are obtained directly together with 1-O-alkyl-2-acyl-*sn*-glycerols or 1,2-diacyl-*sn*-glycerols. The optically pure 1,2-diacyl-glycerols are important messengers in signal chains and activators of protein kinase C. They are also starting products for the synthesis of phospholipids with natural configuration.

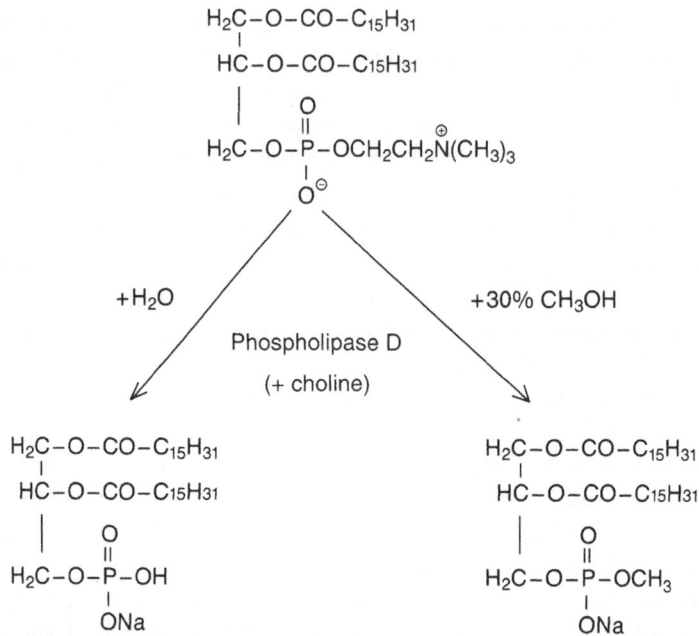

Figure 7. Interconversion of phospholipids with phospholipase D in water or in solutions of primary alcohols.

2.3 Phospholipase D

Phospholipase D (EC 3.1.4.4) is a hydrolytic enzyme discovered in extracts of cabbage leaves (88). At least the enzyme from white cabbage differs from the phospholipases described above. It is less specific. For instance, it accepts as substrates phospholipids with either the natural or the unnatural configuration, and the rates for single-chain and double-chain molecules are not very different (89). It is an ideal enzyme for the interconversion of phospholipids irrespective of their configuration. We prefer to prepare the fresh enzyme from white cabbage according to a fast and simple method (89), which is described in Appendix 4.4.

A wide variety of different products can be obtained by such interconversions of phospholipids. Some examples are summarised in Table 4. In the presence of water, phosphatidic acids are the only products formed. If the amount of alcohol in the incubation mixture is 20–30% (v/v) of the buffer solution, then the respective alkyl esters are obtained by transesterification (Figure 7).

In Table 4, different interconversions of phospholipids are shown. Only primary alcohols are good acceptors for the phosphatidyl residue. If the alcohol

Table 4. Primary alkanols and their derivatives as acceptors for the phosphatidyl residue from egg phosphatidylcholine under standard incubation conditions (2 h, 20 °C).

Acceptor		Alkyl ester of phosphatidic acid (%)	Phosphatidic acid (%)
Alkanols	$x = 1-3$	100	0
HO–$(CH_2)_x$–H	$x = 4$	80	20
	$x = 5$	50	50
	$x = 6$	20	80
	$x = 7-10$	0	100
Alkandiols	$y = 2$ and 3	100	0
HO–$(CH_2)_y$–OH	$y = 4$	90	10
	$y = 5$	50	50
	$y = 6$ and 7	0	100
Unsaturated propanols	HO–CH_2–CH=CH_2	100	0
	HO–CH_2–C≡CH	100	0
Haloethanols	HO–$(CH_2)_2$–F	100	0
	–Cl	65	35
	–Br	30	70
	–I	0	100

molecule has a straight chain containing more than four carbon atoms, the alcohols do not serve as acceptors in the enzyme-catalysed reaction.

3. CONCLUDING REMARKS

Lipases are valuable tools for the synthesis of acyl derivatives of glycerol because they lead to specific products if one position in the glycerol (1 or 3) is already blocked by an alkyl chain. The hydrolysis then leads to a defined product, racemic or enantiomerically pure depending on the history of the starting material. After phosphorylation, racemates can be resolved into optical isomers by hydrolysis with phospholipase C.

Phospholipase A$_2$ is most useful for the preparation of mixed-chain phospholipids, while phospholipase C is important for the resolution of racemic mixtures into their enantiomers. In addition, phospholipase D can be used for the interconversion of phospholipid molecules: since its stereospecificity is low, phospholipids both of natural or unnatural configuration can be converted into phospholipids differing in the polar region of the molecule.

4. APPENDIX

4.1 Assay for lipase activity

Lipase activity can be assayed simply, by pH-stat titration (40). For instance, the substrate triolein (5 mM) is emulsified by sonication in 5% (w/v) gum arabic with 4 mM desoxycholic acid and 0.1 M NaCl, and the pH is adjusted to 8.0. Lipase (0.01–0.1 units) is added to 5 ml substrate emulsion and the mixture is stirred continuously at 30 °C. The oleic acid liberated by lipase action is estimated by continuous titration with 0.02 M NaOH from an automatic burette, while the pH is maintained at 8.0. Titration is continued for 2–10 min.

This method has the advantage that triolein can be replaced by a wide variety of other substrates. There are several other methods for the estimation of the liberated fatty acid. The substrate and the reaction products can be extracted from the incubation mixture by a methanol/chloroform mixture (1:2 v/v) and subsequently separated on thin-layer chromatography plates (hexane/diethyl ether, 1:1 v/v). The products formed are quantified densitometrically after staining with CuSO$_4$ (39).

4.2 Hydrolysis by phospholipase A$_2$

We describe the preparation of 1-palmitoyl-sn-glycero-3-phospho bromo ethyl ester, calcium salt. 1,2-dipalmitoyl-sn-glycero-3-phosphoric acid bromoethyl ester, sodium salt (0.01 mol; 8.0 g) was dissolved in a mixture of 300 ml diethyl ether and 300 ml distilled water that contained CaCl$_2$ (0.016 mol). Palitzsch buffer A (285 ml) was added, followed by 15 ml Palitzsch buffer B. The pH of the resulting emulsion was 7.5. After the addition of 10 mg phospholipase A$_2$, the reaction mixture was stirred for 60 min at 35 °C. As shown by thin-layer chromatography (in chloroform/methanol/acetic acid/water in the ratio 100:60:20:5 v/v), the starting material was transformed completely to the calcium salt of 1-palmitoyl-sn-glycero-3-phosphoric acid bromo ethyl ester and palmitic acid. Stirring was stopped, and the reaction mixture separated in two phases. The upper phase, which contained the product, was mixed with 100 ml toluene and evaporated to dryness (15 torr; 35 °C). The residue formed white crystals after the addition of 100 ml acetone. The yield was 93% (5.5 g).

4.3 Hydrolysis by phospholipase C

We describe the preparation of 1-O-hexadecyl-2-acetyl-*sn*-glycerol and of 3-O-hexadecyl-2-acetyl-*sn*-glycero-1-phosphocholine from the racemate. 1-O-hexadecyl-2-acetyl-*rac*-glycero-3-phosphocholine (0.01 mol; 5 g) was dissolved in a mixture of 100 ml diethyl ether and 100 ml 0.1 M Tris-maleate buffer containing 10 mmol $CaCl_2$ with the pH adjusted to 7.1. After the addition of 0.001 units phospholipase C from *B. cereus*, the mixture was stirred for 60 min at 25 °C. As shown by thin-layer chromatography (in pentane/diethyl ether in the ratio 1:1 v/v), 1-O-hexadecyl-2-acetyl-*sn*-glycerol was formed and extracted into the diethyl ether phase. It was collected by phase separation and isolated by evaporation to dryness. The yield was 83% (1.5 g).

Again, 100 ml diethyl ether was added to the lower phase. Stirring was continued for a further hour, but no product formation was observed, indicating that the water-soluble 3-O-hexadecyl-2-acetyl-*sn*-glycero-1-phosphocholine was pure enantiomeric PAF. It was extracted with chloroform/methanol, evaporated to dryness and precipitated with acetone. The yield was 80% (2.0 g).

4.4 Hydrolysis by phospholipase D

Enzyme preparation: white cabbage (500 g; stored at 5 °C) was cut into small pieces. After the addition of 50 ml distilled water (5 °C), the cabbage was homogenised in a Braun MX 32 household homogeniser. During this operation, the temperature should not exceed 15 °C. The yellow-green slurry was filtered by suction (normal paper filter; 15 torr) and the filtrate was collected. The volume of the extract was 200 ml. It was stirred at 5 °C and used within 48 h.

For the preparation of 1,2-dimyristoyl-*sn*-glycero-3-phosphoethanolamine, 0.01 mol (6.8 g) 1,2-dimyristoyl-*sn*-glycero-3-phosphocholine was dissolved in 150 ml diethyl ether and 50 ml buffer (0.2 M sodium acetate and 0.1 M $CaCl_2$). After the addition of ethanolamine hydrochloride (0.25 mol; 25 g) the pH was adjusted to 5.6 by the dropwise addition of acetic acid. Under continuous stirring, 50 ml enzyme extract (above) was added. Stirring was continued for 10 h. Diethyl ether was then removed by evaporation, and 1,2-dimyristoyl phosphatidyl ethanolamine was extracted with 150 ml $CHCl_3$ and 200 ml CH_3OH. The chloroform extract was evaporated to dryness, and the residue was precipitated by the addition of 300 ml acetone/methanol (1:1 v/v). The yield was 78% (5 g).

REFERENCES

1. Eliel, E.L. (1962). *Stereochemistry of Carbon Compounds*, p. 436. New York: McGraw-Hill.
2. Klibanov, A.M. (1990). Asymmetric transformations catalyzed by enzymes in organic solvents. *Accts Chem. Res.*, 23, 114–120.

3. Jones, J.B. (1986). Enzymes in organic synthesis. *Tetrahedron*, 42, 3351–3403.
4. Chen, C.-S. & Sih, C.J. (1989). Enantioselektive Biokatalyse in organischen Solventien am Beispiel Lipase-katalysierter Reaktionen. *Angew. Chem.*, 101, 711–724.
5. Klibanov, A.M. (1989). Enzymatic catalysis in anhydrous organic solvents. *Trends Biochem. Sci.*, 14, 141–144.
6. Zaks, A. (1990). Enzymes in organic solvents. In *Biocatalysts for Industry*, vol. 8, ed. Dordick, J.S., pp. 161–180. New York: Plenum Press.
7. Zaks, A. & Klibanov, A.M. (1985). Enzyme-catalyzed processes in organic solvents. *Proc. Natl. Acad. Sci. USA*, 82, 3192–3196.
8. Yamada, H. & Shimizu, S. (1988). Microbial and enzymatic processes for the production of biologically and chemically useful compounds. *Angew. Chem. Int. Ed.* 27, 622–642.
9. Jones, J.B. (1987) Some examples of enzymes in organic synthesis. *Annals N.Y. Acad. Sci.*, 501, 119–128.
10. Takahashi, K., Tamaura, Y., Kodera, Y., Mihama, T., Saito, Y. & Inada, Y. (1987). Magnetic Lipase Active in Organic Solvents. *Biochem. Biophys. Res. Commun.*, 142, 291–296.
11. Lavayre, J., Verrier, J. & Baratti, J. (1982). Stereospecific hydrolysis by soluble and immobilized lipases. *Biotechnol. Bioengin.*, 24, 2175–2187.
12. Macrae, A.R. (1983). Lipase-catalysed interesterification of oils and fats. *J. Am. Oil Chem. Soc.*, 60, 291–294.
13. John, V.T. & Abraham, G. (1990). Lipase catalysis and its applications. In *Biocatalysts for Industry*, vol. 10, ed. Dordick, J.S., pp. 193–127. New York, USA: Plenum Press.
14. Nielsen, T. (1985). Industrial application possibilities for lipase. *Fat Sci. Tech.*, 87, 15-19.
15. Iwai, M. & Tsujisaka, Y. (1984). Fungal lipase. In *Lipases*, ed. Borgström, B. & Brockman, H.L., pp 443–470. Amsterdam: Elsevier Science Publishers.
16. Olivecrona, T. & Bengtsson, G. (1984). Lipases in milk. Chap. IIE in *Lipases*, ed. Borgström, B. & Brockman, H.L., pp. 205–262. Amsterdam: Elsevier Science Publishers.
17. Ganong, B.R. (1991). Roles of lipid turnover in transmembrane signal transduction. *Am. J. Med. Sci.*, 302, 304–312.
18. Snyder, F. (1990). Platelet-activating factor and related acetylated lipids as potent biologically active cellular mediators. *Am. J. Physiol.*, 259, C697–C708.
19. Hanahan, D.J. (1986). Platelet activating factor: a biologically active phosphoglyceride. *Annu. Rev. Biochem.*, 55, 483–509.
20. Workman, P. (1991). Platelet activating factor. *Cancer Cells*, 4, 426–432.
21. Eibl, H. & Unger, C. (1990). Hexadecylphosphocholine: a new and selective antitumor drug. *Cancer Treat. Rev.*, 17, 233–242.
22. Kötting, J., Fleer, E.A.M., Unger, C. & Eibl, H. (1988). Synthetische Alkyllysophospholipide als Antitumormittel-Strukturverwandte des 'Platelet Activating Factors'. *Fat Sci. Technol.*, 90, 345–351.
23. Eibl, H., Hilgard, P. & Unger, C. (1992). *Alkylphosphocholines: New Drugs in Cancer Therapy*, (in the series *Progress in Experimental Tumour Research*, vol. 34), Basel: Karger Verlag.
24. Kaufmann-Kolle, P., Unger, C. & Eibl, H. (1992). Hexadecylphosphocholine in liposomal dispersions. In *Alkylphosphocholines: New Drugs in Cancer Therapy*, ed. Eibl, H., Hilgard, P. & Unger, C. (in the series *Progress in Experimental Tumour Research*, vol. 34), p.12–14. Basel: Karger Verlag.
25. Kranich, A., Kaufmann-Kolle, P., Unger, C. & Eibl, H. (1992) Doxorubicin liposomes: the influence of lipid composition on biodistribution in mice. *Lipids*, in the press.
26. Sonnet, P.E. (1989). Glycerol ethers, synthesis, configuration analysis, and a brief review of their lipase-catalyzed reactions. In *Biocatalysis in Agricultural Biotechnology*, ed. Whitaker, J.R. & Sonnet, P.E., pp. 269–277. Washington, DC, USA: American Chemical Society.

27. Kötting, J., Jürgens, D., Schiller, R. & Fehrenbach, F.-J. (1985). Biochemical and biological properties of *Staphylococcus aureus* lipases (EC 3.1.1.3). In *The* Staphylococci, ed. Jelijaszewicz, J., *Zbl. Bakt. Suppl.*, 14, 301–309. Heidelberg: Springer Verlag.
28. Kötting, J., Eibl, H. & Fehrenbach, F.-J. (1988). Substrate specificity of *S. aureus* (Ten5) lipases with isomeric oleoyl-*sn*-glycero ethers as substrates. *Chem. Phys. Lipids*, 47, 117–122.
29. Kötting, J., Jürgens, D. & Huser, H. (1983). Separation and characterisation of two isolated lipases from *Staphylococcus aureus* (Ten5). *J. Chromat.*, 281, 253–261.
30. Benzonana, G. & Esposito, S. (1971). On the positional and chain specificities of *Candida cylindracea* lipase. *Biochim. Biophys. Acta*, 231, 15–22.
31. Vadehra, D.V. (1974). Staphylococcal lipases. *Lipids*, 9, 158–163.
32. Benzonana, G. (1974). Some properties of an exocellular lipase fron *Rhizopus arrhizus*. *Lipids*, 9, 166–172.
33. Okumura, S., Iwai, M. & Tsujisaka, Y. (1981). The effect of reverse action on triglyceride hydrolysis by lipase. *Agric. Biol. Chem.*, 45, 185–189.
34. Sonnet, P.E. (1988). Lipase selectivities. *J. Am. Oil Chem. Soc.*, 65, 900–904.
35. Litchfield, C. (1972). *Analysis of Triglycerides*, pp. 173–174. New York, USA: Academic Press.
36. Paltauf, F., Esfandi, F. & Holasek, A. (1974). Stereospecificity of lipases. enzymatic hydrolysis of enantiomeric alkyl diacylglycerols by lipoprotein lipase, lingual lipase and pancreatic lipase. *FEBS Letters*, 40, 119–123.
37. Smith, L.C. & Pownall, H.J. (1984). Lipoprotein lipase. In *Lipases*, ed. Borgström, B. & Brockman, H.L., pp. 263–306. Amsterdam: Elsevier.
38. Brockerhoff, H. & Jensen, J.G. (1974). *Lipolytic Enzymes*. New York: Academic Press.
39. Kötting, J., Marschner, N.W., Unger, C. & Eibl, H. (1992). Determination of alkylphosphocholines in biological fluids and tissues. In *Alkylphosphocholines: New Drugs in Cancer Therapy*, ed. Eibl, H., Hilgard, P. & Unger, C. (in the series *Progress in Experimental Tumour Research*, vol. 34), pp.12–14. Basel: Karger Verlag.
40. Rick, W. (1969). Kinetischer Test zur Bestimmung der Serumlipaseaktivität. *Z. klin. Chem. klin. Biochem.*, 7, 530–539.
41. Mattson, F.A. & Volpenheim, R.A. (1962). Synthesis and properties of glycerides. *J. Lipid Res.*, 3, 281–296.
42. Eibl, H. & Lands, W.E.M. (1970). A new, spectrophotometric microdetermination of vicinal diols. *Anal. Biochem.*, 33, 58–66.
43. Zaks, A. & Klibanov, A.M. (1984). Enzymatic catalysis in organic media at 100°C. *Science*, 224, 1249–1251.
44. Tsujisaka, Y., Okumura, S. & Iwai, M. (1977). Glyceride synthesis by four kinds of microbial lipase. *Biochim. Biophys. Acta*, 489, 415–422.
45. Brockman, H.L. (1984). General features of lipolysis: reaction scheme, interfacial structure and experimental approaches. In *Lipases*, ed. Borgström, B. & Brockman, H.L., pp. 3–46. Amsterdam: Elsevier Science Publishers.
46. Rogalska, E., Ransac, S. & Verger, R. (1990). Stereoselectivity of lipases. *J. Biol. Chem.*, 265, 20271–20276.
47. Olivecrona, T. & Bengtsson, G. (1984). Lipases in milk. In *Lipases*, ed. Borgström, B. & Brockman, H.L., pp. 205–262. Amsterdam: Elsevier Science Publishers.
48. Goldberg, M., Thomas, D. & Legoy, M.-D. (1990). The control of lipase-catalysed transesterification and esterification reaction rates. *Eur. J. Biochem.*, 190, 603–609.
49. Chen, C.-S., Fujimoto, Y., Girdaukas, G. & Sih, C.J. (1982). Quantitative analyse of biochemical kinetic resolutions of enantiomers. *J. Am. Chem. Soc.*, 104, 7294–7299.
50. Martin, V.S., Woodard, S.S., Katsuki, T., Yamada, Y., Ikeda, M. & Sharpless, K.B. (1981). Kinetic resolution of racemic allylic alcohols by enantioselective epoxidation: a route to sustances of absolute enantiomeric purity. *J. Am. Chem. Soc.*, 103, 6237–6240.

51. Cambou, B. & Klibanov, A.M. (1984). Preparative production of optically active esters and alcohols using esterase-catalyzed stereospecific transesterification in organic media. *J. Am. Chem. Soc.*, 106, 2687–2692.
52. Belan, A., Bolte, J., Fauve, A., Gourcy, J.G. & Veschambre, H. (1987). Use of biological systems for the preparation of chiral molecules. 3. An application in pheromone synthesis: preparation of sulcatol enantiomers. *J. Org. Chem.*, 52, 256–260.
53. Rao, A.S., Paknikar, S.K. & Kirtane, J.G. (1983). Recent advances in the preparation and synthetic applications of oxiranes. *Tetrahedron*, 39, 2323–2367.
54. Kirchner, G., Scollar, M.P. & Klibanov, A.M. (1985). Resolution of racemic mixtures *via* lipase catalysis in organic solvents. *J. Am. Chem. Soc.*, 107, 7072–7076.
55. Verger, R. (1984). Pancreatic lipase. In *Lipases*, ed. Borgström, B. & Brockman, H.L., pp. 83–150. Amsterdam: Elsevier Science Publishers.
56. Theisen, P.D. & Heathcock, C.H. (1988). Improved procedure for preparation of optically active 3-hydroxyglutarate monoesters and 3-hydroxy-5-oxoalkanoic acids. *J. Org. Chem.*, 53, 2374–2378.
57. Stokes, T.M. & Oehlschlager, A.C. (1987). Enzyme reactions in apolar solvents: the resolution of (±)-sulcatol with porcine pancreatic lipase. *Tetrahedron Letters*, 28, 2091–2094.
58. Degueil-Castaing, M., De Jeso, B., Drouillard, S. & Maillard, B. (1987). Enzymatic reactions in organic synthesis: 2-ester interchange of vinyl esters. *Tetrahedron Letters*, 28, 953–954.
59. Wang, Y.-F., Lalonde, J.J., Momogan, M., Bergbreiter, D.E. & Wong, C.-H. (1988). Lipase catalyzed irreversible transesterifications using enol esters as acylating reagents: preparative enantio- and regioselective syntheses of alcohols, glycerol derivatives, sugars, and organometallics. *J. Am. Chem. Soc.*, 110, 7200–7205.
60. Wang, Y.-F., Chen, S.-T., Liu, K.K.-C. & Wong, C.-H. (1989). Lipase-catalyzed irreversible transesterification using enol esters: resolution of cyanohydrins and synthesis of ethyl (*R*)-2-hydroxy-4-phenylbutyrate and (*S*)-propanolol. *Tetrahedron Letters*, 30, 1917–1920.
61. Boaz, N.W. (1989). Enzymatic esterification of 1-ferrocenylethanol: an alternate approach to chiral ferrocenyl bis-phosphines. *Tetrahedron Letters*, 30, 2061–2064.
62. Langrand, G., Secchi, M., Buono, G., Baratti, J. & Triantaphylides, C. (1985). Lipase catalyzed ester formation in organic solvents, an easy preparative resolution of alpha-substituted cyclohexanols. *Tetrahedron Letters*, 26, 1857–1860.
63. Langrand, G., Baratti, J., Buono, G. & Triantaphyles, C. (1986). Lipase catalyzed reactions and strategy for alcohol resolution. *Tetrahedron Letters*, 27, 29–32.
64. Hemmerle, H. & Gais, H.-J. (1987). Asymmetric hydrolysis and esterification catalyzed by esterases from porcine pancreas in the synthesis of both enantiomers of cyclopentanoid building blocks. *Tetrahedron Letters*, 28, 3471–3474.
65. Ladner, W.E. & Whiteside, G.M. (1984). Lipase-catalyzed hydrolysis as a route to esters of chiral epoxy alcohols. *J. Am. Chem. Soc.*, 106, 7250–7251.
66. Mori, K. & Akao, H. (1980). Synthesis of optically active alkynyl alcohols and α-hydroxy esters by microbial asymmetric hydrolysis of their corresponding alcohols. *Tetrahedron*, 36, 91–96.
67. Mori, K. (1991). Preparation of chiral building blocks by biochemical methods. In *Biocatalysis in Agricultural Biotechnology*, vol. 24, ed. Baker, D.R., Fenyes, J.G. & Moberg, W.K. Washington, DC, USA: American Chemical Society.
68. Yamamoto, K., Nishioka, T., Oda, J. & Yamamoto, Y. (1988). Asymmetric ring opening of cyclic acid anhydrides with lipase in organic solvents. *Tetrahedron Letters*, 29, 1717–1720.
69. Gutman, A.L. & Bravdo, T. (1989). Enzyme-catalyzed enantioconvergent lactonization of gamma-hydroxy diesters in organic solvents. *J. Org. Chem.*, 54, 4263–4265.
70. Ngooi, T.K., Scilimati, A., Guo, Z. & Sih, C.J. (1989). A chemoenzymatic route to

(–)-pyrenophorin. *J. Org. Chem.*, 54, 911–914.

71. Sweers, H.M. & Wong, C.-H. (1986). Enzyme-catalyzed regioselective deacylation of protected sugars in carbohydrate synthesis. *J. Am. Chem. Soc.*, 108, 6421–6422.

72. Therisod, M. & Klibanov, A.M. (1986). Facile enzymatic preparation of monoacylated sugars in pyridine. *J. Am. Chem. Soc.*, 108, 5638–5640.

73. Therisod, M. & Klibanov, A.M. (1987). Regioselective acylation of secondary hydroxyl groups in sugars catalyzed by lipases in organic solvents. *J. Am. Chem. Soc.*, 109, 3977–3981.

74. Therisod, M. & Klibanov, A.M. (1987). Regioselective acylation of secondary hydroxyl groups in sugars catalyzed by lipases in organis solvents. *J. Am. Chem. Soc.*, 109, 3977–3981.

75. Holla, W. (1989). Enzymatic synthesis of selectively protected glycals. *Angew. Chem. Int. Ed.*, 28, 220–221.

76. Riva, S., Bovara, R., Ottolina, G., Secundo, F. & Carrea, G. (1989). Regioselective acylation of bile acid derivatives with *Candida cylindracea*. *J. Org. Chem.*, 54, 3161–3164.

77. Riva, S. & Klibanov, A.M. (1988). Enzymochemical regioselective oxidation of steroids without oxidoreductases. *J. Am. Chem. Soc.*, 110, 3291–3295.

78. Hirschmann, H. (1960). The nature of substrate asymmetry in stereoselective reactions. *J. Biol. Chem.*, 235, 2762–2767.

79. IUPAC-IUB (1977). The nomenclature of lipids, recommendations 1976. *Eur. J. Biochem.*, 79, 11–21.

80. Eibl, H. & Woolley, P. (1986). Synthesis of enantiomerically pure glyceryl esters and ethers. I. Methods employing the precursor 1,2-isopropylidene-*sn*-glycerol. *Chem. Phys. Lipids*, 41, 53–63.

81. Eibl, H. & Woolley, P. (1986). Synthesis of enantiomerically pure glyceryl esters and ethers. II. Methods employing the precursor 3,4-isopropylidene-D-mannitol. *Chem. Phys. Lipids*, 47, 47–53.

82. Slotboom, A.J., Verger, R., Verheij, H.M., Baartmans, P.H.M. & van Deenen, L.L.M., (1976). Application of enantiomeric 2-*sn*-phophatidylcholines in interfacial enzyme kinetics of lipolysis. *Chem. Phys. Lipids.*, 17, 128–147.

83. van Deenen, L.L.M. & de Haas, G.H. (1963). The substrate specificity of phospholipase A. *Biochim. Biophys. Acta*, 70, 538–553.

84. Eibl, H. (1980). An efficient synthesis of mixed-acid phospholipids using 1-palmitoyl-*sn*-glycero-3-phosphoric acid bromoalkyl ester. *Chem. Phys. Lipids*, 26, 239–247.

85. Ries, U., Fleer, E.A.M., Unger, C. & Eibl, H. (1992). Synthetic phospholipids as substrates for phospholipase C from *Bacillus cereus*. *Biochim. Biophys. Acta*, 1125, 166–170.

86. Eibl, H. (1978). Phospholipid synthesis: oxazophospholanes and dioxaphospholanes as intermediates. *Proc. Natl. Acad. Sci. USA*, 75, 4074–4077.

87. Eibl, H. & Engel, J. (1992). Synthesis of hexadecylphosphocholine (Miltefosine). In *Alkylphosphocholines: New Drugs in Cancer Therapy*, ed. Eibl, H., Hilgard, P. & Unger, C. (in the series *Progress in Experimental Tumour Research*, vol. 34), pp. 1–5. Basel: Karger Verlag.

88. Hanahan, D.J. & Chaikoff, J.L. (1948). On the nature of the phosphorus-containing lipids of cabbage leaves and their relation to a phospholipid-splitting enzyme contained in these leaves. *J. Biol. Chem.*, 172, 191–198.

89. Eibl, H. & Kovatchev, S. (1981). Preparation of phospholipids and their analogs by phospholipase D. *Methods Enzymol.*, 72, 632–639.

90. Palitzsch, S. (1915). Über die Anwendung von Borax- und Borsäurelösungen bei der colorimetrischen Messung der Wasserstoffkonzentration der Meerwassers. *Biochem. Z.*, 70, 333–343.

15.

Medical aspects of triglyceride lipases

GUNILLA BENGTSSON-OLIVECRONA
and THOMAS OLIVECRONA

Lipases participate in a multitude of biological reactions, many of which relate to human physiology and pathology. The quantitatively dominant pathway is in the use of lipids as fuel. An adult on a typical Western diet eats and absorbs more than 100 g of triglycerides (TG) per day, and transports around 150 g of TG in the bloodstream. It is the lipases involved in these processes that we shall discuss in this chapter. The reader is referred to other reviews for information on lipases involved in other medically important pathways, e.g., the generation of signalling molecules from phospholipids (PL) (1), the provision of substrate for formation of eicosanoids (2), or the degradation of sphingolipids and associated storage diseases (3).

1. THE TRIGLYCERIDE LIPASE SUPERFAMILY

Following the cloning of cDNAs for lipases from different animal species, it became apparent that lipoprotein lipase (LPL), hepatic lipase (HL) and pancreatic lipase are structurally related (for a list of sources on sequences, see reference 4). The main function of pancreatic lipase and of LPL is to make possible the transfer of lipids across cell membranes, while HL may have its main function in the remodelling of lipoproteins (see section 4). The evolutionary relationships between the lipases, based on gene structures and degrees of conservation, have been discussed in several recent articles (4–12). The canine pancreatic lipase gene contains 13 exons, while the human LPL gene has ten exons and the human HL gene has nine. The exon structures of LPL and HL are similar, but LPL contains an additional exon that codes for a long untranslated 3' region. Kirchgessner *et al.* have suggested that pancreatic lipase is more closely related to the putative

---

Abbreviations used: apo CII, apolipoprotein CII; CETP, cholesteryl ester transfer protein; HDL, high-density lipoprotein; HL, hepatic lipase; LCAT, lecithin cholesterol acyl transferase; LDL, low-density lipoprotein; LPL, lipoprotein lipase; LRP, LDL-receptor-related protein; PL, phospholipid; TG, triglyceride; VLDL, very low-density lipoprotein.

ancestral gene than either LPL or HL is (5). These two show a closer structural relationship to each other than to pancreatic lipase. In every species investigated, there is only one variant each of LPL and HL (one gene for each). In contrast, two proteins related to pancreatic lipase were recently cloned from human pancreas (12). They show about 70% sequence identity with human pancreatic lipase. One of them had no catalytic activity, though residues in the active-site region were conserved. The other had catalytic activity, but was less dependent on colipase than the pancreatic lipase previously known.

The lipase family has a fourth member in some yolk proteins from *Drosophila* (literature is collated in reference 13). The regions of maximal similarity between the yolk proteins and the lipases coincide with the region of maximal similarity between the lipases themselves. This region spans about 100 amino-acid residues and contains the active-site structures of the lipases. The serine residue at the active site, found also in other hydrolases (esterases, proteases and other lipases), is not conserved in the yolk proteins. This accords with the idea that the yolk proteins have no known catalytic function. However, the sequence similarity in critical regions and the prevalence of conservative substitutions suggest strongly that all four family members are folded similarly in this region (4, 8, 11). If so, the yolk proteins too may have a hydrophobic pocket structure similar to the active site of the lipases. It is not known whether this pocket is functional. Interestingly, yolk proteins bind fatty-acid–ecdysteroid conjugates and are transported in the hæmolymph to the oocyte, where they are ingested by receptor-mediated endocytosis. Thus, the theme of lipid transport is relevant also for this distant family member. The lipases may have gained (or retained) a catalytic function.

2. PANCREATIC LIPASE

Pancreatic lipase constitutes 1–3% of the protein mass in pancreatic juice (for a review see reference 14). The action of the lipase is very sensitive to the physical state of the lipid substrate. The enzyme is easily competed out from binding to lipid–water interfaces by amphiphiles, *e.g.* bile salts, and by other lipid-binding proteins. Under intestinal conditions, pancreatic lipase is therefore dependent on its activator protein, colipase, both for binding to the lipid–water interface and for subsequent rapid catalysis. Most other pancreatic enzymes are secreted as inactive pro-forms, zymogens, but the lipase is secreted in its active form. Colipase is secreted as a pro-form, pro-colipase. This is activated by tryptic cleavage, resulting in the release of an amino-terminal pentapeptide (15). The cleavage does not change markedly the affinity of colipase for the lipase, but it appears to be important for proper binding of the complex to the interface under certain conditions. Interestingly, the pentapeptide enters the bloodstream and may act in the brain as a satiety signal (15).

Figure 1. Three-dimensional structure of pancreatic lipase. Drawing by Hans Nilsson representing an artist's interpretation of the three-dimensional figure in reference 16. Stippled segments show the four β sheets that are conserved in both the lipases and the insect yolk proteins. The hatched region corresponds to the segment postulated to be heparin-binding in LPL. The arrow with asterisk marks the active-site Ser152. N and C indicate the corresponding termini. From reference 8, by permission.

2.1 Molecular structure

Important steps in the understanding of structure–function relationships in this protein family were taken by Winkler *et al.* (reference 16; Chapter 7), who resolved the crystal structure of pancreatic lipase, and recently by van Tilbeurgh *et al.* (17) who analysed crystals of the lipase–pro-colipase complex. The data show that human pancreatic lipase (449 amino-acid residues) is composed of two independently folded domains (Figure 1). The amino-terminal domain spans residues 1–335 and contains the residues of the active-site triad (Ser152, Asp176 and His263). The active site is covered by a short loop between the disulphide-linked residues 237 and 261. Similar loops are found in some fungal lipases (18, 19). The loop has to be moved to allow the entry of substrate molecules into the active site. It has been suggested that this is the molecular basis for the so-called interfacial activation of lipases. The catalytic efficiency of pancreatic lipase increases by orders of magnitude when the enzyme binds to lipid–water interfaces. The loop structure has a hydrophobic side which in water is turned to face the active site. On interfacial binding, the loop may be repositioned so that its inside becomes part of the hydrophobic, lipid-binding region around the entrance of the active site. The conformational changes induced by active-site inhibitors have been studied for the fungal lipases (20, 21) and are probably representative also for members of the triglyceride lipase superfamily.

One point still not understood in detail is the mechanism by which the activator protein (colipase) increases the catalytic efficiency of pancreatic lipase. Pro-colipase has a flattened structure, consisting of three finger-shaped regions which are held together by disulphide bridges (17). The overall structure is similar to that of a snake-venom toxin. The lipid-binding site of colipase is probably formed by the tips of the fingers, while interaction with the lipase occurs at the base of the fingers on the opposite side of the colipase molecule. The colipase-binding site in the lipase is in the carboxy-terminal domain and is contained in the β sheet formed by residues 352–403. The binding of colipase does not induce conformational changes in the lipase molecule, but the additional lipid-binding capacity of the complex indirectly allows interfacial activation involving loop-opening. Its is not known whether any contacts are made between the opened loop and colipase.

3. LIPOPROTEIN LIPASE AND HEPATIC LIPASE

3.1 Molecular structure

Amino-acid and cDNA sequences are known for LPL from five species (human, cow, mouse, guinea-pig and chicken; for literature, see reference 4) and for HL from four species (human, rat, rabbit and mouse; references 4, 22). The gene structures are known for human and guinea-pig LPL (5, 23), and for human HL (24). The LPL gene (human and mouse) has been mapped to chromosome 8, and the human HL gene is on chromosome 15. Human LPL consists of 448 amino-acid residues. It shows about 50% sequence similarity with human HL and about 30% similarity with human pancreatic lipase. In a central region, spanning residues 125–230 in pancreatic lipase, there is 33% residue identity among all known lipases of this protein family, whereas in neighbouring segments on either side the degree of identity is only 15%. On the basis of structure-prediction analyses it has been concluded that this region is folded similarly in all members of the family (4, 8, 11) and that it contains the active site. Natural and synthetic mutants affecting Ser132, Asp156 or His241 in LPL (25–29) or Ser147 in rat HL (30) have no catalytic activity, which supports the suggestion that these residues are in the catalytic triad, by analogy with corresponding residues in pancreatic lipase. Emmerich *et al.* (26) showed that other functions (binding of heparin or lipid) were largely unaffected in the active-site mutants, indicating that the overall structure of the LPL molecule was not seriously disturbed. For several other mutations, especially those involving exchange of a non-polar residue for a polar one in the densely-packed central region, it can be expected that the main effect on the catalytic activity will be caused by hindrance to folding and/or to assembly into dimers, rather than by direct effects on the catalytic site (11, 31, 32). Therefore, no direct conclusions about the function of the mutated residues can be drawn. In the three exons coding for the carboxy-terminal folding region, only few mutants are known, and their consequences for LPL function are unclear (31).

In contrast to the strong similarities in sequence between family members in the central, homologous region, there is not a single residue identity between any of the three lipases in the loop that covers the active site in pancreatic lipase (8). However, the disulphide bond that closes the loop is conserved. In a recent study by Faustinella *et al.* (27), the loop in LPL was partly or fully replaced by the loop sequence from HL. This lowered the LPL activity by 40% at most. In contrast, a chimera of LPL with the loop sequence of pancreatic lipase was completely inactive. The same result was obtained when a part of the loop sequence was deleted. The predicted secondary structure of the loop is similar for LPL and HL, and these enzymes may therefore be functionally almost equivalent. It was not obvious from secondary-structure predictions why the pancreatic lipase loop inactivated LPL. It is of interest that the tip of the loop in LPL is very sensitive to trypsin. Cleavage occurs in bovine LPL at Arg230, indicating that this bond is particularly exposed (33). The nicked enzyme retains activity against soluble esters and short-chain triglycerides (tributyrin), while its activity against emulsified long-chain triglycerides is lowered dramatically (34). Binding of the LPL to lipid emulsions and to heparin was not affected by the nick, indicating that the overall structure of the lipase was retained. The cleavage at this site probably disturbs the finely-tuned motions undergone by the putative loop on binding to the lipid–water interface.

Another area of special interest in the LPL molecule is the site for binding to heparin (Figure 1). Its physiological function is to anchor the lipase to the glycosaminoglycan chains of heparan sulphate proteoglycans (Figure 2). The heparin-binding site has been localised to the region that in pancreatic lipase forms a surface loop between the two folding units on the back side of the molecule (8). Within this segment (residues 260–306 in human LPL and 281–330 in human pancreatic lipase according to the alignment in reference 4) there are 14 positively

VLDL/chylo-receptor   LRD-receptor

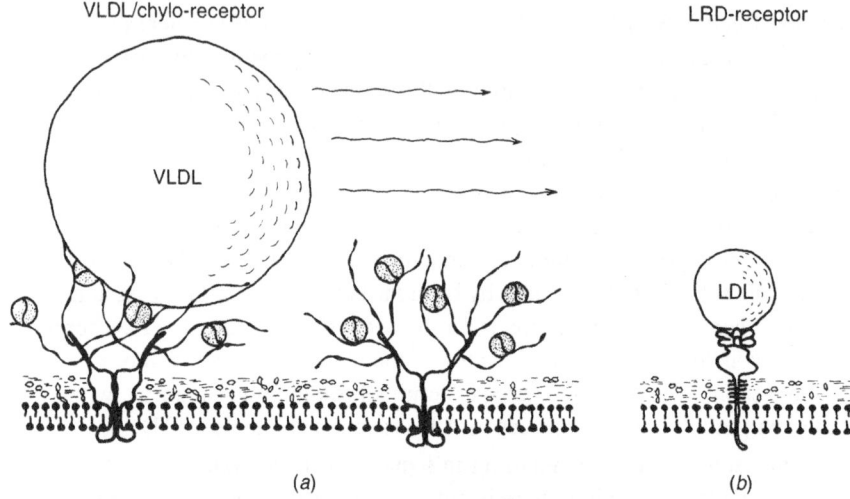

(a)   (b)

Figure 2. Endothelial lipolysis sites. (a) Heparan sulphate proteoglycans are intercalated in the plasma membrane of an endothelial cell. Each core protein carries several polysaccharide chains. The attached LPL molecules (small spheres) are in a position where they can freely interact with lipoproteins, illustrated here by a VLDL particle. Chylomicrons are much larger than VLDL and probably interact simultaneously with several heparan sulphate proteoglycans. (b) The low-density-lipoprotein (LDL) receptor also has its ligand-binding domain outside the glycocalyx. This is accomplished by a domain with O-linked glycan chains, which spans the glycocalyx. From reference 48, by permission.

charged residues in LPL, nine in HL and four in pancreatic lipase. The segment of human LPL contains the sequence KVRAKRSSK, which carries a high density of positive charges (five out of nine residues) and is similar to the consensus sequence found in other heparin-binding proteins such as apolipoproteins B and E. It is also likely that the three-dimensional arrangement of the charges is important for the binding affinity. Active LPL is a homodimer (35). Conformational changes and loss of catalytic activity occur in the molecule on monomerisation. Both monomeric, inactive LPL and active HL bind with lower affinity to heparin than the LPL dimer does (36, 37). The stronger binding of the dimer may occur because heparin-binding sites on both subunits co-operate (38), or because the dimer has a more favourable arrangement of charges in the heparin-binding region. Some mutations in regions that are not considered to be involved in heparin binding lower the heparin affinity (31). This is probably due to structural rearrangements in the molecule, possibly leading to monomerisation. It is not known what parts of LPL are involved in the subunit interaction.

The carboxy-terminal domains of members of the lipase family show little sequence similarity. Of 114 residues only ten are conserved. However, structure-prediction calculations lead, to the view that these three lipases are arranged similarly (in a β-sandwich) in this region, as well as elsewhere (11). One piece of evidence for this is that LPL is readily cleaved by chymotrypsin at Phe390 and Trp392 (A. Lookene and G. Bengtsson-Olivecrona, unpublished results). In comparison with the pancreatic lipase structure, this site would be in the middle of the putative carboxy-terminal folding unit in a surface loop connecting the β5 and β6 strands. In pancreatic lipase, binding of colipase occurs to a pre-formed site in this region (17). A similar binding of apolipoprotein C-II (apo C-II) to LPL is less likely, since studies of LPL–HL chimeras in two different laboratories (31, 39) indicate that binding of apo C-II occurs within the 314 amino-terminal residues, *i.e.,* to the amino-terminal folding unit. Furthermore, the chymotrypsin-truncated LPL has the same affinity for apo C-II as the intact LPL has (A. Lookene and G. Bengtsson-Olivecrona, unpublished results). Monoclonal antibodies against the carboxy-terminal region of LPL inhibit the activity toward emulsified lipid substrates (31, 39) and the chymotryptic cleavage made LPL less active toward emulsified lipid substrates (33). These data indicate that the carboxy-terminal region of LPL contains structures of importance for proper arrangement of the enzyme at certain interfaces (a secondary lipid-binding site).

3.2 Activator proteins

Both LPL and pancreatic lipase are dependent upon activator proteins (apo C-II and colipase, respectively) for function in their physiological environment. For HL, no activator is known. The main rôle of colipase is to assist pancreatic lipase in binding to lipid–water interfaces that are covered by bile salts and proteins. This can be understood in terms of a mediator that binds to the lipase on the one hand and to lipid on the other. LPL binds to most interfaces, even in the absence of apo C-II. Therefore, the activator has its main effect on LPL that is already bound to the interface. It can only be assumed that the interaction causes structural changes in LPL that make the enzyme more efficient in catalysis. A logical mechanism would be to facilitate the movement of the loop covering the active site. This, however, is less likely, since the LPL–HL chimera with the loop sequence from HL was stimulated by apo C-II (27). No cross-reactivity between the activator proteins has ever been reported. This is understandable, since there is no sequence homology or structural homology between the two molecules, even though they are similar in size (101 residues for colipase, 79 for apo C-II). Colipase contains five disulphide bonds that stabilise three finger-like structures (see section 2.1). Apo C-II contains no cysteine and is a member of the apolipoprotein family, which contains flexible regions that bind to lipid by forming amphiphilic helices (for a review, see reference 40). The lipid-binding region of apo C-II has been localised to the amino-terminal two-thirds of the molecule, while the structures involved in binding to LPL and activation are in the carboxy-terminal third. From studies with synthetic fragments of apo C-II, it was concluded that the lipid-binding regions were not necessary for activation of the lipase. This may not be true for synthetic interfaces at high surface pressures (41) or with biologically packaged lipids, such as chylomicrons (G. Bengtsson-Olivecrona and U. Beisiegel, unpublished results). Patients with apo C-II deficiency have symptoms similar to patients with LPL deficiency (31). All mutations identified up to now in patients with clinically manifest apo C-II deficiency lead to disturbed expression or to expression of prematurely truncated forms of the protein. As yet, no example is known of a point mutation of an individual amino-acid residue leading to loss of function.

4. REACTIONS CATALYSED IN THE PHYSIOLOGICAL ENVIRONMENT

4.1 Lipid digestion

In the intestine, the colipase-dependent pancreatic lipase hydrolyses efficiently the bulk of dietary TGs and ensures their efficient absorption. The action of this enzyme is supported by several other factors.

1. The lipase appears to have rather strict requirements for the physical properties of the substrate–water interface. Bile components help to provide a suitable environment. One of their rôles is to solubilise the lipolytic products. Another is that bile salts open the way for the lipase by assisting the desorption of other proteins from the surface of the lipid droplets.

2. Colipase ensures that the lipase can still become adsorbed to the droplets.

3. Gastric lipase is a 50-kDa glycoprotein (42) which is stable and able to act over a wide range of pH. This lipase can attack virtually any lipid droplet. Its action results in hydrolysis of some of the TG to diglycerides and fatty acids. This facilitates emulsification of the lipids and prepares them for subsequent intestinal digestion.

4. Carboxyl ester lipase is a 100-kDa glycoprotein (43) which has a wide substrate specificity and hydrolyses monoglycerides, cholesteryl esters, retinyl esters and other lipid esters. Hence, it complements the narrow substrate specificity of the colipase-dependent lipase. Carboxyl ester lipase is secreted by the pancreas, but also by the lactating mammary gland. The lipase in milk ensures efficient digestion and absorption of the milk lipids by the new-born. This is an important safeguard, since pancreatic function is sometimes poorly developed in new-born children. This lipase is active only in the presence of certain bile salts. Hence, it is inactive in the milk (and pancreas) and becomes

(a)

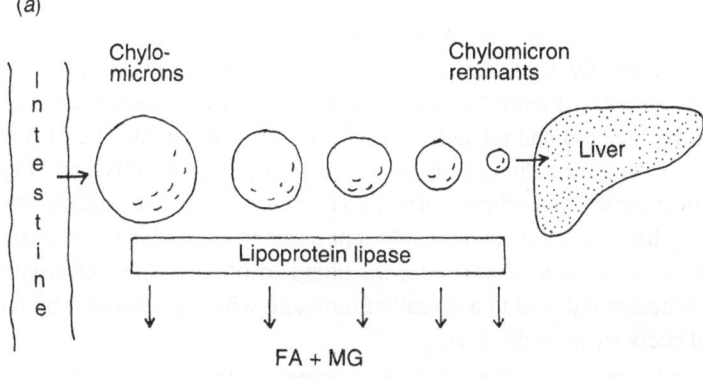

(b)

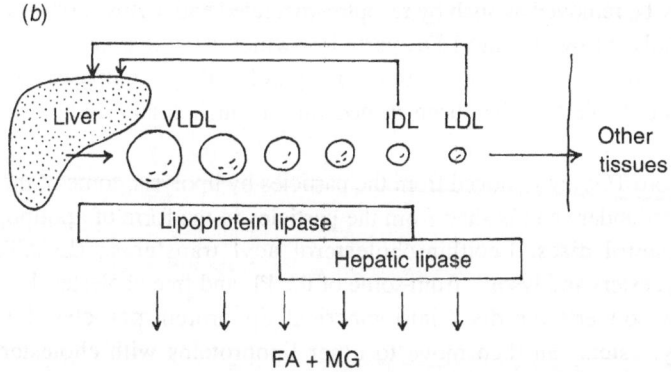

Figure 3. De-lipidation of TG-rich lipoproteins by the lipases. (a) Exogenous pathway; (b), endogenous pathway.

active only after it has arrived in the intestine. It is sometimes called the bile-salt-stimulated lipase, BSSL.

5. Pancreatic phospholipase A_2, which takes care of the dietary PL, is discussed in Chapter 6.

Fat malabsorption is a fairly common clinical problem. Many cases of this are due to diseases of the intestinal mucosa, or to defective bile secretion, but some are due to insufficient lipase secretion. This is seen in diseases of the pancreas, *e.g.* cystic fibrosis, and can be alleviated by peroral administration of lipase preparations.

4.2 Lipoprotein metabolism

As a background for our discussion of the physiological rôles of LPL and HL, we must outline briefly some features of lipoprotein metabolism (44, 45). In the exogenous pathway (Figure 3), dietary lipids are assembled into chylomicrons by the intestinal mucosa and released to the blood *via* the lymph. The TGs are then unloaded through the action of LPL at vascular binding-lipolysis sites (Figure 2), mainly in muscles and adipose tissue. The fatty acids and 2-monoglycerides released by lipolysis move into the adjacent tissue to be used in cellular metabolic reactions, or they may be carried with blood to other tissues. Ultimately, the particle becomes reduced to a so-called remnant, which is removed by receptor-mediated endocytosis in the liver.

In the endogenous pathway, TGs are packaged by the liver with PL, cholesterol and apolipoproteins and released in VLDL particles. These undergo a lipolytic cascade, similar to that described for chylomicrons, to form VLDL remnants. These may be removed as such by receptor-mediated endocytosis, or they may be further lipolysed by HL into LDL particles, which have lost almost all TG but retain the cholesteryl esters that cannot be hydrolysed by the lipases. Hence, the TG-rich VLDL particle has been turned into a much smaller, cholesterol-rich particle.

When core TGs are removed from the particles by lipolysis, some of the surface becomes redundant and is shed from the particles in the form of apolipoprotein–PL–cholesterol discs. Lecithin-cholesterol acyl transferase (LCAT) forms cholesteryl esters and lysoPL from some of the PL and free cholesterol molecules (45). This converts the discs into spherical lipoprotein particles, HDL. The cholesteryl esters can then move to other lipoproteins with cholesteryl ester transfer protein CETP (45).

LDLs typically circulate in the blood for two days before they are catabolised through receptor-mediated endocytosis, and HDLs have even longer lifetimes. In contrast, de-lipidation of a chylomicron is usually accomplished within ten minutes and that of a VLDL in less than two hours. Hence, TG transport, the bulk process, is rapid and the concentrations of TG-rich lipoproteins in plasma are relatively low. The process leaves behind long-lived LDL and HDL particles, which are the main lipoproteins in the blood of fasting humans.

4.3 Regulation and action of LPL

The activity of LPL in individual tissues is under hormonal control (46). For instance, the activity is relatively low in adipose tissue in the fasted state, but rises rapidly after a meal and directs an increased fraction of the blood lipids to adipose tissue for storage. Conversely, the activity in heart muscle is higher in the fasted than in the fed state. Another example is provided by the mammary gland (47).

(a) Liposome   (b) Droplet

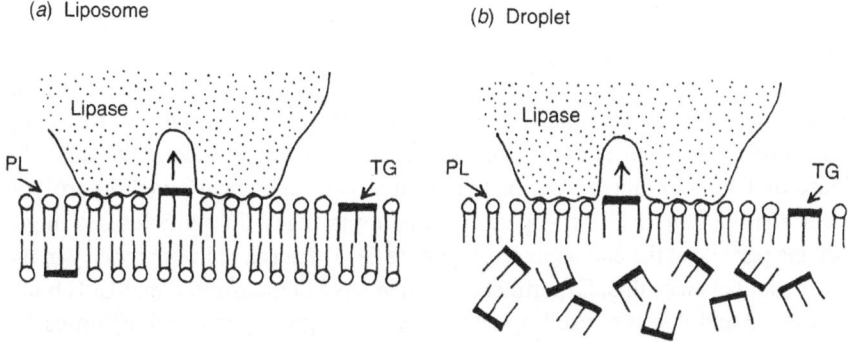

Figure 4. Entry of substrate molecules into the active site of lipases. The figure illustrates the different situation with (a) mixed liposomes of PL and TG and (b) PL-enveloped TG emulsion droplets.

Here LPL activity is low, but it rises dramatically on parturition and then stays high throughout lactation. This channels a large fraction of blood lipids to the gland for production of milk fat.

LPL is an efficient enzyme. Its turnover number is about 1000 s^{-1} for TG hydrolysis under 'physiological conditions' (48). This raises the questions of how substrate molecules enter and products leave the active site. To approach this, Rojas *et al.* compared the hydrolysis of TG in liposomes and in emulsion droplets, made up to give the same proportion of PL and TG in the surface layer (49). The relation between rate and surface area was similar for liposomes and emulsion droplets, indicating that the lipid-binding sites on the lipase did not detect any marked difference between the surfaces of the two types of particles. If each hydrolytic event is followed by dissociation of the lipase into the aqueous phase, one would expect to see the same ratio of PL to TG hydrolysis and similar kinetics with liposomes and emulsion droplets. The ratios of TG to PL hydrolysis were, however, quite different. With liposomes, the enzyme hydrolysed more PL than it did TG, whereas with emulsion droplets TG was the preferred substrate. Furthermore, the maximal rate of TG hydrolysis was more than 40 times higher with the emulsion droplets than with the liposomes. These results indicate that when LPL binds to the substrate particle, it stays for several rounds of lipolysis, *i.e.*, the action is processive. The results furthermore suggest that when products leave the active site it is easier for a new substrate molecule to come in from below (from the core) than from the side (from the surface), as Figure 4 shows. This would seem to fit with the X-ray structure of pancreatic lipase, which shows the active site at the bottom of a hydrophobic pocket (16).

This picture of lipolysis fits the action of LPL (and HL) *in vivo*. Degradation of chylomicrons is a rapid process, in which many TGs and a few PL molecules are hydrolysed. A typical chylomicron may contain two million TG molecules. Hydrolysis of 90% of this TG to form a remnant particle requires 3.6 million hydrolytic events. Hence, several lipase molecules must act simultaneously on the particle, and the lipase molecules must stay attached to the particle for many rounds of lipolysis. It is clear that this process cannot be accomplished if the lipase molecule migrates from one chylomicron to another for each hydrolytic event. On the other hand, co-operative binding to many lipase molecules might lock the particle at the endothelial site (Figure 2). There is a built-in property in the lipase that balances this. Experiments *in vitro* have demonstrated that LPL binds fatty acids and that the resulting complexes have greatly reduced affinities for lipid–water interfaces, for apo C-II and for heparan sulphate (50, 51). Saxena *et al.* have shown that physiological concentrations of free fatty acids dissociate LPL from cultured endothelial cells (52). Thus, the following sequence of events is suggested (53, 54): fatty acids are sometimes released by LPL at endothelial sites more rapidly than the underlying tissue can transport and metabolise the fatty acids, which, in turn, results in spillage of fatty acids into the blood, and to the formation of LPL–fatty-acid complexes with inhibition of continued hydrolysis and dissociation of LPL from endothelial heparan sulphate. The implication is that LPL may sometimes, often or even usually be present in excess, while the limiting factor is transport and metabolism of fatty acids by the underlying tissue.

It is generally assumed that LPL is bound at the endothelial cell surface by interaction with the glycosaminoglycan chains of heparan sulphate proteoglycans (Figure 2). This model demonstrates two separate aspects of LPL action; the lipase attaches the lipoprotein to the cell surface, and it hydrolyses the TG of the lipoprotein. Interest has focussed mainly on the hydrolytic function. More recently, it has been demonstrated that LPL can mediate the binding of lipoproteins to other cells (55, 56), and that this can lead to the transfer of components between the cells and the lipoproteins (57), to enhanced transfer of the particles to other receptors (55), and to endocytosis of the lipase-lipoprotein complexes (58). These actions occur with all types of lipoproteins (55, 56) and even with liposomes prepared from non-hydrolysable PL analogues (59). Hence the ligating function by which LPL links lipoprotein particles to heparan sulphate on the cell surface appears to be independent of the hydrolytic function of the enzyme. Beisiegel *et al.* demonstrated recently that LPL also mediates the binding of lipoproteins to the putative receptor for chylomicron remnants, the low-density-lipoprotein-related protein, LRP (60).

LPL thus appears to be a multifunctional protein, with properties of importance for several aspects of lipoprotein catabolism. The lipase is required for rapid unloading of TG from chylomicrons at the site of endothelial binding and

Table 1. Lipase activities in post-heparin plasma of some species.

Source	LPL	HL
Man	350	370
Rat	440	700
Dog	630	170
Guinea-pig	790	170
Calf	530	< 5

Heparin (100 IU/kg body weight, about 0.65 mg) was injected intravenously, and a plasma sample was taken 10–15 min later. Hepatic lipase was assayed by using a gum-arabic-stabilised triglyceride emulsion in the presence of 1 M NaCl to suppress LPL activity. LPL was assayed with a phosphatidylcholine-stabilised triglyceride emulsion with rat serum as a source of activator. For assay of LPL activity in human and rat plasma, HL was suppressed by immunoinhibition with anti-HL serum. For dog, guinea-pig and calf plasma, LPL represents the difference in activity between samples treated respectively with anti-LPL or with control serum. Values are in nmol/(ml·min). (From reference 86, by permission of Elsevier Science Publishers BV.)

lipolysis. Lipase–fatty-acid interactions may underlie the mechanism by which the metabolic state of adjacent cells determines how much unloading is to be allowed. Lipase molecules adsorbed to the leaving particle may signal that it is ready for receptor-mediated removal mechanisms to come into action.

4.4 LPL and HL – what are the functional differences?

While the rôle of LPL in de-lipidation of TG-rich lipoproteins is fairly well defined, the rôle of HL has remained somewhat enigmatic. In HL deficiency in humans, VLDL remnants and TG-rich HDL accumulate in the plasma (61, 62). Studies on lipoprotein kinetics in patients with HL deficiency (63), and studies on the effects of addition of purified HL to plasma (64), also indicate that HL acts upon HDL and LDL, and that it may be involved in the terminal stages of de-lipidation of chylomicron and VLDL remnants. Hence, LPL and HL appear to have partially overlapping rôles in lipoprotein metabolism (Figure 3). Table 1 shows lipase activities in post-heparin plasma from a number of species. These data indicate that LPL activity is high in all species. In contrast, HL activity varies widely between species (65). It is high in humans and in rats, moderately high in dogs, low in guinea-pigs and virtually non-existent in calves. Similar observations have been made for other factors involved in metabolism of the cholesterol-rich lipoproteins (66). For instance, lipid-transfer activity in plasma differs widely between species. The concentrations of LDL and HDL in plasma also vary widely between animals. For instance, both LDL and HDL concentrations are relatively high in humans, whereas dogs have low concentrations of LDL and high concentrations of HDL, and guinea-pigs have high concentrations of LDL and low concentrations of HDL. Our view is that LPL carries out an indispensable initial reaction in lipoprotein metabolism and is consistently high in animal species, whereas there is more latitude in how animals are disposed to handle metabolism of remnant particles and of LDL and HDL.

That LPL and HL do in fact select different types of lipoproteins as their substrate is evident from a number of observations *in vitro* and *in vivo*. For instance, when the lipases are added to whole plasma, LPL hydrolyses lipids mainly in VLDL, whereas HL acts mainly on HDL (64). This particle selection probably reflects differences in binding affinity for the lipoprotein particles, *i.e.*, differences in the lipid-binding or interface-recognition sites on the two lipases. A direct demonstration of substrate selection has come from competition experiments. For instance, in a study by Bengtsson *et al.*, HL was found to hydrolyse efficiently TG in PL-stabilised emulsion droplets when these were the only lipid particles in the system (67). When HDL was added, hydrolysis of the emulsified TG was suppressed and the enzyme acted on the HDL instead. In contrast, addition of HDL to such a system does not impede the action of LPL.

Another aspect of substrate selection is the question of how the enzymes seek out and hydrolyse individual lipid molecules at the interface. LPL and HL are relatively non-specific enzymes. They hydrolyse TG, diglycerides, monoglycerides, PL and a variety of model substrates such as *p*-nitrophenyl butyrate (48). An important exception is that they do not hydrolyse cholesteryl esters. In lipoproteins, the two major substrates are therefore TG and PL. TGs are largely confined to the core of the lipoproteins. Hamilton and Small have shown, by experiments with liposomes, that the molar ratio of PL to TG in the surface film is more than 30 (68). Scow and Egelrud were the first to show that LPL hydrolyses not only TG but also PL in rat chylomicrons (69). Subsequent studies, mainly in model systems, have led to the view that HL is a more potent phospholipase than is LPL. This was tested directly in a recent study on the action of the two lipases on PL and TG in isolated lipoproteins (70). For each type of lipoprotein used, HL always hydrolysed more PL for a given amount of TG hydrolysis, demonstrating that there are in fact differences between the two enzymes in how they select substrate molecules at the interface.

To obtain a model system in which substrate selection could be studied in more detail. Rojas *et al.* incorporated trace amounts of triolein in liposomes of egg-yolk phosphatidylcholine (49). In this model system, the lipase would be expected to recognise a few TG molecules exposed among many PL molecules at the surface of the liposomes (see Figure 4). These liposomes behaved as stable substrates for lipase action, that is, the hydrolysis of triolein and of phosphatidylcholine was linearly related to the time of incubation and to the amount of enzyme added. Curves for reaction velocity *versus* amount of substrate added were parallel for triolein hydrolysis and also for phosphatidylcholine hydrolysis. That is, substrate saturation was reached at the same amount of particles, irrespective of which hydrolysis was monitored. Using identical liposomes as substrate for LPL and HL, Rojas *et al.* found that HL hydrolysed more phosphatidylcholine molecules for each triolein molecule than LPL did. This is the same result as had been obtained with lipoproteins, and it reinforces the conclusion that there are differences in how the enzymes select substrate molecules at the interface. There are several other studies in the literature that lead to similar conclusions. For instance, Ikeda *et al.*, using LPL and HL purified from human post-heparin plasma, compared product profiles during hydrolysis of TG in a TG/PL emulsion by the two enzymes (71). As has been demonstrated before, LPL hydrolysed TG mainly to monoglycerides and free fatty acids. On the other hand, HL under identical conditions hydrolysed more monoglycerides, yielding primarily free fatty acids as products. Another piece of evidence comes from observations by Åke Nilsson and his co-workers (72). They noted that LPL has relatively low activity toward ester bonds involving fatty acids with double bonds close to the carbonyl group. This is probably due to steric hindrance, as has been shown for pancreatic lipase. Under identical conditions, HL shows a higher activity toward these ester bonds. In fact, the addition of HL to reaction mixtures with LPL will drive the hydrolysis of such glycerides towards completion in a mixture of natural TGs.

4.5 Is LPL rate-limiting for the metabolism of TG-rich lipoproteins?

It is clear that LPL is needed for normal catabolism of TG-rich lipoproteins. The massive hypertriglyceridæmia seen in patients deficient in LPL or in its activator, apolipoprotein C-II, attest to this (31). However, it is not so clear whether LPL is rate-limiting in normal individuals.

Post-heparin LPL activity is assumed to give a measure of the LPL available at endothelial sites. A recent study showed an activity of 483 (± 180) nmol fatty acid produced per minute and millilitre plasma for a group of middle-aged, healthy, normolipidæmic men (54). This assay was with a PL-stabilised TG emulsion at 25 °C and pH 8.5. The activity would be higher at 37 °C, but lower at pH 7.4; these factors roughly cancel out. The activity is somewhat higher with chylomicrons than with the synthetic emulsion. None the less, the value should give an estimate of capacity for TG hydrolysis *in vivo*. For each TG molecule, two ester bonds are cleaved by the enzyme. Therefore, the activity corresponds to the clearing of roughly 250 nmol TG per minute and millilitre plasma, or about 40 g of TG per hour if the plasma volume is 3 litres. This is well above rates of TG transport after normal meals, suggesting that other steps, *e.g.* lipid absorption, are rate-limiting for the entire process. Even after peroral fat loads, this LPL activity should be enough for efficient clearing, but the margin is not large. Hence, we may expect to see a correlation between clearing capacity and post-heparin LPL activity (or LPL activity measured in tissue biopsies). Several studies have shown an inverse relation between LPL activity and the magnitude of postprandial lipæmia in healthy subjects (73, 74), but the correlations have been weak. A recent study with heterozygotes for LPL deficiency showed pronounced lipæmia in response to a peroral fat load (76). In patients with coronary artery disease, on the other hand, LPL activity did not allow the lipæmia to be predicted (54, 77). Taken together, these studies indicate that low activity of LPL can be a limiting factor in the clearing of large peroral fat loads, but is probably not limiting under normal conditions. The implication is that the amount of LPL, as a restriction, can be overridden by other factors. One possibility that has been raised is that TG transport may be limited more by the ability of the tissues to assimilate the products of lipolysis than by the capacity of the organism for lipolysis as such (53, 80). Molecular coupling could occur by way of the interaction of fatty acids with LPL (see section 4.3).

A related question is whether LPL is required for the assimilation of lipids into extrahepatic tissues. Humans have limited capacity for synthesis of fatty acids in adipose tissue. Hence, transport in lipoprotein TG and uptake through LPL is thought to be the major pathway by which adipose tissue acquires fatty acids. In rats, a strong correlation has been demonstrated between LPL activity in adipose tissue and the uptake of fatty acids from radiolabelled chylomicrons (80). However, Brun et al. (81) found no difference in total body fat content, or in the distribution of subcutaneous fat tissue, between patients with LPL deficiency and normal controls. The implication is that the assimilation of fatty acids by adipose tissue depends less on the pathway of transport than on metabolic factors.

Another question is what impact LPL activity has on the concentration and composition of lipoproteins in the post-absorptive and fasting states. Most studies on the relation between LPL activity and basal levels of TG in plasma have yielded weak or non-significant relations (78), even in individuals heterozygotic for LPL deficiency (76, 79). This is not surprising, since rates of TG transport are much lower for VLDL than for chylomicrons (only a few grammes per hour). The most consistent correlation found has been a positive relation between LPL activity and HDL cholesterol, particularly HDL2 cholesterol (82). The molecular mechanisms behind this relation have been discussed extensively. One line of thought is based on the fact that surface components are transferred from TG-rich lipoproteins to HDL as a result of LPL-mediated hydrolysis of core TG. Hence, efficient de-lipidation, as opposed to early receptor-mediated particle removal, would channel more PL, cholesterol and apolipoproteins to HDL (82). Another line of thought stresses the rôle of CETP, which catalyses homo- and heteroexchange of cholesteryl esters and TG between lipoproteins (44, 66, 83). Increased levels of TG-rich lipoproteins (basal or postprandial) would cause increased flow of cholesteryl esters from LDL and HDL into the TG-rich lipoproteins. Conversely, TG transferred to LDL and HDL would be susceptible to hydrolysis by HL. The result would be a preponderance of small HDL (HDL3) and small LDL (pattern B). Recent studies have in fact demonstrated an association between LPL activity and small dense LDL (76, 84).

Many studies have demonstrated an inverse relation between HL activity and HDL cholesterol levels (82). The underlying mechanisms suggested follow the same lines of thought as for LPL. On the basis of the observation that HL acts on HDL, it has been argued that high activity of HL leads to a generally enhanced catabolism of the particles (82). Alternatively, it has been suggested that the major impact of high HL activity is to enhance hydrolysis of TG transferred into the particles with CETP, and hence drive depletion of HDL core lipids through exchange with TG-rich lipoproteins (83, 85).

Whatever the mechanisms are, many studies now indicate that low LPL activity and high HL activity alone, or in combination, can drive the lipoproteins towards an 'atherogenic profile', probably imparting increased risk of cardiovascular disease for the individual. Both enzymes respond profoundly to physiological conditions such as hormones, diet and exercise. It is becoming increasingly probable that imbalances in the lipases are at the root of many of the common derangements of lipoprotein composition and concentration.

REFERENCES

1. Vance, D.E. (1991). Phospholipid metabolism and cell signalling in eucaryotes. In *Biochemistry of Lipids, Lipoproteins and Membranes*, ed. Vance, D.E. & Vance, J., pp. 205–240. Amsterdam: Elsevier.
2. Smith, W.L., Borgeat, P. & Fitzpatrick, F.A. (1991). The eicosanoids: cyclooxygenase, lipoxygenase, and epoxygenase pathways. In *Biochemistry of Lipids, Lipoproteins and Membranes*, ed. Vance, D.E. & Vance, J., pp. 297–325. Amsterdam: Elsevier.
3. Sweeley, C.C. (1991). Sphingolipids. In *Biochemistry of Lipids, Lipoproteins and Membranes*, ed. Vance, D.E. & Vance, J., pp. 327–361. Amsterdam: Elsevier.
4. Hide, W.A., Chan, L. & Li, W.-H. (1992). Structure and evolution of the lipase superfamily. *J. Lipid Res.*, 33, 167–178.
5. Kirchgessner, T.G., Chaut, J.C., Heinzmann, C., Etienne, J., Guilhot, S., Svenson, K., Ameis, D., Pilon, C., D'Auriol, L., Andalibi, A., Schotz, M.C., Galibert, F. & Lusis, A.J. (1989). Organization of the human lipoprotein lipase gene and evolution of the lipase gene family. *Proc. Natl. Acad. Sci. USA*, 86, 9647–9651.
6. Persson, B., Bengtsson-Olivecrona, G., Enerbäck, S., Olivecrona, T. & Jörnvall, H. (1989). Structural features of lipoprotein lipase. Lipase family relationships, binding interactions, non-equivalence of lipase cofactors, vitellogenin similarities, and functional subdivisions of lipoprotein lipase. *Eur. J. Biochem.*, 179, 39–45.

7. Mickel, F.S., Weidenbach, F., Swarovsky, B., LaForge, K.S. & Scheele, G.A. (1989). Structure of the canine pancreatic lipase gene. *J. Biol. Chem.*, 264, 12895–12901.

8. Persson, B., Jörnvall, H., Olivecrona, T. & Bengtsson-Olivecrona, G. (1991). Lipoprotein lipases and vitellogenins in relation to the known three-dimensional structure of pancreatic lipase. *FEBS Letters*, 288, 33–36.

9. Kern, P.A. (1991). Lipoprotein lipase and hepatic lipase. *Curr. Opin. Lipidol.*, 2, 162–169.

10. Bensadoun, A. (1991). Lipoprotein lipase. *Annu. Rev. Nutr.*, 11, 217–237.

11. Derewenda, Z.S. & Cambillau, C. (1991). Effects of gene mutations in lipoprotein and hepatic lipases as interpreted by a molecular model of the pancreatic triglyceride lipase. *J. Biol. Chem.*, 266, 23112–23119.

12. Giller, T., Buchwald, P., Blum-Kaelin, D. & Hunziker, W. (1992). Two novel human pancreatic lipase related proteins, hPLRP1 and hPLRP2. Differences in colipase dependence and in lipase activity. *J. Biol. Chem.*, 267, 16509–16516.

13. Bownes, M. (1992). Why is there sequence similarity between insect yolk proteins and vertebrate lipases. *J. Lipid Res.*, 33, 777–790.

14. Chapus, C., Rovery, M., Sarda, L. & Verger, R. (1988). Minireview on pancreatic lipase and colipase. *Biochimie*, 70, 1223–1234.

15. Erlanson-Albertsson, C. (1992). Pancreatic colipase. Structural and physiological aspects. *Biochim. Biophys. Acta*, 1125, 1–7.

16. Winkler, F.K., D'Arcy, A. & Hunziker, W. (1990). Structure of human pancreatic lipase. *Nature*, 343, 771–774.

17. van Tilbeurgh, H., Sarda, L., Verger, R. & Cambillau, C. (1992). Structure of the pancreatic lipase–colipase complex. *Nature*, 359, 159–162.

18. Brady, L., Brzozowski, A.M., Derewenda, Z.S., Dodson, E., Dodson, G., Tolley, S., Turkenburg, J.P., Christiansen, L., Huge-Jensen, B., Norskov, L., Thim, L. & Menge, U. (1990). A serine protease triad forms the catalytic centre of a triacylglycerol lipase. *Nature*, 343, 767–770.

19. Schrag, J.D., Li, Y., Wu, S. & Cygler, M. (1991). Ser-His-Glu triad forms the catalytic site of the lipase from *Geotrichum candidum*. *Nature*, 351, 761–764.

20. Brzozowski, A.M., Derewenda, U., Derewenda, Z.S., Dodson, G.G., Lawson, D.M., Turkenburg, J.P., Björkling, F., Huge-Jensen, B., Patkar, S.A. & Thim, L. (1991). Model for interfacial activation in lipases from the structure of fungal lipase–inhibitor complex. *Nature*, 351, 491–494.

21. Derewenda, U., Brzozowski, A.M., Lawson, D.M. & Derewenda, Z.S. (1992). Catalysis at the interface: the anatomy of a conformational change in a triglyceride lipase. *Biochemistry*, 31, 1532–1541.

22. Oka, K., Nakano, T., Tkalcevic, G.T., Scow, R.O. & Brown, W.V. (1991). Molecular cloning of mouse hepatic triacylglycerol lipase: gene expression in combined lipase-deficient (cld/cld) mice. *Biochim. Biophys. Acta*, 1089, 13–20.

23. Enerbäck, S. & Bjursell, G. (1989). Genomic organization of the region encoding guinea-pig lipoprotein lipase: evidence for exon fusion and unconventional splicing. *Gene*, 84, 391–397.

24. Cai, S.-J., Wong, D.M., Chen, S.-H. & Chan, L. (1989). Structure of the human hepatic triglyceride lipase gene. *Biochemistry*, 28, 8966–8971.

25. Faustinella, F., Smith, L.C., Semenkovich, C.F. & Chan, L. (1991). Structural and functional roles of highly conserved serines in human lipoprotein lipase. Evidence that serine 132 is essential for enzyme catalysis. *J. Biol. Chem.*, 266, 9481–9485.

26. Emmerich, J., Beg, O.U., Peterson, J., Previato, L., Brunzell, J.D., Brewer, H.B. Jr & Santamarina-Fojo, S. (1992). Human lipoprotein lipase. Analysis of the catalytic triad by site-directed mutagenesis of Ser132, Asp156, and His241. *J. Biol. Chem.*, 267, 4161–4165.

27. Faustinella, F., Smith, L.C. & Chan, L. (1992). Functional topology of a surface loop shielding the catalytic center in lipoprotein lipase. *Biochemistry*, 31, 7219–7223.
28. Faustinella, F., Chang, A., Van Biervliet, J.P., Rosseneu, M., Vinaimont, N., Smith, L.C., Chen, S.-H. & Chan, L. (1991). Catalytic triad residue mutation (Asp156–Gly) causing familial lipoprotein lipase deficiency. Co-inheritance with a nonsense mutation (Ser447–Ter) in a Turkish family. *J. Biol. Chem.*, 266, 14418–14424.
29. Ma, Y., Bruin, T., Tuzgol, S., Wilson, B.I., Roederer, G., Liu, M.-S., Davignon, J., Kastelein, J.J.P., Brunzell, J.D. & Hayden, M.R. (1992). Two naturally occurring mutations at the first and second bases of codon aspartic acid 156 in the proposed catalytic triad of human lipoprotein lipase. In vivo evidence that aspartic acid 156 is essential for catalysis. *J. Biol. Chem.*, 267, 1918–1923.
30. Davis, R.C., Stahnke, G., Wong, H., Doolittle, M.H., Ameis, D., Will, H. & Schotz, M.C. (1990). Hepatic lipase: site directed mutagenesis of a serine residue important for catalytic activity. *J. Biol. Chem.*, 265, 6291–6295.
31. Fojo, S.S. (1992). Genetic dyslipoproteinemias: role of lipoprotein lipase and apolipoprotein C-II. *Curr. Opin. Lipidol.*, 3, 186–195.
32. Lalouel, J.-M., Wilson, D.E. & Iverius, P.-H. (1992). Lipoprotein lipase and hepatic triglyceride lipase: molecular and genetic aspects. *Curr. Opin. Lipidol.*, 3, 86–95.
33. Bengtsson-Olivecrona, G., Olivecrona, T. & Jörnvall, H. (1986). Lipoprotein lipase from cow, guinea-pig and man. Structural characterization and identification of protease-sensitive internal regions. *Eur. J. Biochem.*, 161, 281–288.
34. Bengtsson, G. & Olivecrona, T. (1981). Lipoprotein lipase: modification of its kinetic properties by mild tryptic digestion. *Eur. J. Biochem.*, 113, 547–554.
35. Osborne, J.C., Jr, Bengtsson-Olivecrona, G., Lee, N. & Olivecrona, T. (1985). Studies on inactivation of lipoprotein lipase. Role of dimer to monomer dissociation. *Biochemistry*, 24, 5606–5611.
36. Bengtsson-Olivecrona, G. & Olivecrona, T. (1985). Binding of active and inactive forms of lipoprotein lipase to heparin: effects of pH. *Biochem. J.*, 226, 409–413.
37. Liu, G., Bengtsson-Olivecrona, G. & Olivecrona, T. (1993). Assembly of lipoprotein lipase in perfused guinea-pig hearts. *Biochem. J.*, 292, 277–282.
38. Clarke, A.R., Luscombe, M. & Holbrook, J.J. (1983). The effect of the chain length of heparin on its interaction with lipoprotein lipase. *Biochim. Biophys. Acta*, 747, 130–137.
39. Wong, H., Davies, R.C., Nikazy, J., Seebart, K.E. & Schotz, M.C. (1991). Domain exchange: characterization of a chimeric lipase of hepatic lipase and lipoprotein lipase. *Proc. Natl. Acad. Sci. USA*, 88, 11290–11294.
40. Wang, C.-S., Hartsuck, J. & McConathy, W.J. (1992). Structure and functional properties of lipoprotein lipase. *Biochim. Biophys. Acta*, 1123, 1–17.
41. Jackson, R.L., Balasubramaniam, A., Murphy, R.F. & Demel, R.A. (1986). Interaction of synthetic peptides of apolipoprotein C-II and lipoprotein lipase at monomolecular films. *Biochim. Biophys. Acta*, 875, 203–210.
42. Gargouri, Y., Moreau, H. & Verger, R. (1989). Gastric lipases: biochemical and physiological studies. *Biochim. Biophys. Acta*, 1006, 255–271.
43. Nilsson, J., Bläckberg, L., Carlsson, P., Enerbäck, S., Hernell, O. & Bjursell, G. (1990). cDNA cloning of human-milk bile-salt-stimulated lipase and evidence for its identity to pancreatic carboxylic ester hydrolase. *Eur. J. Biochem.*, 192, 543–550.
44. Eisenberg, S. (1990). Metabolism of apolipoproteins and lipoproteins. *Curr. Opin. Lipidol.*, 1, 205–221.
45. Fielding, P.E. & Fielding, C.J. (1991). Dynamics of lipoprotein transport in the circulatory system. In *Biochemistry of Lipids, Lipoproteins and Membranes*, ed. Vance, D.E. & Vance, J., pp. 427–459. Amsterdam: Elsevier.
46. Eckel, R.H. (1989). Lipoprotein lipase: a multifunctional enzyme relevant to common metabolic diseases. *New Engl. J. Med.*, 320, 1060–1068.

47. Scow, R.O. & Chernick, S.S. (1987). Role of lipoprotein lipase during lactation. In *Lipoprotein Lipase*, ed. Borensztajn, J., pp. 149–185. Chicago: Evener Publishers.
48. Olivecrona, T. & Bengtsson-Olivecrona, G. (1987). Lipoprotein lipase from milk – the model enzyme in lipoprotein lipase research. In *Lipoprotein lipase*, ed. Borensztajn, J., pp. 15–58. Chicago: Evener.
49. Rojas, C., Olivecrona, T. & Bengtsson-Olivecrona, G. (1991). Comparison of the action of lipoprotein lipase on triacylglycerols and phospholipids when presented in mixed liposomes or in emulsion particles. *Eur. J. Biochem.*, 197, 315–321.
50. Bengtsson, G. & Olivecrona, T. (1980). Lipoprotein lipase. Mechanism of product inhibition. *Eur. J. Biochem.*, 106, 557–562.
51. Saxena, U. & Goldberg, I.J. (1990). Interaction of lipoprotein lipase with glycosaminoglycans and apolipoprotein C-II: effects of free fatty acids. *Biochim. Biophys. Acta*, 1043, 161–168.
52. Saxena, U., Witte, L.D. & Goldberg, I.J. (1989). Release of endothelial cell lipoprotein lipase by plasma lipoproteins and free fatty acids. *J. Biol. Chem.*, 264, 4349–4355.
53. Peterson, J., Bihain, B.E., Bengtsson-Olivecrona, G., Deckelbaum, R.J., Carpentier, Y.A. & Olivecrona, T. (1990). Fatty acid control of lipoprotein lipase: a link between energy metabolism and lipid transport. *Proc. Natl. Acad. Sci. USA*, 87, 909–913.
54. Karpe, F., Olivecrona, T., Walldius, G. & Hamsten, A. (1992). Lipoprotein lipase in plasma after an oral fat load: relation to free fatty acids. *J. Lipid Res.*, 33, 975–984.
55. Eisenberg, S., Sehayek, E., Olivecrona, T. & Vlodavsky, I. (1992). Lipoprotein lipase enhances binding of lipoproteins to heparan sulfate on cell surfaces and extracellular matrix. *J. Clin. Invest.*, 90, 2013–2021.
56. Mulder, M., Lombardi, P., Jansen, H., Van Berkel, T.J.C., Frants, R.R. & Havekes, L.M. (1992). Heparan sulphate proteoglycans are involved in the lipoprotein lipase-mediated enhancement of the cellular binding of very low density and low density lipoproteins. *Biochem. Biophys. Res. Commun.*, 185, 582–587.
57. Friedman, G., Chajek-Shaul, T., Stein, O., Olivecrona, T. & Stein, Y. (1981). The role of lipoprotein lipase in the assimilation of cholesteryl linoleyl ether by cultured cells incubated with labeled chylomicrons. *Biochim. Biophys. Acta*, 666, 156–164.
58. Williams, K.J., Fless, G.M., Petrie, K.A., Snyder, M.L., Brocia, R.W. & Swenson, T.L. (1992). Mechanisms by which lipoprotein lipase alters cellular metabolism of lipoprotein (a), low density lipoprotein, and nascent lipoproteins. Roles for low density lipoprotein receptors and heparan sulfate proteoglycans. *J. Biol. Chem.*, 267, 13284–13292.
59. Stein, O., Halperin, G., Leitersdorf, E., Olivecrona, T. & Stein, Y. (1984). Lipoprotein lipase mediated uptake of non-degradable ether analogues of phosphatidylcholine and cholesteryl ester by cultured cells. *Biochim. Biophys. Acta*, 795, 47–59.
60. Beisiegel, U., Weber, W. & Bengtsson-Olivecrona, G. (1991). Lipoprotein lipase enhances the binding of chylomicrons to low density lipoprotein receptor-related protein. *Proc. Natl. Acad. Sci. USA*, 88, 8342–8346.
61. Breckenridge, W.C. (1987). Deficiencies of plasma lipolytic activities. *Am. Heart J.*, 113, 567–573.
62. Auwerx, J.H., Marzetta, C.A., Hokanson, J.E. & Brunzell, J.D. (1989). Large buoyant LDL-like particles in hepatic lipase deficiency. *Arteriosclerosis*, 9, 319–325.
63. Demant, T., Carlson, L.A., Holmquist, L., Karpe, F., Nilsson-Ehle, P., Packard, C.J. & Shepherd, J. (1988). Lipoprotein metabolism in hepatic lipase deficiency: studies on the turnover of apolipoprotein B and on the effect of hepatic lipase on high density lipoprotein. *J. Lipid Res.*, 29, 1603–1611.
64. Clay, M.A., Hopkins, G.J., Ehnholm, C. & Barter, P.J. (1989). The rabbit as an animal model of hepatic lipase deficiency. *Biochim. Biophys. Acta*, 1002, 173–181.
65. Jansen, H. & Hülsmann, W.C. (1985). Enzymology and physiological role of hepatic lipase. *Biochem. Soc. Trans.*, 13, 24–26.

66. Brown, M.L., Hesler, C. & Tall, A.R. (1990). Plasma enzymes and transfer proteins in cholesterol metabolism. *Curr. Opin. Lipidol.*, 1, 122–127.
67. Bengtsson, G. & Olivecrona, T. (1980). The hepatic heparin-releasable lipase binds to high density lipoproteins. *FEBS Letters*, 119, 290–292.
68. Hamilton, J.A. & Small, D.M. (1981). Solubilization and localization of triolein in phosphatidylcholine bilayers: a carbon-13 NMR study. *Proc. Natl. Acad. Sci. USA*, 78, 6878–6882.
69. Scow, R.O. & Egelrud, T. (1976). Hydrolysis of chylomicron phosphatidylcholine in vitro by lipoprotein lipase, phospholipase A_2 and phospholipase C. *Biochim. Biophys. Acta*, 431, 538–549.
70. Deckelbaum, R.J., Ramakrishnan, R., Eisenberg, S., Olivecrona, T. & Bengtsson-Olivecrona, G. (1992). Triglyceride and phospholipid hydrolysis in human plasma lipoproteins: role of lipoprotein and hepatic lipase. *Biochemistry*, 31, 8544–8551.
71. Ikeda, Y., Takagi, A. & Yamamoto, A. (1989). Purification and characterization of lipoprotein lipase and hepatic triglyceride lipase from human postheparin plasma: production of monospecific antibody to the individual lipase. *Biochim. Biophys. Acta*, 1003, 254–269.
72. Nilsson, Å., Landin, B. & Schotz, M.C. (1987). Hydrolysis of chylomicron arachidonate and linoleate ester bonds by lipoprotein lipase and hepatic lipase. *J. Lipid Res.*, 28, 510–517.
73. Weintraub, M.S., Eisenberg, S. & Breslow, J.L. (1987). Different patterns of postprandial lipoprotein metabolism in normal, type IIa, type III, and type IV hyperlipoproteinemic individuals. *J. Clin. Invest.*, 79, 1110–1119.
74. Patsch, J.R., Prasad, S., Gotto, A.M. Jr & Patsch, W. (1987). High density lipoprotein 2. Relationship of the plasma levels of this lipoprotein species to its composition, to the magnitude of postprandial lipemia, and to the activities of lipoprotein lipase and hepatic lipase. *J. Clin. Invest.*, 80, 341–347.
75. Miesenböck, G., Hölzl, B., Föger, B., Brandstätter, E., Paulweber, B., Sandhofer, F. & Patsch, J.R. (1993). Heterozygous lipoprotein lipase deficiency due to a missense mutation as the cause of triglyceride intolerance with multiple lipoprotein abnormalities. *J. Clin. Invest.*, in the press.
76. Groot, P.H.E., van Stiphout, W.A.H.J., Krauss, X.H., Jansen, H., Van Tol, A., Van Ramhorst, E., Chin-On, S., Hofman, A., Cresswell, S.R. & Havekes, L.M. (1991). Postprandial lipoprotein metabolism in normolipidemic men with and without coronary artery disease. *Arterioscler. Thromb.*, 11, 653–662.
77. Taskinen, M.R. (1987). Lipoprotein lipase in hypertriglyceridemias. In *Lipoprotein Lipase*, ed. Borensztajn, J., pp. 201–228. Chicago: Evener Publishers.
78. Wilson, D.E., Emi, M., Iverius, P.-H., Hata, A., Wu, L.L., Hillas, E., Williams, R.R. & Lalouel, J.-M. (1990). Phenotypic expression of heterozygous lipoprotein lipase deficiency in the extended pedigree of a proband homozygous for a missense mutation. *J. Clin. Invest.*, 86, 735–750.
79. Sniderman, A., Baldo, A. & Cianflone, K. (1992). The potential role of acylation stimulating protein as a determinant of plasma triglyceride clearance and intracellular triglyceride synthesis. *Curr. Opin. Lipidol.*, 3, 202–207.
80. Cryer, A.S., Riley, S.E., Williams, E.R. & Robinson, D.S. (1976). Effect of nutritional status on rat adipose tissue, muscle and post-heparin plasma clearing factor lipase activities: their relationship to triglyceride fatty acid uptake by fat cells and to plsma insulin concentrations. *Clin. Sci. Mol. Med.*, 50, 213–221.
81. Brun, L.-D., Gagné, C., Julien, P., Tremblay, A., Moorjani, S., Bouchard, C. & Lupien, P.-J. (1989). Familial lipoprotein lipase-activity deficiency: Study of total body fatness and subcutaneous fat tissue distribution. *Metabolism*, 38, 1005–1009.
82. Taskinen, M.R. & Kuusi, T. (1987). Enzymes involved in triglyceride hydrolysis. In

Clinical Endocrinology and Metabolism. Lipoprotein Metabolism, ed. Shepherd, J., pp. 639–666. London: W.B. Saunders.

83. Miesenböck, G. & Patsch, J.R. (1992). Postprandial hyperlipemia: the search for the atherogenic lipoprotein. *Curr. Opin. Lipidol.*, 3, 196–201.

84. Karpe, F., Tornvall, P., Olivecrona, T., Steiner, G., Carlson, L.A. & Hamsten, A. (1993). Composition of human low density lipoproteins. Effects of postprandial triglyceride-rich lipoproteins, lipoprotein lipase, hepatic lipase and cholesterol ester transfer protein. *Atherosclerosis*, 98, 33–49.

85. Deckelbaum, R.J., Granot, E., Oschry, Y., Rose, I. & Eisenberg, S. (1984). Plasma triglyceride determines structure-composition in low and high density lipoproteins. *Arteriosclerosis*, 4, 225–231.

86. Olivecrona, T., Bengtsson-Olivecrona, G. & Hultin, M. (1991). Lipases in lipoprotein metabolism. In *Lipoproteins and the Pathogenesis of Atherosclerosis*, ed. Shepherd, J., Packard, C.J. & Brownlie, S.M., pp. 51–58. Amsterdam: Elsevier.

16.

Protein engineering of lipases

RICK BOTT, JOHN W. SHIELD and A.J. POULOSE

The study of lipases has been stimulated by several factors. These include the availability of techniques offered by recombinant-DNA technology to produce large quantities of several representative lipases, an increasing interest in the industrial application of enzymes and the availability of three-dimensional structures determined at high resolution by X-ray crystallography. In this chapter we focus on the current state of and the future opportunities for the application of protein engineering to enzymes of this class.

1. PROTEIN ENGINEERING

Recombinant DNA technology makes it possible to alter the DNA sequence at specific sites of a lipase enzyme in order to alter the amino-acid sequence and to produce a variant lipase. The central dogma of molecular biology is that the deoxynucleotide sequence determines the amino-acid sequence of a protein, and this sequence in turn determines the three-dimensional folding of the polypeptide chain to give a single stable tertiary structure. This tertiary structure brings together various amino-acid side-chains, including a catalytic triad composed of a serine, an aspartic or glutamic acid and a histidine that are widely dispersed in the linear amino-acid sequence; the juxtaposition of these confers catalytic properties on the lipase to hydrolyse ester linkages. In addition to the catalytic centre formed by the tertiary folding of the polypeptide backbone, the overall surface topology often contributes to the binding and subsequent hydrolysis of particular substrates that interact favourably with this surface. Thus the lipase is not only a catalyst, but very often a rather selective one, hydrolysing specific ester linkages associated with particular fatty-acid side-chains. Recombinant DNA technology allows the deliberate alteration of these functional properties by modification of the DNA sequence encoding the amino-acid sequence of a protein and in turn the three-dimensional structure of the enzyme. When successful, this technique is referred to as protein engineering.

Protein engineering can be performed before or after translation. Pre-translational modifications are accomplished by employing genetic engineering techniques to introduce changes in the DNA sequence coding for the amino-acid residues comprising the primary structure of the protein. Post-translational modifications involve changes to amino-acid side-chains or to glycosylation after the protein has been synthesized. In this chapter, we look at primarily pre-translational modifications to the gene leading to changes in the amino-acid sequence of the protein.

1.1 Cloning and expression of natural lipases

Recombinant DNA technology has now made it possible to identify and transfer the DNA encoding the amino-acid sequence of any particular lipase into a heterologous host and to alter the regulation of this gene so that it is produced in large quantities. If a lipase is to be prepared economically for industrial applications, it is likely that the lipase will have to be produced in a heterologous host. Consequently, the wild-type gene will have to be cloned into an organism well suited for the production of commercial quantities of industrial enzymes. The requirements made of the host organism are numerous and include: that the organism be safe to handle and to manipulate genetically, that the organism secrete the target enzyme, and that production of the enzyme with the organism be economic. *Bacillus subtilis* and *Aspergillus niger* are two common host organism systems used in the production of lipase enzymes (1, 2).

1.2 Current industrial interest in lipase

Lipases have evolved to be efficient catalysts for lipolytic reactions involving the hydrolysis of ester linkages of mono-, di- and triglycerides in aqueous emulsions. There are a number of industrial applications where hydrolysis of these bonds is important: the generation of fatty acids from natural oils for the production of soaps; the removal of oils and fats from fabrics, machinery, hides and waste water; and the production of mono- and diglycerides for food emulsifiers. Lipases can also be used to conduct transesterification reactions. The selectivity of lipases towards the length of the fatty acids or the number and location of unsaturations in the fatty acids is employed to produce high-value fats or oils (3). The high stereospecificity of lipases can also be exploited to synthesize specific compounds, in particular enantiomerically pure compounds containing ester bonds. For thermodynamic reasons, it is often necessary for this type of synthesis reaction to be conducted in environments containing minimal amounts of water. Consequently, there have been numerous reports of the ability of lipases in organic solvents to synthesize a variety of compounds (4), including precursors for biologically active therapeutics, herbicides, or pesticides.

2. LIPASE STRUCTURE

2.1 Sequence homology

Many lipase amino-acid sequences have been deduced from the sequences of cloned lipases. Comparison of human (5) bovine (6) and murine (7) lipoprotein lipases revealed more than 87% identity (8), while the sequence of chicken lipoprotein lipase (9), with a lower degree of identity, still shows clear homology. The amino-acid sequences of pancreatic lipases from man (10), pig (11), dog (12), rat (13, 14) and mouse (15) have been determined and they also show considerable sequence homology. Recently, two human proteins related to pancreatic lipase have been cloned; these showed more than 65% sequence identity with human pancreatic lipase (16). Similarly, hepatic lipase from humans (17) and rats (18) also show significant sequence homology with the pancreatic and lipoprotein lipases.

It is evident from the known sequences that lipoprotein lipase, pancreatic lipase and hepatic lipase are homologous enzymes, belonging to a lipase gene family. However, even though there are striking similarities between the three-dimensional structures of microbial and mammalian lipases (19, 20), there is very little overall sequence identity, except for the sequences around the active-site residues. Comparison of sequences around the amino-acid residues of the catalytic triad in proteases, lipases, cutinases and esterases show the presence of GxSxG consensus sequence (Table 1) at the active serine site. The only known exception to this consensus sequence is the substitution of alanine for the second glycine in subtilisins. No such consensus sequences are identified around the catalytic histidine or aspartic (glutamic) acid residues, except in similar enzymes from closely related species.

A recent analysis of the conformation of the pentapeptide GxSxG in trypsin, subtilisin, lipase and esterase enzymes showed that in the case of proteases, the invariant glycine residues are conserved because of conformational requirements (32). Furthermore, as will be discussed below, the stereochemistry of the active serine loop is different in subtilisin when this is compared with trypsin; these two serine protease families are not regarded as homologous. In the case of lipases and esterases, the invariant glycines are conserved because of the steric restrictions imposed by the stereochemistry of packing. This study also indicates that the active-serine loop in lipases and esterases is a β-ε Ser-α structural motif, and thus these enzymes are evolutionarily linked. The same study also concluded that there is no evolutionary link between serine proteases and other serine hydrolases. However, the existence of the consensus amino-acid sequence GxSxG among all known serine hydrolases from a wide variety of organisms is intriguing.

Table 1. Comparison of sequences around active serine.

Enzyme class	Enzyme	Local sequence	Reference
		* * **	
Proteases	Trypsin (bovine)	Q - G - D - S - G - G - P	(21)
	Elastase (pig)	Q - G - D - S - G - G - P	(21)
	Plasmin (human)	Q - G - D - S - A - G - G	(21)
	Protease-A (*S. griseus*)	P - G - D - S - G - G - S	(21)
Lipases	Pancreatic lipase (human)	I - G - H - S - L - G - A	(10)
	Lipase (*R. miehei*)	T - G - H - S - L - G - G	(22)
	Lipase (*G. candidum*)	F - G - E - S - A - G - A	(23)
	Lipoprotein lipase (human)	L - G - Y - S - L - G - A	(5)
	Lingual lipase (rat)	V - G - H - S - Q - G - T	(24)
Cutinases	Cutinase (*P. mendocina*)	S - G - H - S - Q - G - G	(25)
	Cutinase (*F. solani*)	G - G - Y - S - Q - G - A	(26)
	Cutinase (*C. capsici*)	G - G - Y - S - Q - G - T	(27)
Esterases	Butyrylcholine esterase (human)	F - G - E - S - A - G - A	(28)
	Carboxylesterase (murine)	F - G - E - S - S - G - G	(29)
	Acetylcholine esterase (*T. californica*)	F - G - E - S - A - G - G	(30)
	Thioesterase (duck)	F - G - H - S - F - G - S	(31)

The conserved residues of the consensus sequence GxSxG are indicated by asterisks.

2.2 Three-dimensional structure

With the publication of the first two X-ray-crystallographic, three-dimensional structures of lipases, the triacylglycerol lipase from *Rhizomucor miehei* (19) and the human pancreatic lipase (20), it immediately became apparent that the catalytic centres of these enzyme are composed of a Ser...His...Asp triad, as in the serine proteases trypsin and subtilisin. It was also obvious that the lipase active sites were not evolutionarily related to the serine proteases. There are two classes of serine proteases: the trypsin-like class and subtilisin-like class. In these two classes of serine proteases, the catalytic-triad amino acids are arranged in a different linear order: Asp32, His64 and Ser221 in the amino-acid sequence of subtilisin BPN′ and His57, Asp102 and Ser195 in the amino-acid sequence of

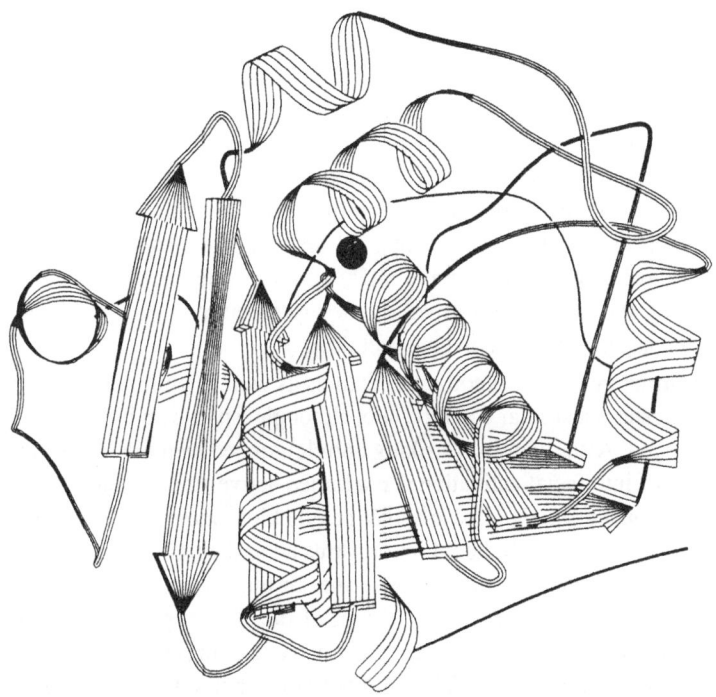

Figure 1. The secondary structure and folding of *Rhizomucor miehei* lipase (19). The location of the active site is indicated by a dot below the 'lid', which is formed by a helical segment consisting of residues 81–85.

bovine pancreatic trypsin. The lipase structures have a catalytic triad composed of serine, histidine and aspartic acid arranged in a third linear order: Ser126, Asp176 and His206 in the sequence of *Pseudomonas mendocina* lipase. As such, the lipases represent a third example of convergent evolution toward structures forming a serine catalytic triad. The lipase triad is also different in another manner, in that the three-dimensional placing of the three catalytic residues is the mirror-image of the corresponding placing in catalytic triads of proteases. The active sites of two lipases are inaccessible for solvent: in *Rhizomucor miehei* lipase the active site is buried under the head of a long loop or 'lid' comprising residues 84–91 (Figure 1), while in human pancreatic lipase three segments block access to the catalytic triad; these are a 'flap' formed by residues in a surface loop between residues 237 and 261, and neighbouring segments 75–79 and 212–216.

Two additional lipase structures have been reported for the lipase from *Geotrichum candidum* (33) and the cutinase from *Fusarium solani* (34). The *Geotrichum* lipase structure produced a surprise, as the catalytic triad of this lipase is formed by serine, histidine and a glutamic acid rather than aspartic acid. The *Geotrichum* enzyme also established a strong link by sequence and three-dimensional structural homology to acetylcholinesterase, a representative enzyme of a growing class of related structures known as the α/β hydrolases. The α/β hydrolases represent a class of related protein structures that includes several enzymes having quite diverse catalytic functions while sharing a common tertiary fold (35). The structure of the cutinase from *Fusarium* is also noteworthy in that unlike the previous lipase structures, all sharing the property of interfacial activation and having buried active sites, cutinase does not exhibit interfacial activation and the active site formed by the more common catalytic triad serine, histidine and aspartic acid is solvent accessible.

In these structures, several common features as well as points of divergence are apparent. All share a similar catalytic site, although aspartic and glutamic acid can be interchanged. Also, there is a striking correlation between the lipases that share the property of interfacial activation and a buried active site. An inhibitor complex between *Rhizomucor miehei* lipase and *N*-hexyl chlorophosphonate ethyl ester has been reported (36) in which the inhibitor forms a tetrahedral intermediate between the phosphonate group and the active serine. In this structure, a conformational change is observed involving a movement of the 'lid' segment responsible for burying the active site in the native enzyme to give a conformation that renders the active site more accessible. This movement results in the exposure of a hydrophobic binding site and surface. This result, along with the similar findings for phospholipase A_2 discussed below and the accessibility of the active site in *Fusarium* cutinase, confirmed the existence of a correlation between a buried active site on the one hand and interfacial activation involving a conformational change to facilitate substrate binding on the other.

2.3 Lipases are α/β hydrolases

Just as there is variation in the identity of the acid moiety in the catalytic triad, it appears that there may be two variations in the tertiary folding found in the three-dimensional structures of lipases. All bear some or considerable structural similarity with a recently discovered class of enzymes known as the α/β hydrolases. Three of the four reported lipase structures share the 'hydrolase fold' (Figure 2) first seen in the structure of dienelactone hydrolase (37), and later reported for wheat-germ serine carboxypeptidase (38), acetylcholine esterase (39) and haloalkane dehalogenase (40). These enzymes catalyse a similar hydrolytic action on a wide variety of substrates ranging from proteins, acetylcholine and halogenated compounds and lipids. This is by far the widest-ranging family of enzymes found to share a common tertiary fold. Only the lipase from *Rhizomucor miehei* differs in that, while it shares a common carboxy-terminal half with the hydrolases above, the N-terminal half of this enzyme has a different alignment of

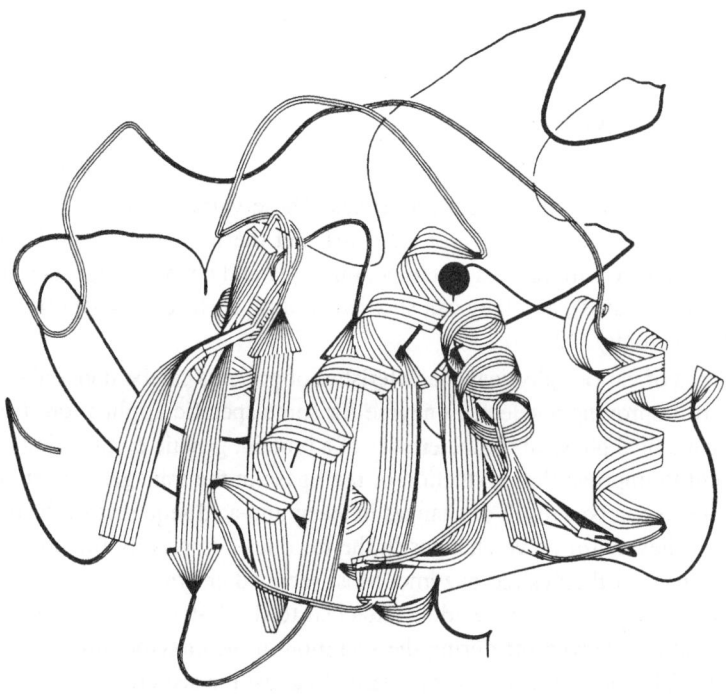

Figure 2. The secondary structure and folding of the α/β hydrolase fold. This example is taken from the three-dimensional structure of wheat-germ carboxypeptidase II (38), where features of secondary structure that are unique to this enzyme have been suppressed. The location of the active site is marked by a dot and the diagram is oriented to facilitate comparison with Figure 1.

parallel β-sheet strands. Recently the structure of a *Humicola* enzyme has been determined that shares a common tertiary folding pattern with the *Rhizomucor miehei* lipase (41). Thus the lipases may have the property of accepting a much broader range of substrates than previously described families of enzymes, such as the serine proteases of the subtilisin and trypsin classes, which have been the objects of extensive protein engineering. All these hydrolases share a common catalytic triad that is a mirror image of the catalytic triad of serine proteases and papain.

This structural homology indicates that the entire structural database relating structure and function in different lipases can be utilised in designing experiments to engineer a particular lipase. This homology in the three-dimensional structures of very closely related enzymes has been exploited for other enzymes as a guide to selection of what sites to modify and what substitutions to make in order to alter the properties of one enzyme in order to make them resemble more closely those of another.

3. PROTEIN ENGINEERING

Protein engineering can be employed in enzyme systems to produce variant enzymes with altered amino-acid sequences. These variants can have modified specific activity, increased k_{cat}, altered pH and thermal activity profiles and increased stability with respect to temperature, pH and chemical agents (such as oxidants or proteases), and they can show altered pI, surface hydrophobicity and substrate specificity (42, 43).

Alteration of various physical characteristics of proteins can be undertaken with the goal of improving a selected enzyme for some specific application. Besides changing the final physical characteristics of a protein, genetic engineering can be carried out to improve the production of the lipase in the transgenic host. These improvements can be a result of changes in the amino-acid sequence of the desired protein, or they can involve changes in the residues comprising the pre- and/or pro- sequences of the desired enzyme. Most secreted enzymes contain a pre- or signal sequence, a series of residues attached to the N-terminus of the mature enzyme that are cleaved off during the secretion of the enzyme through the cell membrane. This pre-sequence is considered to assist the secretion of the enzyme by influencing the enzyme's configuration and its interaction with the secretion apparatus of the cell. Changes in the amino-acid sequence of this pre- region may alter the rate of production of the desired enzyme. However, this type of engineering does not alter any of the characteristics of the product enzyme and will not be discussed further in this chapter.

3.1 Engineering of phospholipase A₂

Phospholipases A_2 are enzymes that hydrolyse specifically the secondary acyl ester linkage of glycerophospholipids. Considerable progress has been made in the understanding of structure–function relationships of the phospholipase A_2 enzymes (44) and in engineering different properties into them. These results may serve as a paradigm for the engineering of lipases. It is therefore worth noting the similarities and differences between the phospholipases and triglyceride lipases. The molecular weights of phospholipases A_2 are more closely clustered and are lower (12–14 kDa) than those of the triglyceride lipases (25–55 kDa). Both classes of enzymes are selective for ester bonds and include enzymes manifesting the property of interfacial activation in the presence of a lipid bilayer. The active sites of the two enzymes share a histidine and either an aspartic or a glutamic acid. The serine found in the catalytic triad of the lipases is absent from the phospholipases A_2, being replaced by an ordered solvent molecule occupying the position of the serine hydroxyl oxygen atom in the structure of lipase active sites. Efforts to understand the mechanism of interfacial activation through analysis of the three-dimensional structure of complexes with transition-state analogues have also highlighted a number of similarities and some differences. Both classes of enzymes with the property of interfacial activation display in their complexes with transition-state analogues conformational changes that expose more hydrophobic surface. In the phospholipases A_2 complexes, the conformational change involves changes in the conformations of side-chains of different amino-acid residues that line the active site, while in the only similar complex with a triglyceride lipase (36), the folding of the main chain of the backbone of surface loops undergoes conformational changes to expose an enlarged hydrophobic surface and the active site that previously was buried. The enzymes found in the phospholipase A_2 class are divided into three groups. These groups are selected by probable evolutionary homology and structural similarities. Many examples of phospholipases are known from the venom of poisonous snakes and insects. As might be expected, the two largest groups separate into phospholipase A_2 coming either from snakes such as the rattlesnake, found in the western hemisphere, or from the cobra, found in the eastern hemisphere. The third, more distantly related class of phospholipase A_2 includes the enzyme found in bee venom.

The structures of porcine pancreatic phospholipase A_2 (45) and bovine pancreatic phospholipase A_2 (46) are quite similar. The snake-venom phospholipases A_2 from rattlesnake (47) and cobra (48) are homologous to these examples of pancreatic phospholipase A_2. All share a folding pattern that is not at all similar to that found in the known structures of triglyceride lipases. The three-dimensional structures of cobra- and bee-venom phospholipases A_2 in complexes with transition-state analogues have been published (48, 49). The bee-venom phospholipase A_2 retains the same catalytic mechanism as the phospholipases A_2 from pancreas and snake, although the sequence of the bee enzyme has diverged strongly in comparison with the others. The active sites of the phospholipases A_2 are conserved and also are quite different from the active sites of the lipases, in that serine is replaced by a bound solvent molecule that serves as the nucleophile in the hydrolysis reaction.

One of the earliest, ground-breaking experiments in protein engineering was to delete part of a surface loop, residues 62–72, found in the structures of bovine and porcine pancreatic phospholipase A_2 (50). Part of this loop is absent from the structure of the phospholipase A_2 from rattlesnake (47), which is more active than the pancreatic phospholipases A_2 and structurally is otherwise homologous to the pancreatic phospholipases A_2 (45, 46). A variant porcine pancreatic phospholipase A_2 having a five-residue deletion (residues 62–66) and the substitutions D59S, S60G and N67Y showed increased enzymic activity (50).

The native porcine phospholipase A_2 has a turnover number (k_{cat}) for micellar substrates that is lower by a factor of 9–160 compaared with that of snake-venom phospholipase A_2. The variant enzyme shows intermediate increases in enzymic activity, both on monomeric and on micellar substrates, by factors ranging from 2 to 16. The three-dimensional structure of this variant has been determined at 2.1 Å resolution (51). All three phospholipases A_2, the porcine pancreatic native enzyme, the porcine pancreatic variant enzyme and the rattlesnake enzyme, have identical active sites. The major differences are found in the region of the loop deletion, where the variant enzyme adopts a loop conformation that is intermediate between those of the native pancreatic enzyme and the rattlesnake enzyme. The deletion of residues 62–66 in the variant enzyme smoothes a bulge seen in the porcine enzyme, and the variant enzyme's binding site for lipid aggregates forms a smooth plane (51). This work represents the first attempt to 'sculpt' an enzyme by deleting and replacing amino-acid residues on the enzyme surface, on the basis of detailed comparison between related enzymes. This result corroborates the details of enzyme–micelle interaction that have been deduced from X-ray crystallographic studies with cobra and bee-venom phospholipase A_2 transition-state analogue complexes (48). This example parallels earlier experiments in protein engineering with the serine protease subtilisin (52). In that study, the amino-acid sequences of subtilisin BPN′ and subtilisin Carlsberg were compared in the region of the active site. This pinpointed differences likely to confer upon subtilisin Carlsberg its rather different specificity and its greater turnover number. The introduction into subtilisin BPN′ of three of the 88 possible differences resulted in a variant of subtilisin BPN′ that had 'recruited' the altered substrate specificity and the increased k_{cat} of subtilisin Carlsberg.

Additional examples of engineered phospholipase A_2 variants bearing single or double site-specific substitutions that led to improved enzymic properties have been reported by a number of laboratories. Replacing lysine 56 with methionine in the amino-acid sequence of bovine pancreatic phospholipase A_2 resulted in a variant enzyme with increased k_{cat} and decreased apparent K_m for a series of different micellar substrates (53). Replacing the tyrosine at position 69 with phenylalanine in porcine pancreatic phospholipase A_2 altered the stereospecificity of the variant enzyme (54). The native enzyme shows no activity on D-lecithin and D-glycerosulphate, whereas the variant Y69F shows low but significant activity on these substrates (54). Leucine is conserved in all pancreatic phospholipases A_2 at position 31, while in snake-venom phospholipases it is quite variable. The different residues found at position 31 in the amino-acid sequences of different snake-venom phospholipases A_2 were introduced into porcine pancreatic phospholipase A_2. One phospholipase A_2 variant having tryptophan at position 31 showed increased activity for monomeric substrates and increased affinity for the micellar interface (55).

In addition to altering the enzymic properties of different phospholipases A_2, increased stability has also been achieved by altering the pattern of charged residues near the N terminus of helix 5 in porcine pancreatic phospholipase A_2. Two variants with aspartic acid in place of asparagine at position 89 (N89D) both showed increased stability relative to the native enzyme (56), as did a double variant having the same substitution at position 89 along with lysine instead of glutamic acid at position 92 (N89D/E92K) (56). It seems clear that the phospholipases A_2 are as amenable to the alteration of their functional properties as the subtilisin and trypsin classes of serine proteases are (42, 43).

3.2 Engineering of lipases

The lipases share a catalytic-site topology with the serine proteases, for which it was possible to alter overall enzymic activity, substrate specificity, activity–pH profiles and stability toward various chemical oxidants and denaturing factors. Lipases also share a number of functional properties such as interfacial activation, stereospecificity and ester group selectivity with the phospholipases A_2, for which there are examples of engineered variants showing changes in these properties. Given the immediate industrial applications of lipases, any one of these properties or a combination of them in an engineered lipase could have considerable impact.

However, in spite of the numerous potential physical properties of lipase enzymes that could be altered by protein engineering, only a few references can be found in the literature concerning efforts to engineer lipases. Compared with other classes of hydrolases, lipases tend to be harder to work with, primarily because lipases are interfacially active and relatively difficult to characterise and assay. The most significant work regarding protein engineering of lipase enzymes has been conducted in industrial laboratories, in attempts to make industrially significant improvements in potentially high-volume markets.

Lipase enzymes have a Ser...His...Asp catalytic triad, in which the active serine is often located in an elongated, hydrophobic binding-pocket. This region is referred to as the lipid contact zone, and it is probably separated from the bulk aqueous environment by a small hydrophobic flap on the enzyme. It might be possible to change, *via* protein engineering, the accessibility of the active site and the residues that participate in the interaction with the substrate.

At present, the largest market for lipase enzymes is due to their use as ingredients in heavy-duty laundry detergents. Lipase enzymes hydrolyse natural fat and oil stains that are difficult to remove at low wash temperatures with economical surfactants (57). However, a component in many laundry detergents is protease, and it is possible that the proteases will hydrolyse any added lipase, reducing the efficacy of the lipase. Consequently, workers at Unilever (58) have identified the sites within a *Pseudomonas* species lipase that are hydrolysed by commercially available proteases used in detergents, and have engineered the lipase so as to increase its resistance to proteolysis. The engineering was accomplished by substituting or inserting amino-acid residues at the site of proteolytic attack or in its vicinity, resulting in a substantial increase in resistance of the lipase to proteolysis. A number of different approaches were shown to be effective in increasing stability toward proteolytic attack. Specifically, these were: (a) the amino-acid residue that upon cleavage would become the N terminus of the cleaved fragment is replaced by proline; (b) an amino-acid residue two or four positions away from the cleavage site is replaced by a charged or more polar residue; (c) the electrostatic potential of the cleavage site is modified by insertion or substitution of positively charged residues, or by removal of negatively charged residues; (d) 1, 2 or 3 residues from the proteolytic cleavage site are deleted.

An important characteristic for enzymes destined for use in laundry detergents is the ability of the enzyme to withstand oxidative degradation caused by bleach components in the detergent formulation. Primarily, methionine residues in detergent enzymes are subject to oxidative attack, and oxidative stability can be increased by removing methionine residues that are susceptible to oxidative attack on the enzyme, or by replacing them with oxidatively stable residues. This approach to improve oxidative stability has been successfully applied in detergent protease enzymes (58, 59), and has also been applied to potential detergent lipase enzymes (60).

Another important characteristic of enzymes used in laundry detergents is the need for the enzyme to have an alkaline pH optimum, or to have an alkaline pI. Protein engineering can be used to increase the pI of the enzyme by increasing its nett positive charge. This is achieved by the deletion of negatively charged residues or their replacement by neutral residues, by substituting positively charged residues for neutral or negatively charged residues, or by the insertion of positively charged residues.

At Genencor International we have worked extensively with a lipase from *Pseudomonas mendocina*. This enzyme, which originally was identified and characterised as a cutinase (61, 62), hydrolyses cutin, a plant polyester that envelopes all plant leaves. This enzyme is very efficient in the hydrolysis of triglyceride substrates. The gene encoding this lipase was cloned and expressed in *Escherichia coli*, and its amino-acid sequence was deduced from the gene sequence (63). The *E. coli* production system produced substantial quantities of the lipase, making this enzyme a convenient model for protein engineering studies on lipases.

Before the crystal structure of this enzyme became available, it was possible to identify the functionally critical regions in the enzyme by chemical modification and site-specific mutagenesis (25). Diethyl *p*-nitrophenyl phosphate inactivated the enzyme by modification of a single serine at position 126. Replacement of serine in the native enzyme with alanine, by site-specific mutagenesis, resulted in an inactive protein. These results suggested that this lipase contains an active serine similar to the other hydrolases containing catalytic triads. Identification of the histidine and aspartic acid residues of the catalytic triad was achieved by site-specific mutagenesis. Each histidine in the lipase was replaced by a glutamine, and the resulting enzymes were analysed for lipase activity. It was found that the enzyme activity was lost only when His206 was replaced with glutamine, suggesting that His206 is part of the catalytic triad. Similarly, aspartic acid at position 176 was identified as the third amino-acid residue in the catalytic triad.

In order to alter the catalytic properties of this lipase, we decided to perturb the active site by making amino-acid substitutions around the catalytic-triad residues. We examined the catalytic constants k_{cat} and K_m for synthetic p-nitrophenyl esters (25). Significant alteration in K_m for p-nitrophenyl butyrate was observed when amino-acid substitutions were made around the active serine and histidine. K_m was decreased by a factor of 30 when serine was substituted for glutamine at position 127; an increase in K_m was observed when serine was replaced by glutamic acid at position 205. Improvements in k_{cat} were observed when substitutions were made near the active histidine, whereas dramatic decreases in the value of k_{cat} were observed when substitutions were made in the active-serine region. It is worth noting that substitution of asparagine for glutamine at position 127 resulted in an almost inactive enzyme.

Additionally, we observed that with substitutions in the vicinity of the catalytic triad residues, one can generate a series of enzymes that will have a wide range of substrate specificity. A measure of this specificity is the ratio of the specificity constants for two synthetic substrates. In Table 2, this ratio is given for several variants of the lipase enzyme. Single amino-acid substitution near the catalytic triad produced enzymes where this ratio is changed by a factor of up to 17.

The nucleophile specificity of this lipase was also altered by amino-acid substitutions in the regions around the catalytic-triad residues. Hydrolysis reactions proceed with an initial formation of acyl enzyme by the transfer of acyl group from the ester substrate to the active serine on the enzyme. Subsequent nucleophilic attack upon the acyl enzyme by water releases the acid from the enzyme and makes the serine available for further acylation. When alcohols are present in the reaction medium, they can act as nucleophiles to form esters instead of acids. When both water and alcohol are present, the enzyme generates both acids and esters. The relative amounts of each will depend on the specificity of the enzyme towards the nucleophiles present in the reaction medium. We have observed that substitution of amino acids around active serine influences the specificity of the enzyme towards methanol. A twofold increase in methanolysis/hydrolysis ratio was observed to be induced by a single amino-acid substitution at position 127 (64).

Table 2. Altered substrate specificity of *Pseudomonas mendocina* lipase by single amino-acid substitutions.

Enzyme	$\dfrac{k_{cat}/K_m \text{ (} p\text{-nitrophenyl butyrate)}}{k_{cat}/K_m \text{ (} p\text{-nitrophenyl caprylate)}}$
Ser205 → Arg	77
Phe207 → Ala	56
Phe207 → His	28
Wild type	14
Ser205 → Thr	9
Gln127 → Leu	3
Gln127 → Arg	1
Gln127 → Thr	0.8

The putative catalytic triad for this enzyme is Ser126...His206...Asp176 (see text).

Similarly, hydrogen peroxide can act as the nucleophile, allowing lipases to generate per-acids from triglycerides in an aqueous reaction medium (65, 66). The hydrogen peroxide attacks the acyl-enzyme resulting in per-acid formation. Preference for hydrogen peroxide as a nucleophile can be improved by amino-acid substitutions near either the active serine or the active histidine (66, 67). Substitution of serine with asparagine at position 205, resulted in a twofold improvement in the perhydrolysis/hydrolysis ratio (Table 3). Similar improvement was obtained by the replacement of phenylalanine with threonine at position 207. When threonine or serine was substituted for glutamine at position 127, a fourfold improvement in the perhydrolysis/hydrolysis ratio was obtained.

It is clear that amino-acid substitutions around the catalytic-triad amino acids of a lipase can influence the catalytic properties such as K_m, k_{cat}, substrate specificity and nucleophile specificity. The magnitude of the changes observed by random substitutions is substantial. Since the crystal structure of the enzyme is not complete, it is impossible even to attempt to rationalise any of the effects due to these amino-acid substitutions. We hope that we can explain some of them through molecular modelling. We anticipate that, once we have the structural information, we shall be able to build further improvements into this lipase.

4. CONCLUDING REMARKS

The stage has been set for lipase engineering. A database is being accumulated relating structure and function for different members of this class of enzymes. In

Table 3. Alteration of perhydrolysis/hydrolysis ratio of *Pseudomonas mendocina* lipase by site-specific mutagenesis.

Enzyme	Perhydrolysis/hydrolysis ratio (trioctanoin substrate)
Wild type	0.14
Asn 205	0.25
Thr 207	0.25
Ser 127	0.56
Thr 127	0.56

addition, initial engineering results point to the similarities with engineering efforts in related hydrolases, such as the serine proteases and phospholipases A_2. In the case of lipase, we can expect to draw on an even more extended database from the α/β hydrolases, which have a very broad substrate range. While none of the future changes that can be introduced in lipase can be predicted with any certainty, this class of enzyme seems predestined to become surveyed exhaustively in the search for commercially interesting engineered enzymes.

REFERENCES

1. Van Kimmenade, A. & Power, S.D. (1992). Secretion of heterologous proteins by *Bacillus subtilis*: A systematic structural approach. Presented at the American Chemical Society meeting, April 1992, San Francisco. *Abstr. Pap. Am. Chem. Soc.*, 203, 1–3.
2. Boel, E. & Huge-Jensen, I.B. (1989). Recombinant *Humicola* lipase and process for the production of recombinant *Humicola* lipases. European patent application EP 0305216.
3. Macrae, A.R. (1983). Lipase catalyzed interesterification of oils and fats. *J. Am. Oil Chem. Soc.*, 60, 291–294.
4. Boland, W., Frobel, C. & Lorenz, M. (1991). Esterolytic and lipolytic enzymes in organic synthesis. Synthesis, *J. Synthetic Org. Chem.*, 12, 1049–1072.
5. Wion, K.L., Kirchgessner, T.D., Lusis, A.J., Schotz, M.C. & Lawn, R.M. (1987). Human lipoprotein lipase complementary DNA Sequence. *Science*, 235, 1638–1641.
6. Senda, M., Oka, K., Brown, W.V., Qasba, P.Q. & Furuichi, J. (1987). Molecular cloning and sequence of a cDNA coding for bovine lipoprotein lipase. *Proc. Natl. Acad. Sci. USA*, 84, 4369–4373.
7. Kirchgessner, T.G., Svenson, K.L., Lusis, A.J. & Schotz, M.C. (1987). The sequence of cDNA encoding lipoprotein lipase. *J. Biol. Chem.*, 262, 8463–8466.
8. Wang, C.S., Hartsuck, J. & McConathy, W.J. (1992). Structure and functional properties of lipoprotein lipase. *Biochim. Biophys. Acta*, 1123, 1–17.

9. Cooper, D.A., Stein, J.C., Strieleman, P.J. & Bensadoun, A. (1989). Avian adipose lipoprotein lipase: cDNA sequence and reciprocal regulation of mRNA levels in adipose and heart. *Biochim. Biophys. Acta*, 1008, 92–101.

10. Lowe, M.E., Rosenblum, J.L. & Strauss, A.W. (1989). Cloning and characterization of human pancreatic lipase cDNA. *J. Biol. Chem.*, 264, 20042–20048.

11. De Caro, J., Boudouard, M., Bonicel, J., Guidoni, A., Desnuelle, P. & Rovery, M. (1981). Porcine pancreatic lipase: Completion of the primary structure. *Biochim. Biophys. Acta*, 671, 129–138.

12. Kerfelec, B., LaForge, K.S., Puigserver, A. & Scheele, G. (1986). Primary structures of canine pancreatic lipase and phospholipase A_2 messenger RNA. *Pancreas*, 1, 430–437.

13. Sims, H.F., Strauss, A.W. & Lowe, M.E. (1991). GenBank accession number M58369.

14. Wicker-Planquart, C. & Puigserver, A. (1992). Primary structure of rat pancreatic lipase mRNA. *FEBS Letters*, 296, 61–66.

15. Grusby, M.J., Nabavi, N., Wong, H., Dick, R.F., Bluestone, J.A., Schotz, M.C. & Glimcher, L.H. (1990). Cloning of an interleukin-4 inducible gene from cytotoxic T lymphocytes and its identification as a lipase. *Cell*, 60, 451–459.

16. Giller, T., Buchwald, P., Blum-Kaclin, D. & Hunziker, W. (1992). Two novel human pancreatic lipase related proteins, hPLRP1 and hPLRP2. *J. Biol. Chem.*, 267, 16509–16516.

17. Datta, S., Luo, C.C., Li, W.H., Van Tuinen, P., Ledbetter, D.H., Brown, M. A., Chen, S.H., Liu, S.W. & Chan, L. (1988). Human hepatic lipase: cloned cDNA sequence, restriction fragment length polymorphisms, chromosomal localization and evolutionary relationships with lipoprotein lipase and pancreatic lipase. *J. Biol. Chem.*, 263, 1107–1110.

18. Komaromy, M.C. & Schotz, M.C. (1987). Cloning of rat hepatic lipase cDNA: evidence for a lipase gene family. *Proc. Natl. Acad. Sci. USA*, 84, 1526–1530.

19. Brady, L., Brzozowski, A.M., Derewenda, Z.S., Dodson, E., Dodson, G., Tolley, S., Turkenburg, J. P., Christiansen, L., Huge-Jensen, B., Nørskov, L., Thim, L. & Menge, U. (1990). A serine protease triad forms the catalytic centre of a triacyl glycerol lipase. *Nature*, 343, 767–770.

20. Winkler, F.K., D'Arcy, A. & Hunziker, W. (1990). Structure of human pancreatic lipase. *Nature*, 343, 771–774.

21. Dayhoff, M.O., Barker, W.C. & Hardman, J. K. (1972). In: *Atlas of Protein Sequence and Structure*, vol. 5, ed. Dayhoff, M.O., pp. 53–66. Washington, DC, USA: National Biomedical Research Foundation.

22. Boel, E., Huge-Jensen, B., Christiansen, M., Lars, T. & Fil, N.P. (1988). *Rhizomucor miehei* triglyceride lipase is synthesized as a precursor. *Lipids*, 23, 701–706.

23. Shimada, Y., Sugihara, A., Tominaga, J., Iizmi, T. & Tsunasawa, S. (1989). cDNA molecular cloning of *Geotrichum candidum* lipase. *J. Biochem.*, 106, 383–388.

24. Docherty, A.J.P., Bodmer, M.W., Angal, S., Verger, R., Rivière, C., Lowe, P.A., Lyons, A., Emptage, J.S. & Harris, T.J.R. (1985). Molecular cloning and nucleotide sequence of rat lingual lipase cDNA. *Nucl. Acids Res.*, 13, 1891–1903.

25. Poulose, A.J., Van Beilen, J., Power, S., Shew, B., Gray, G. & Norton, S. (1990). Protein engineering of a lipase from *Pseudomonas* sp. Presented at the 81st. annual meeting of the American Oil Chemists Society, May 1990, Baltimore. *Inform*, 1 (4), 318.

26. Soliday, C.L. & Kolattukudy, P.E. (1983). Primary structure of the active site region of fungal cutinase, an enzyme involved in phytopathogenesis. *Biochem. Biophys. Res. Commun.*, 114, 1017–1022.

27. Ettinger, W.F., Thukral, S.K. & Kolattukudy, P.E. (1987). Structure of cutinase gene, cDNA and the derived amino acid sequence from phytopathogenic fungi. *Biochemistry*, 26, 7883–7892.

28. Arpagaus, M., Kott, M., Vatsis, K.P., Bartels, C.F., Bert, N.L.D. & Lockridge, O. (1990). Structure of the gene for human butyryl choline esterase: evidence for a single copy. *Biochemistry*, 29, 124–131.

29. Ovnic, M., Tepperman, K., Medda, S., Elliot, R.W., Stephenson, D.A., Grant, S.G. & Ganschow, R.E. (1991). Characterization of a murine cDNA encoding a member of the carboxylase multigene family. *Genomics*, 9, 344–354.
30. Schumacher, M., Camp, S., Maulet, Y., Newton, M., MacPhee-Quigley, K., Taylor, S.S., Friedman, T. & Taylor, P. (1986). Primary structure of *Torpedo californica* acetylcholine esterase deduced from its cDNA sequence. *Nature*, 319, 407–409.
31. Poulose, A.J., Rogers, L., Cheesbrough, T.M. & Kolattukudy, P.E. (1985). Cloning and sequencing of the cDNA for S-acyl fatty acid synthetase thioesterase from the uropygial gland of mallard duck. *J. Biol. Chem.*, 260, 15953–15958.
32. Derewenda, Z.S. & Derewenda, U. (1991). Relationships among serine hydrolases: evidence for a common structural motif in triacyl-glyceride lipases and esterases. *Biochem. Cell Biol.*, 69, 842–851.
33. Schrag, J.D., Li, Y., Wu, S. & Cygler, M. (1991). Ser-His-Glu triad forms the catalytic site of the lipase from *Geotrichum candidum*. *Nature*, 351, 761–764.
34. Martinez, C., DeGeus, P., Lauwereys, M., Matthyssens, G. & Cambillau, C. (1992). *Fusarium solani* cutinase is a lipolytic enzyme with a catalytic serine accessible to solvent. *Nature*, 356, 615–618.
35. Ollis, D.L., Cheah, E., Cygler, M., Dijkstra, B., Frolow, F., Franken, S. M., Harel, M., Remington, S.J., Silman, I., Schrag, J., Sussman, J.L., Verschueren, K.H.G. & Goldman, A. (1992). The α/β hydrolase fold. *Prot. Engin.*, 5, 197–211.
36. Brzozowski, A. M., Derewenda, U., Derewenda, Z.S., Dodson, G.G., Lawson, D.M., Turkenburg, J. P., Björkling, F., Huge-Jensen, B., Patkar, S.A. & Thim, L. (1991). A model for interfacial activation in lipases from the structure of a fungal lipase–inhibitor complex. *Nature*, 351, 491–494.
37. Pathak, D. & Ollis, D. (1990). Refined structure of dienelactone hydrolase at 1.8 Å. *J. Mol. Biol.*, 214, 497–525.
38. Liao, D.-L. & Remington, S.J. (1990). Structure of wheat serine carboxypeptidase II at 3.5 Å resolution. *J. Biol. Chem.*, 265, 6528–6531.
39. Sussman, J. L., Harel, M., Frolow, F., Ofner, C., Goldman, A., Toker, L. & Silman, I. (1991). Atomic structure of acetylcholinesterase from *Torpedo californica*: a prototypic acetylcholine-binding protein. *Science*, 253, 872–879.
40. Franken, S.Y., Rozeboom, H.J., Kalk K.H. & Dijkstra, B.W. (1991). Crystal structure of haloalkane dehalogenase: an enzyme to detoxify haloalkanes. *EMBO J.*, 10, 1297–1302.
41. Derewenda, Z. Personal communication.
42. Wells, J.A. & Estell, D.A. (1988). Subtilisin – an enzyme designed to be engineered. *Trends Biochem. Sci.*, 13, 291–297.
43. Corey, D.R. & Craik, C.S. (1992). Trypsin: a model for the introduction of novel properties into proteins. In *Innovations on Proteins and Their Inhibitors*, ed. Avilles, F., (in the press). Berlin: Walter DeGruyter.
44. Waite, M. (1987). *The Phospholipases*. Handbook Lipid Res. vol. 5, New York: Plenum.
45. Dijkstra, B.W., Renetseder, R., Kalk, K.H., Hol, W.G.J. & Drenth, J. (1983). Structure of porcine pancreatic phospholipase A_2 at 2.6 Å and comparison with bovine phospholipase A_2. *J. Mol. Biol.*, 168, 163–179.
46. Dijkstra, B.W., Van Ness, G.J.H., Kalk, K.H., Brandenburg, N.P., Hol, W. G. J. & Drenth, J. (1982). The structure of bovine pancreatic phospholipase A_2 at 3.0 Å resolution. *Acta Crystallogr. B*, 38, 793–799.
47. Brunie, S., Bolin, J., Gewirth, P. & Sigler, P.B. (1985). The refined structure of dimeric phospholipase A_2 at 2.5 Å. Access to a shielded catalytic center. *J. Biol. Chem.*, 260, 9742–9747.
48. White, S.P., Scott, D.L., Otwinowski, Z., Gelb, M.H. & Sigler, P.B. (1990). Crystal structure of cobra-venom phospholipase A_2 in a complex with a transition-state analog. *Science*, 250, 1560–1563.

49. Scott, D.L., Otwinowski, Z., Gelb, M.H. & Sigler, P.B. (1990). Crystal structure of bee-venom phospholipase A_2 in a complex with a transition-state analog. *Science*, 250, 1563–1566.

50. Kuipers, O.P., Thunnissen, M.M.G.M., DeGues, P., Dijkstra, B.W., Drenth, J., Verheij, H.M. & DeHaas, G.H. (1989). Enhanced activity and altered specificity of phospholipase A_2 by deletion of a surface loop. *Science*, 244, 82–85.

51. Thunnissen, M.M.G.M., Kalk, K.H., Drenth, J. & Dijkstra B.W. (1990). Structure of an engineered porcine phospholipase A_2 with enhanced activity at 2.1 Å resolution: comparison with the wild-type porcine and *Crotalus atrox* phospholipase A_2. *J. Mol. Biol.*, 216, 425–439.

52. Wells, J.A., Cunningham, B.C., Graycar, T.P. & Estell, D.A. (1987). Recruitment of substrate specificity properties of one enzyme into a related one by protein engineering. *Proc. Nat. Acad. Sci. USA*, 84, 5167–5171.

53. Noel, J.P., Bingman, C.A., Deng, T., Dupureur, C.M., Hamilton, K.J., Jiang, R.T., Kwak, J.G., Sekharudu, C., Sundaralingam, M. & Tsai, M.D. (1991). Phospholipase A_2 engineering. X-ray structural and functional evidence for the interaction of lysine 56 with substrates. *Biochemistry*, 30, 11801–11811.

54. Kuipers, O. P., Dijkman, R., Pals, C.E.G.M., Verheij, H.M. & de Haas, G.H. (1989). Evidence for the involvement of tyrosine-69 in the control of stereospecificity of porcine pancreatic phospholipase A_2. *Prot. Engin.*, 2, 467–471.

55. Kuipers, O. P., Kerver, J., van Meersbergen, J., Vis, R., Dijkman, R., Verheij, H.M. & de Haas, G.H. (1990). Influence of size and polarity of residue 31 in porcine pancreatic phospholipase A_2 on catalytic properties. *Prot. Engin.*, 3, 599–603.

56. Pickersgill, R.W., Sumner, I.G., Collins, M.E., Warwicker, J., Perry, B., Bhat, K.M. & Goodenough, P.W. (1991). Modification of the stability of phospholipase A_2 by charge engineering. *FEBS Letters*, 281, 219–222.

57. Gormsen, E. & Malmos, H. (1991). A new lipase for the detergent industry. *Household and Personal Products Industry (happi)*, October 1991. pp. 122–125.

58. Estell, D.A., Graycar, T.P. & Wells, J.A. (1985). Engineering an enzyme to be resistant to chemical oxidation. *J. Biol. Chem.*, 260, 6518–6521.

59. Estell, D.A. & Wells, J.A. (1988). Modified enzymes and methods for making same. US Patent 4760025.

60. Batenburg, A.M., Egmond, M.R., Frenken, L.G.J. & Verrips, C.T. (1991). Enzymes and enzymatic detergent compositions. European patent application EP 0407225.

61. Sebastian, J., Chandra, A.K. & Kolattukudy, P.E. (1987). Discovery of a cutinase-producing *Pseudomonas* sp. cohabiting with an apparently nitrogen-fixing *Corynebacterium* sp. in the phyllosphere. *J. Bacteriol.*, 169, 131–136.

62. Sebastian, J. & Kolattukudy, P.E. (1988). Purification and characterization of cutinase from fluorescent *Pseudomonas putida* bacterial strain isolated from phyllosphere. *Arch. Biochem. Biophys.*, 263, 77–85.

63. Gray, G., Poulose, A.J. & Power, S. (1988). Novel hydrolase and method of production. European Patent application EP 0268452.

64. Arbige, M.A., Estell, D.A., Pepsin, M.J. & Poulose, A.J. (1988). Preparation of enzymes having altered activity. European patent application. EP 0260105.

65. Weirsma, R.J., Stanislowski, A. G., Gray, G., Poulose, A.J. & Power, S. (1991). Enzymatic peracid bleaching system. US Patent 5030240.

66. Poulose, A.J., Ashizawa, E., Power, S., Shew, B., Gray, G. & Norton, S. (1992). Enzyme engineering of a lipase for improved enzymatic peracid bleaching system. Presented at the American Chemical Society meeting April 1992, San Francisco. *Abstr. Pap. Am. Chem. Soc.*, 203, 1–3.

67. Poulose, A.J. & Anderson, S.A. (1992). Enzymatic peracid bleaching system with modified enzyme. US Patent 5108457.

Index